MW01001746

A Search for Structure

An aggregate of bubbles illustrating the structure of metals on an atomic scale and serving as a visual metaphor for the hierarchy of interactions of all kinds. Note the formation of regions of order that tolerate a few internal local anomalies but conflict on a larger scale to produce linear boundaries of connected disorder.

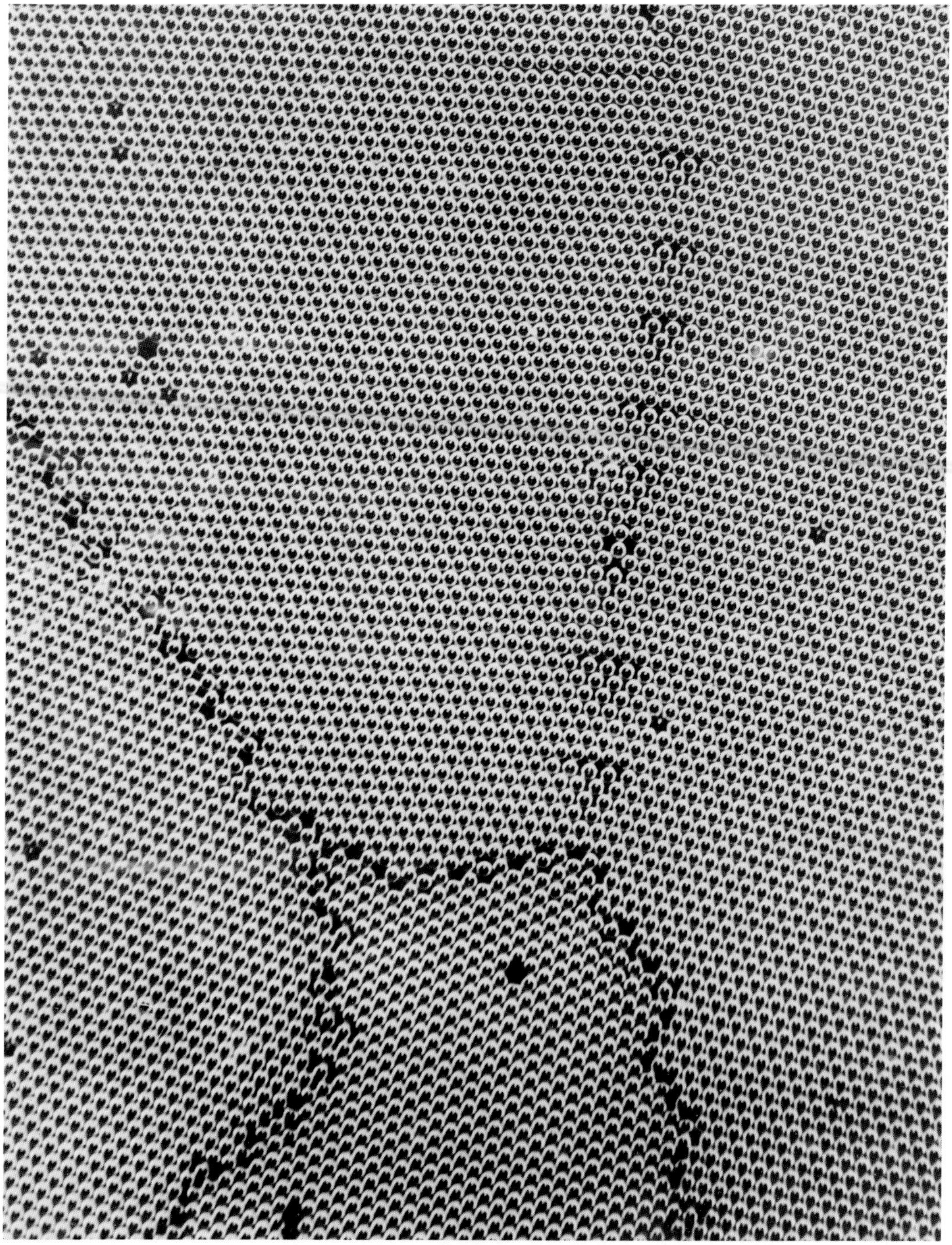

A Search for Structure
Selected Essays on Science, Art, and History

Cyril Stanley Smith

The MIT Press
Cambridge, Massachusetts, and London, England

Publication of this volume has been aided by grants from the National Endowment for the Humanities and Mr. Bern Dibner.

The author gratefully acknowledges permission from the copyright holders to republish the essays of this collection. A detailed listing of the histories of the essays will be found in the section entitled "Sources," which is hereby incorporated onto the copyright page by reference.

© 1981 by
The Massachusetts Institute of Technology

All rights reserved. No part of this book may be reproduced in any form or by any means, electronic or mechanical, including photocopying, recording, or by any information storage and retrieval system, without permission in writing from the publisher.

This book was set in VIP Bembo by Achorn Graphic Services, Inc., and printed and bound by Halliday Lithograph in the United States of America.

**Library of Congress
Cataloging in Publication Data**

Smith, Cyril Stanley.
A search for structure.

Bibliography: p.
Includes index.
Contents: Grain shapes and other metallurgical applications of topology—The discovery of carbon in steel—Structure, substructure, and superstructure.—[etc.]
1. Science—Addresses, essays, lectures.
2. Art—Addresses, essays, lectures. 3. History—Addresses, essays, lectures. I. Title.
Q171.S618 500 81-822
ISBN 0-262-19191-1 AACR2

Contents

Apologia

It was an interesting experience to skim through the nearly two hundred papers of mine that have been published since 1926 to find which, if any, might be of general enough interest to merit inclusion in this collection. Not one has survived from the period of professional work as an industrial metallurgist, the job for which I was trained; only one is included from the time when I turned to more or less pure science in an attempt to provide an explanation of the structure of metals and alloys; and there is no hint of wartime experiences or service on Washington committees. Limitations on space have also excluded an article written in collaboration with the eminent historian of science, Melvin Kranzberg, that summarizes my view on the present state of my profession. Number 195 in the bibliography, it might be noted by anyone particularly concerned with the history and present state of the field of materials science and engineering.

What is left is largely the product of interests peripheral to my profession—history, art, and what impinges on philosophy. In all these fields I can claim only the standing of a rank amateur, but it was my professional familiarity with materials that provided a bridge for conversations with experts in other fields, in rather the same way as materials themselves have interacted with mainstream history, for they are the stuff on which virtually all human activities are based. It was the search for the historical origins of down-to-earth metallurgy that introduced me to the techniques of the artist, and it was noting the similarities between the hierarchy of concepts in art and the structural hierarchy in materials that turned me into a philomorph,* even at times into a rudimentary philosopher.

The exclusion of scientific papers has left little in this collection that has to be believed. The essays are just that: experiments in the development of a viewpoint. They are meant to provoke thinking about the topics discussed rather than to stop it by suggesting that truth has been found. They are interdisciplinary, perhaps at times even nondisciplinary. My ignorance of many of the works familiar to any graduate student in history or art history will be obvious. I deal mainly with details, not with the integral beauty of works of art, and with only one thread in the whole fabric of human history.

Though my conclusions are related to "structuralism" in anthropology, linguistics, and psychology, I arrived at them by a totally different route. I started with the scientific analysis of atomic and microscopically visible structures in solids, but I had been drawn into the study of these forms because I enjoyed looking at them. (To this day I experience a moment of aesthetic pleasure whenever a properly prepared specimen comes into focus under my microscope.) Eventually my brush with finer art led me to see that the understanding of such structures required more than calculation, and I have learned not to be ashamed of the intuitive-pictorial approach which as a young scientist I was taught to abhor. Almost all fields today are concerned in one way or another

*Perhaps it should be mentioned that the suffix of this word "philomorph" was taken from the Greek *morphe*, meaning form, and has nothing to do with the slumbering son of Somnus! The label Philomorphs (which I coined) was adopted in 1965 by a group of people in the Cambridge area who were meeting informally to discuss morphology from many different viewpoints.

with hierarchical structure, and a theory, or perhaps more usefully a metaphor, common to all may emerge if the features of many are compared. Though the units in different fields are different, in all of them meaning comes through communication: patterns of communication are common to all, with aggregation leading to diversity or unity, and the clumps of unity themselves serving in turn as units in larger structures based upon more complex but still direct communication.

Objects seem to provide the common meeting ground for all professions, while verbal structures tend to separate them. I had been studying the history of metals for many years before I fully realized that the best sources for the early period were not the conventional written documents of the historian but the material artifacts in art museums. I had traced many changes in the history of science and technology before I saw the close analogy between them and the detailed structural mechanisms of phase transformation such as occurs in the hardening and softening of steel.

Newton picked up the pebbles on his metaphoric beach with an intellectual objective in mind, but his ancestor in paleolithic times picked up real minerals because he enjoyed looking at them: quite inadvertently he started the chain of practice and craftsmanship and thought that led to the diversity of specialized materials and generalized theory today. More like the early *Homo sapiens* than the sixteenth-century intellectual giant I have enjoyed a life of rather undisciplined wandering and search. How grateful I am to those great institutions, the University of Chicago and the Massachusetts Institute of Technology, that have given me freedom to wander in return for a minimum of service to formal classroom teaching!

Probably the article "Metallurgical footnotes to the history of art" (chapter 9 in this volume) and the article on hierarchical structure (chapter 14) will be of greatest interest to the general reader, but I have included some articles in which laboratory analysis is used to "read" the internal records in artifacts in order to provide a glimpse of the laboratory experience without which the more general viewpoints could not have been attained. One paper examining some specific historical events in detail is here to support my general thesis that aesthetic curiosity has been a common driving force in technological invention: this idea is developed at some length in "Porcelain and plutonism" and summarized in "On art, invention, and technology." The "Highly personal view of science and its history" is an attempt to find balance between what I regard as the current overemphasis by historians on the social environment of science and the earlier overemphasis on science as an isolated intellectual activity by showing the importance of technology in the development of individuals and societies alike: with the postscript it also comes as close to an autobiography as I am ever likely to write.

For all my intellectual truancies, I remain a scientist and I believe profoundly that the scientific method with its interlocking mixture of observation, experiment, theory, and open-minded criticism is the best means of approaching "truth." Yet the world is far too complex to be described scientifically, explicitly. The very strength of science lies in the recognition of boundary conditions for any interaction. Science *must* be simple, yet the human brain has a structure that gives it the capacity for relating to the world in its undivided complexity in ways that are not logical, though they are effective. Aesthetic interest aroused by observation and half-formed perception seems usually, perhaps always, to precede exact analysis, and the extension of exact understanding of local interaction to larger areas of complexity inevitably leads to trouble.

In both the structure of matter itself and in that of human understanding of matter there has been an evolutionary interplay between simple predictable replication and the accidents that allow and sometimes force the discovery of new structures. The parts may have an internal structure that tells them what they must do if . . . , but they cannot anticipate what the environment will be or whether local discomfort is temporary or permanent—whether it will be erased or will serve as the nucleus for the growth of a new, happier structure.

The integrity of any aggregate or organism depends on the fact that the units are internally modified by association with their neighbors: there is a continuity or reflection of external symmetry across the interface and through the body. It is the continuation of duality across units that defines the larger structure. Uncertainty and indeterminacy are not a monopoly of the subatomic level; most things are internally variable unless or until they have resonantly communicated with their environment.

Democritus spoke truly when he said that everything existing is the fruit of chance and necessity. Biologists have told how their interplay has given rise to organisms and environments. Acting in history this interplay has somehow produced the association of body and human brain as a mechanism to extend the same mixed principles to a larger world of thought patterns—one that is capable of far more rapid evolution than was possible with material interactions alone. The great success of the logical analytical reductionist approach to understanding over the last four centuries and the utility of the application of its simple principles has not negated the evidence of history that the sensual-emotional-aesthetic capabilities of the human being also have validity. The problem is to find the proper nonexclusive role of each. These essays are the partial record of one man's attempt to reach out from a base in a rather practical science toward what has come to be regarded as the other world, urged on by a belief, or at least hope, that a sympathetic study might engender some reciprocal understanding. In the past an advance in understanding within a field—art, science, history, religion, politics, or whatever—has usually come from the intense study of a single level of existence and interaction, with necessary (but temporary) exclusion of problems beyond a certain boundary. Now that the units of most levels of material structure have been disclosed, the future offers great opportunities for the exploration of principles of interaction *between* levels. This approach will be dualistic, not holistic. The real hierarchy of things must be appreciated: it is neither a hierarchy of command from on high, nor a structure entirely predetermined by the behavior of parts with their high concentrations of energy; rather, it involves the democratic interaction of both types of organization through history to yield individual assemblies.

The principles of pattern formation, aggregation, and transformation seem to be the same in matter and in the human brain, and if properly formulated they may provide a kind of visual metaphor that will serve to join and mutually illuminate physics on the one hand and geological, biological, and social history on the other—with art in between.

What I seem to have been reaching toward is not a logical philosophy, not a system of words to be communicated and refined by discussion, but rather a system of patterns to be experienced visually and turned into meaning by the sensual finding of a shared duality of the external relations with those of the patterns of and on and in an individual brain. Things take meaning by communication, and levels in matter or thought or experience have limits that are transcended by exploration and the discovery of new possibilities for closure, just as a group of polygons become a polyhedron by their joining, with no change in level on further aggregation as long as the internal vertices continue to share the same dual relationship with their environment.

These papers are probably to be called interdisciplinary—an "in" word these days—but any value they may have derives from the fact that the author started with a rather deep immersion in a single discipline. One cannot hope to understand the nature of interaction between impinging areas without a firm knowledge of at least one of them. Only on such a basis can one appreciate when or where a given body of understanding has ceased to be fully applicable. Interdisciplinary activity is as dangerous for the undergraduate as it is essential for the mature professional in any field.

Despite inevitable disagreement with some of the things I have written, the papers are, save where noted, reprinted without change except for the correction of a few errors of the sort called typographical, bringing some references up to date, and eliminating duplication among the illustrations. The order is chronological.

Cyril Stanley Smith
Cambridge, Massachusetts
4 October 1979

A Search for Structure

1
Grain Shapes and Other Metallurgical Applications of Topology

". . . . peut-être trouverat'on que c'est de cette figure des grains et de leur arrangement que dépendant la ductilité des Métaux et celles de quelques autre matières."

—R. A. F. DE RÉAUMUR, 1724

The structure of matter, on both an atomic and macroscopic scale, is a result of the interplay between the requirements of the physical forces operating between the individual parts and the mathematical requirements of space filling. Unlike the underlying forces, the shapes of objects and their components are immediately apparent, and for this reason the observation and cataloguing of shapes is usually the first stage in the development of any branch of science. As a science matures, attention is focused more on forces, and shape tends to be taken for granted. Day-to-day contact with the limitations of space filling makes us accept them without further thought. Nevertheless, it is desirable from time to time to examine the background and to attempt to deduce relationships of greater complexity and utility than those that are immediately obvious. The present paper is concerned mainly with the role of topology in metallurgical matters, but practitioners in other fields will notice close analogies with the structures and spatial relationships discussed.

The microstructure of a piece of metal consisting of several grains must conform to space-filling requirements. It is the intent of the present paper to call attention to some elementary laws of topology and to show their possible application to metallurgical problems. For proofs of the fundamental relations used, reference should be made to mathematical works,[1] though these rarely contain relations expressed in a form directly applicable to the present type of problem—or, it must be admitted, in terms immediately comprehensible to the average nonmathematician.

Workers in other fields (particularly biologists interested in cell structure) have concerned themselves with related problems and have performed experimental statistical studies on the shapes of cells in unspecialized biological tissue,[2] in soap froth,[3] and the polyhedra formed by compression of lead shot in contact with each other in various modes of stacking.[4] Much of this has been done empirically, although Lewis invoked topological principles to good effect. D'Arcy Thompson, in his truly admirable book, *On Growth and Form,* dealt in general with the relation between shape and function in plants and animals, and included a delightful discussion of topology, geometry, and surface tension as a basis to the understanding of cell shapes.[5]

Desch, in 1919, made a detailed study of the shape of metal grains and showed the relation of these to cells in a soap froth and to the ideal space-filling bodies of Archimedes and Kelvin, but he did not invoke any topological principles.[6] Harker and Parker first showed the importance of grain corner angles in relation to grain growth in a metal,[7] and the present writer has pointed out the significance of this in relation to space filling, and has extended the concept to three-dimensional grains.[8] Scheil, Johnson, Rutherford et al., and others have considered the relation between grain size and grain boundary area, assuming various simple shapes for the grains.[9]

No attempt is made in the present paper to deal with the subject of crystallography, although some of the general relations that are deduced have obvious applications in this field.

Two-Dimensional Relations

Consider first a two-dimensional surface which is to be subdivided into separate areas. The simplest operation is to draw a single closed boundary line of any shape (e.g., figure 1.1A) separating the space within from that outside and forming a single cell in two-dimensional space. This may be further subdivided by other lines joining in various ways and producing an array of polygons, for example as in figure 1.1B. It is at once obvious that certain limitations apply. A line, if it is not endless, cannot have more or less than two ends, and it cannot be shared by more than two polygons. A point may be the junction of any number of polygons or any number of lines, but a network cannot be indiscriminately constructed of points, lines, and polygons without regard to each other. Except at the boundary of the array a line forming an edge of one cell must also always form an edge of a neighboring cell, and an apex of one polygon is similarly an apex of at least one other cell. Finally, two-dimensional space cannot be divided into separate cells except by the introduction of vertices at which at least three cells meet.[10]

As an example of the effects of these restrictions, consider the ways in which space can be subdivided into uniform polygons. It is obvious that, for geometrical reasons, contiguous circles cannot by themselves fill two-dimensional space. Polygons with three or four concave sides are also needed (figure 1.2A). Similarly, although space can be filled with uniform triangles, squares, or hexagons by themselves (figures 1.2B–D), for topological reasons this can be done only if the internal corners at which the various cells meet are, respectively, six-, four-, and three-rayed. Instinctively one feels there must exist a relation between the number of polygons and the number of edges (polygon sides) and corners (vertices). The relation is actually very simple; namely

$$P - E + C = 1, \qquad (1)$$

in which P, E, and C are the number of polygons, edges, and corners respectively (i.e., the number of two-, one-, and zero-dimensional cells).[11]

This simple relation is rigorously true for any simply connected network of lines in two-dimensional space and is not changed by any topologically continuous distortion. As examples, consider figures 1.1B and 1.1C or the more complicated network of figure 1.3 (based on a Picasso painting) which has 35 polygons, 86 edges, and 52 corners, in conformity with equation 1.

Grains and Bubbles in Two Dimensions

In the grains of a metallographic specimen, as well as in a froth of soap bubbles and in the cells of many biological tissues, surface tension eliminates the corners at which (in two dimensions) more than three cells meet and some interesting simplifications result therefrom. Since three edges meet at each and every corner, and since each edge joins two corners (i.e., has two "ends"), we can write

$$2E = 3C. \qquad (2)$$

Also, since each polygon with n sides contributes $n/2$ edges except for those at the boundary of the array, where the edge belongs to only one polygon, it follows that

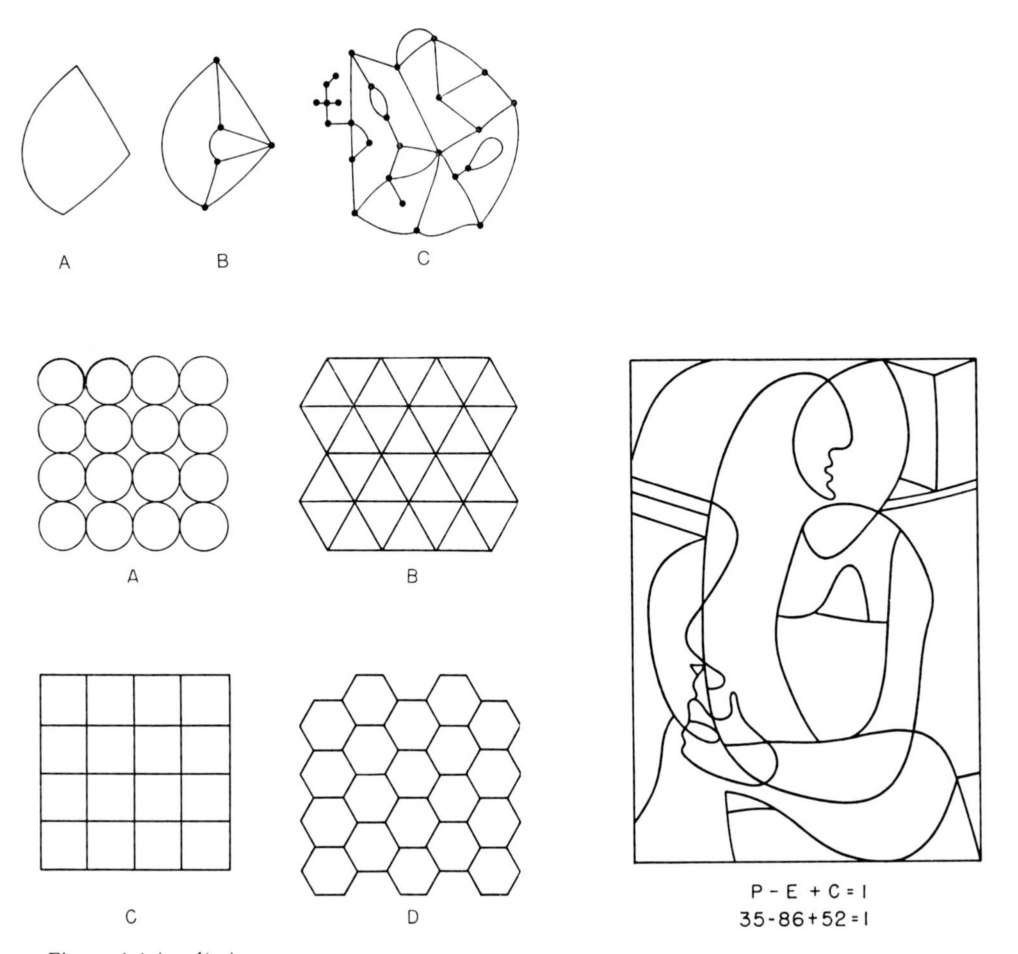

Figure 1.1 (*top line*)
Subdivision of space in two dimensions.

Figure 1.2
Space filling by uniform polygons.

Figure 1.3 (*right column*)
Space filling by irregular polygons.

$$E = \Sigma nP_n/2 + E_b/2, \quad (3)$$

where P_n is the number of polygons of n sides, and E_b is the total number of edges at the boundary. Substituting these values in equation 1 gives

$$\Sigma P_n - (\Sigma nP_n + E_b)/2 + (\Sigma nP_n + E_b)/3 = 1. \quad (4)$$

Simplifying we get

$$\Sigma(6 - n)P_n - E_b = 6 \quad (5)$$

or

$$4P_2 + 3P_3 + 2P_4 + 1P_5 \pm 0P_6 - 1P_7 - 2P_8 - \cdots - (n - 6)P_n - E_b = 6. \quad (6)$$

The relations are easily seen to be true in figures 1.4 and 1.5, which show various simple arrays of bubbles and illustrate the genesis of a typical polygonal grain structure. (These are actual photographs of bubbles blown with a hypodermic syringe between parallel glass plates spaced about 5 millimeters apart. All the bubbles had approximately the same initial volume, except for the intentionally nonuniform cells in figure 1.5.) Note that with 3 cells or more, alternate configurations are possible, but all of these must conform to the above relation.[12]

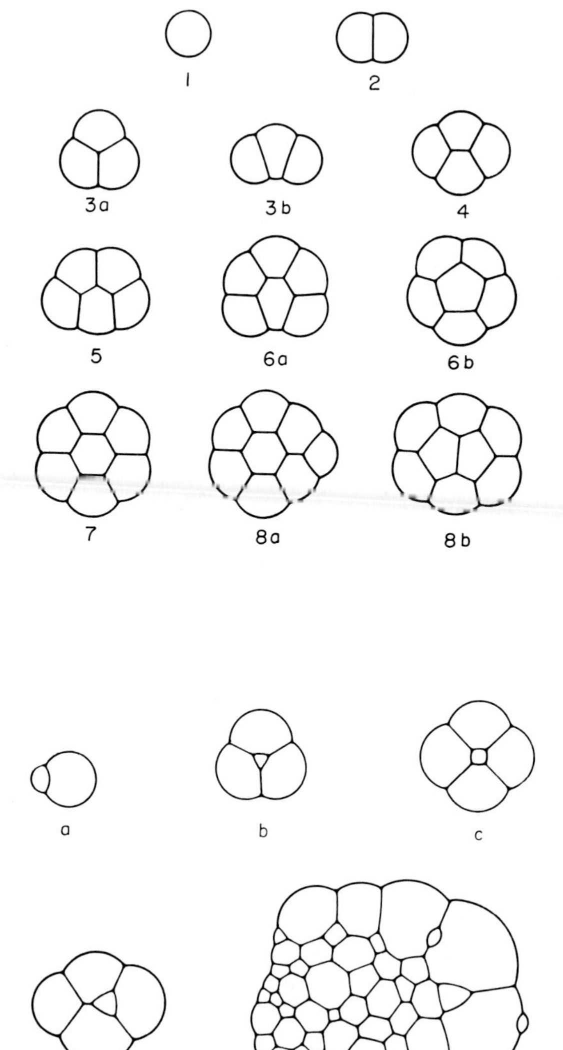

Figure 1.4
Aggregates of various numbers of bubbles of uniform size.

Figure 1.5
Aggregates of nonuniform bubbles.

Sampling of Grains in Two-Dimensional Metallography

The metallurgist may make use of these principles, for they provide a useful criterion for sampling in grain shape or size distribution studies. Equation 5 applies immediately if all of the grains in a given sample are counted, including those at its periphery, at the time of annealing. Alternatively, one may make an isolated system with nothing but three-ray corners by selecting a limited sample of grains and rounding off the grain apices that do not belong to at least two of the grains in the actual network. Thus the sample of grains of alpha iron shown in figure 1.6A can be treated as in figure 1.6B, when it becomes similar to an isolated array of soap bubbles and can be seen to conform to equation 5.[13] This simple relation means that in an array of grains so numerous that those at the boundary become negligible, the average number of sides per grain must be six; the average grain is a hexagon.

An alternate treatment can be devised that does not require either an infinite net of grains or the rounding off of the peripheral grains in the array. It is to be preferred for general use. Equation 5 is a special case of the more fundamental relationship

$$\sum[2r - n(r-2)]P_n + 2E_o - (r-2)E_b = 2r, \quad (7)$$

where P_n is the number of polygons with n sides, E_b the number of sides to the polygon formed by the closed boundary, E_o the number of severed connections reaching out from the vertices of the closed boundary, and r the vertex valence, which can be of any value (the univalent vertices associated with E_o are not counted). This follows from the fact that each vertex is shared by r polygons and each side by two polygons, whence the total number of vertices in the array is $(\sum nP_n + E_b + E_o)/r$ and the number of lines $(\sum nP_n + E_b)/2 + E_o$. Substituting these relations in Euler's relation, $C - E + P = 1$, and simplifying gives equation 7.

Returning to simple foams with only trivalent vertices, the relation becomes

$$\sum(6-n)P_n + 2E_o - E_b = 6. \quad (8)$$

If $r = 4$, then

$$\sum(4-n)P_n + E_o - E_b = 4. \quad (9)$$

Self-duality is possible only when both n and r have the value 4.

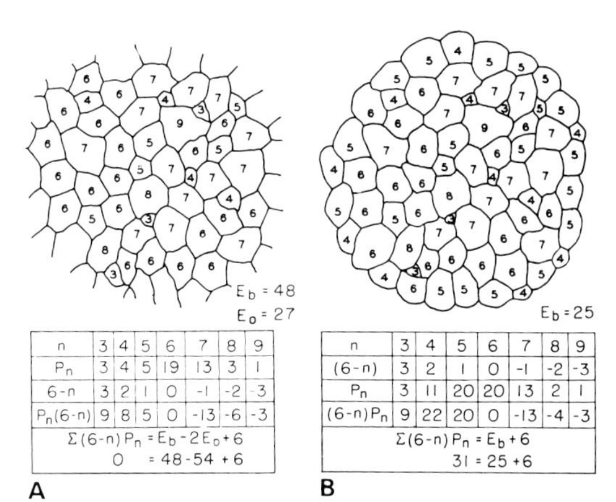

n	3	4	5	6	7	8	9
P_n	3	4	5	19	13	3	1
6-n	3	2	1	0	-1	-2	-3
$P_n(6-n)$	9	8	5	0	-13	-6	-3
$\Sigma(6-n)P_n = E_b - 2E_o + 6$							
0 = 48-54+6							

n	3	4	5	6	7	8	9
(6-n)	3	2	1	0	-1	-2	-3
P_n	3	11	20	20	13	2	1
$(6-n)P_n$	9	22	20	0	-13	-4	-3
$\Sigma(6-n)P_n = E_b + 6$							
31 = 25+6							

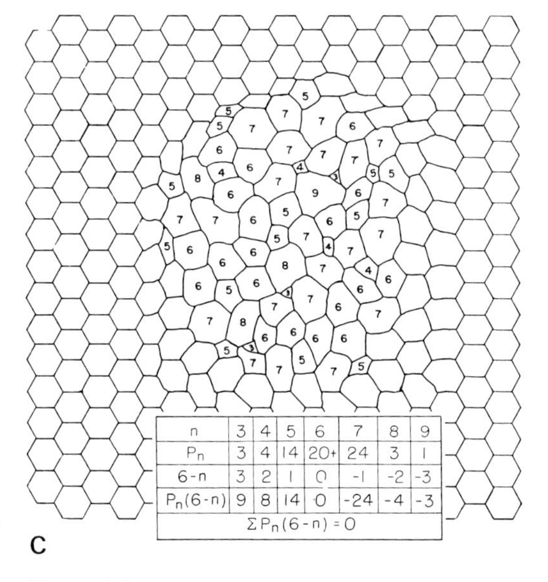

n	3	4	5	6	7	8	9
P_n	3	4	14	20+	24	3	1
6-n	3	2	1	0	-1	-2	-3
$P_n(6-n)$	9	8	14	0	-24	-4	-3
$\Sigma P_n(6-n) = 0$							

Figure 1.6
(A) Microstructure of polycrystalline alpha iron considered as a network of polygons. (Traced from ASTM grain size standard no. 3.) (B) Same network as part A but with corners rounded. (C) Grains of part A inserted in a hexagonal net.

In equations 7–9 all vertices have been assumed to be of constant valence. The relations apply to other cases if the average value $\bar{r} = \Sigma rC_r/C$ is used, but $\bar{n}$ cannot be so used. If n and r both vary, their departures from the dual value are interlocked in such a way as to make the sum of their respective average values equal to 8 after appropriate corrections for the periphery have been made. The basic equation is

$$\Sigma(4 - n)P_n + \Sigma(4 - r)C_r - E_b - 2E_o = 4. \quad (10a)$$

This follows from the fact that the sharing of vertices by edges and polygons makes $2E$ equal to ΣrC_r and also to $\Sigma nP_n + E_b + 2E_o$. Substituting the sum of these in Euler's relation, $4E = 4(P + C - 1)$, and simplifying gives equation 10. This applies to any net of connections. It is often convenient to distinguish E_o, the one-dimensional connections which do not participate in the closure, from the others. Since each one terminates on a univalent vertex, their contribution to $\Sigma(4 - r)C_r$ can be replaced by $3E_o$. If we omit the univalent vertices, then, the equation becomes

$$\Sigma(4 - n)P_n + \Sigma(4 - r)C_r + E_o - E_b = 4. \quad (10b)$$

Quadrilateral grids with their diagonals orthogonal to each other can be superimposed at some scale on *any* extended space-filling tesselation, and within each of them $\Sigma(4 - n)P_n$ must be equal to $\Sigma(4 - r)C_r$. The only symmetry requirement is one between diagonally opposite corners, real and dual, in a set of four rectangles joining to form a new common center, regardless of what other connections may intervene. For one example see figure 1.7A. The arrangement is that of the quincunx, the importance of which was pointed out to English readers by Sir Thomas Browne in his *Garden of Cyrus* (London, 1658). It was well known in classical times in connection with the famed Gardens of Babylon, whose designers wanted to maximize both the size of the flower beds and their accessibility to light and to visitors' admiring eyes.

The connections within a net of any dimensionality can be represented in two-dimensional projection. All continuity in extended systems can be reduced to that of quincunxial nets of vertices having valence twice the effective dimensionality. Individuality within a net involves the presence of a closed boundary within which one or more pairs of external misfits are connected internally in opposition to each other. A new hierarchical level is signaled by the discovery of a new duality based on the quincunxial combination of structures whose boundaries were incompatible on a lower level. [The last four paragraphs, including the derivation of equations 7–10, have been completely rewritten from the original publication.]

The boundary can tell much about what is inside. In the simplest nets with only trivalent vertices to which equation 8 applies, if the boundary is such that the criterion $E_o - 2E_b = 6$ is satisfied, then $\Sigma(6 - n)P_n$ becomes zero, and the average number of sides to the polygons in the net must be exactly six. Conversely, if the average cell is a hexagon, then the peripheral corners that belong to only one cell must outnumber those shared by two cells by exactly six.

An alternate form of equation 8, involving simpler counting, is

$$\Sigma(6 - n)P_n + E_o - E_i = 6, \quad (11)$$

where E_i and E_o are the number of edges joining a closed boundary from the inside and outside, respectively. A simpler form of

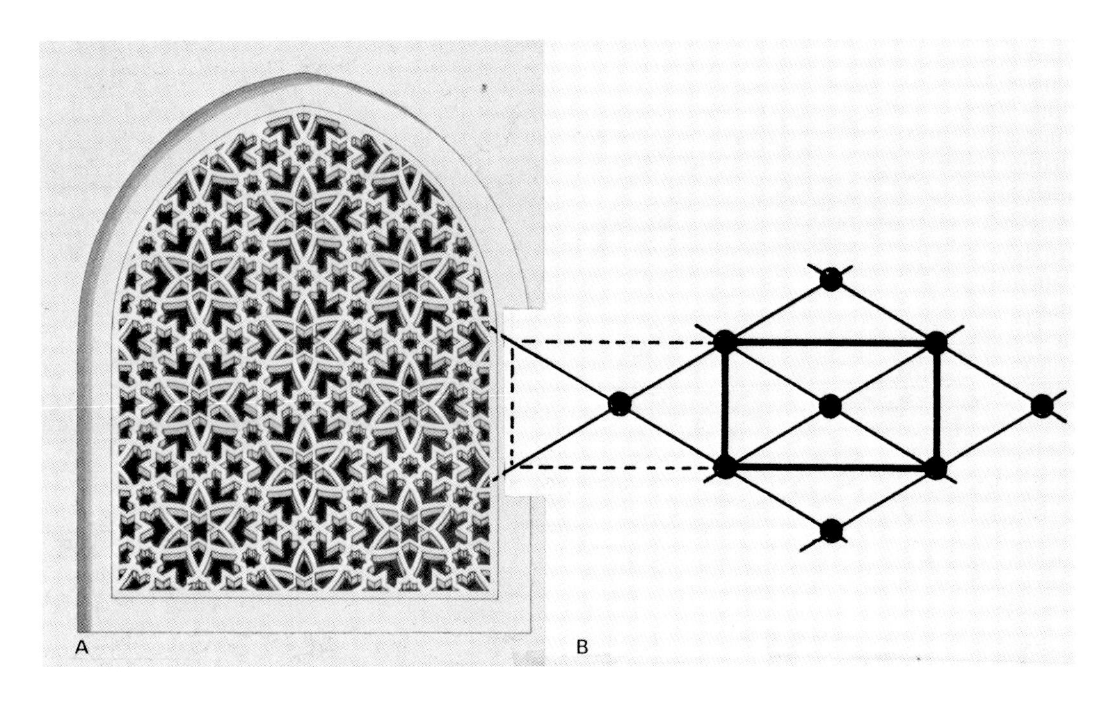

this equation was deduced by Graustein by different methods over 50 years ago.[14]

The metallurgist considering an array of grains in a photomicrograph can use equations 8 or 11, for only when the boundary junction counts are equal can a group of grains be regarded as a representative sample of the structure, capable of being extended by mere duplication to occupy any desired area without the introduction of any new cells to obtain boundary conformity. The grains in figure 1.6 were selected on this basis. The metallurgist is generally interested in the size of grains and not directly in their topological properties, but it is obvious that the larger grains will usually be those of many sides, and the smaller ones those with few sides. Correct sampling for polygon distribution will insure better sampling for size also.[15]

Figure 1.7
Arabic open latticework, from J. Bourgoin's *Arabic Geometrical Pattern and Design* (Paris, 1879; plates reprinted, New York, 1973). Note that a tracing of the dual framework of part B can be superimposed on any vertex or polygon center in part A and the symmetry of both the peripheral and diagonal connections between the vertices will be maintained. The geometric basis of such patterns has been much discussed, most completely by Keith Critchlow in his *Islamic Patterns: An Analytical and Cosmological Approach* (London, 1976), but the more basic topological relationships between connections that are summarized in our equation 10 have not been noted.

The Basis for Grain Growth

Grain growth results from the interaction between the topological requirements discussed above and the geometrical needs of surface-tension equilibrium. The topological relations are rigorously true for any network of lines and polygons, regardless of actual shape. They are completely unaffected by any amount of distortion, however irregular, provided only that the surface on which they are drawn remains continuous. The relation depends on the number of edges, corners, and polygons, and not on their size, straightness, or angular relations. For example, compare with the previous grain and bubble patterns the photograph, figure 1.8, of the pattern of cracks in the glazed surface of a ceramic objet d'art. Since a crack in the brittle glaze relieves the tension stresses due to differential thermal contraction only in a direction normal to itself, late cracks must meet earlier ones at about 90 degrees, and will rarely cross them. However, each point of intersection separates three polygons and topologically the pattern is identical with a network of bubbles, despite the striking difference in geometry. Similarly, the regular patterns formed by rectangular bricks laid in simple heading or stretching bond become topologically identical with a hexagonal tiled floor.

In the specific case of the metallurgist's grains the boundaries between the polygons have associated with them a surface or interfacial tension and, if local equilibrium is allowed to occur, the angles will adjust until the angle becomes 120 degrees, if the energies are equal.[16] The apex angle of a regular straight-sided polygon is $\theta = 180(1 - 2/n)$, where n is the number of sides. Since, therefore, only a hexagon among such polygons

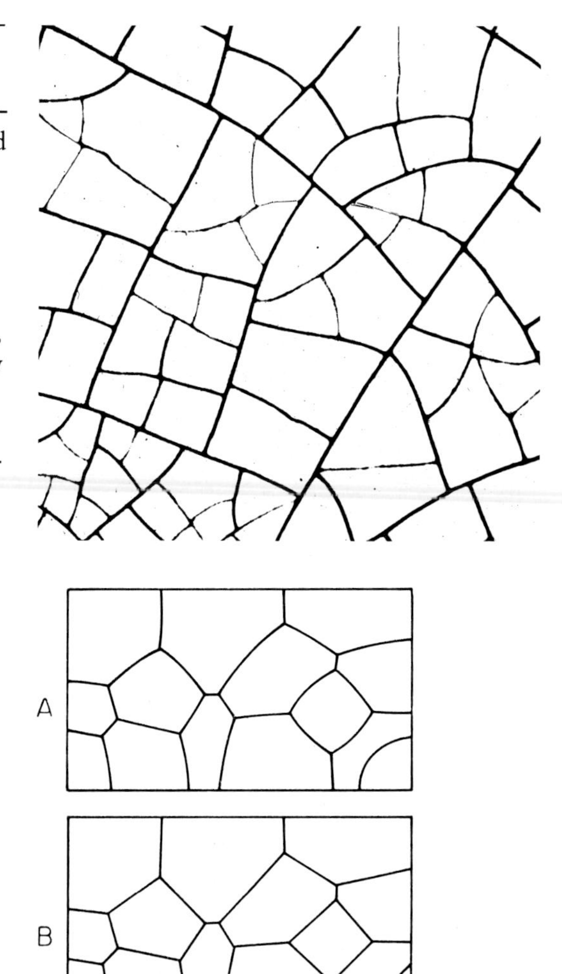

Figure 1.8
Cracks in the surface glaze on a ceramic object. Natural size.

Figure 1.9
(A) Soap bubbles in a rectangular cell; all angles are 120 degrees except where films join rigid cell walls, where the angle is 90 degrees. (B) Polygons with straight sides meeting at the same points as in part A.

can have a corner angle of 120 degrees, the sides of grains in a random array must be curved to a greater or lesser degree in order to reconcile the surface-tension requirements with the varying number of sides to the polygons. Figures 1.9A and 1.9B illustrate this point. The first is a tracing of a group of soap bubbles with surface-tension junctions, while the latter shows straight-sided polygons that meet at the same vertices. Grain growth is possible—indeed inevitable—because of the unavoidability of curvature when surface tension operates in anything but an array of regular hexagons. The importance of such curvatures was first realized by Harker and Parker in their important paper on grain growth.[7] However, since they had not considered the topological requirements, they assumed that grain growth would eventually stop with the achievement of a geometry that would satisfy both local 120-degree surface-tension requirements and long-range absence of curvature. Actually, even in two dimensions, a random array of grains with 120-degree corners is perpetually unstable, and grain growth stops only because of blocking due to inclusions or to variations of grain boundary energy with orientation which permit equilibrium to occur at angles other than 120 degrees. Quite contrary to Harker and Parker, the writer predicts that grain growth will slow and stop (in the absence of inclusions) only when grain corner angles can *depart* from 120 degrees instead of when they *approach* it, as they suggest.

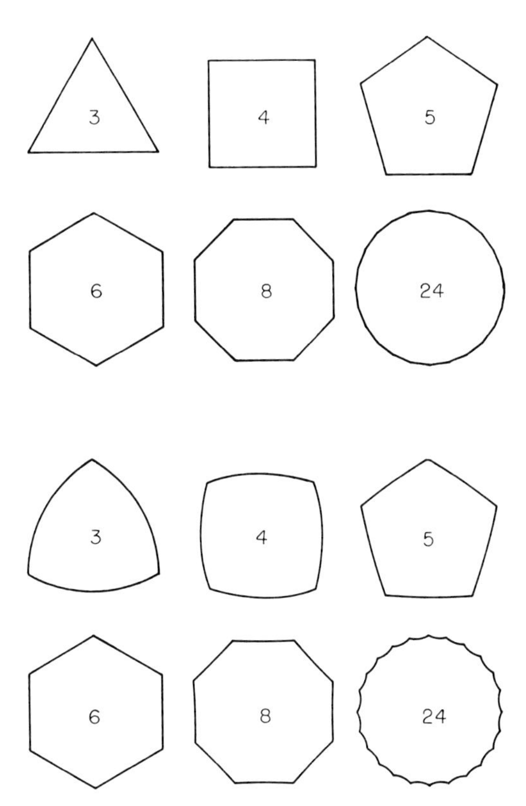

Figure 1.10
Six regular polygons with straight sides.

Figure 1.11
Polygons with curved sides meeting at 120 degrees.

Figure 1.10 shows several regular polygons with various numbers of sides, while figure 1.11 shows similar polygons with the sides formed by arcs of circles intersecting at 120 degrees. It is at once obvious that a hexagon under such conditions has plane sides, while a polygon with more than six sides is

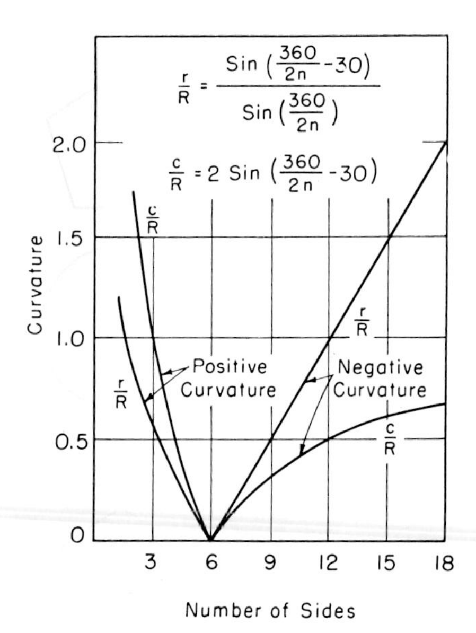

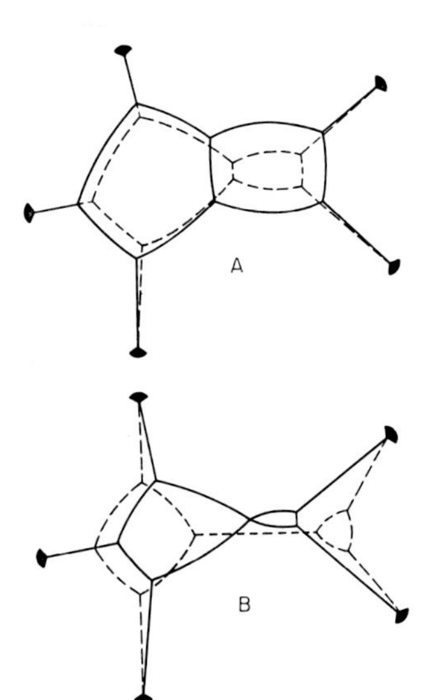

Figure 1.12
Curvature of polygon sides needed to give 120-degree apex angles, as a function of number of sides.

Figure 1.13
Stages in the readjustment of bubble shape as the size diminishes.

concave, and with less than six, convex. Figure 1.12 shows the curvature ($1/R$) of the polygon sides in terms of radius (r) or the chord (c) of the circumscribing circle.

A Dynamic Model of Grain Growth

A random network of polygons must, according to equation 5, have some with both more and less than six sides; consequently there will be both concave and convex cells if the 120-degree criterion is to be satisfied. If, for example, a mass of soap bubbles is brought together irregularly as in figure 1.5, the boundaries of the cells will readjust, compressing some and expanding others until the pressure differences are such that the curvatures precisely agree with 120-degree angles at the corners. The actual pressure difference is equal to $2\gamma/R$, where γ is the surface tension of the soap solution forming the walls of the bubbles. The geometry of the simplest case with two bubbles was elegantly treated by Plateau in the nineteenth century.[17] It should be noted that a froth is stable only because of the existence of these pressure differences between adjacent bubbles, and the structure will change if diffusion is possible under this pressure difference. While a regular hexagonal network in two dimensions would be stable even if diffusion could occur, for it is without pressure differences, such a network is extremely unlikely to appear in practice, and in any case it would be stable only if the films at the boundary were anchored in the correct positions or if they met a rigid framework at an angle of 90 degrees without disturbing the 120-degree relations inside. A single departure from regularity anywhere throughout the network would result in pressure differences which if relieved by diffusion would cause eventual disappearance of all of the cells.

It has usually been supposed, following Lord Kelvin, that a froth of bubbles in three dimensions must be composed of minimum-area tetrakaidecahedra.[18] Although a perfect array of these would be stable even in the absence of gas inside the cells, in practice such regularity can never occur, and the real shape of bubbles is a result of pressure differences balanced by surface tension in accordance with Plateau's rules. A froth cannot be produced in a vacuum, and it is stable only if diffusion of the contained gas cannot occur through the cell walls. If diffusion is possible, the small cells, being under high pressure, will slowly lose gas to the neighboring large ones. If the cell is two- or three-sided, it can disappear symmetrically with continued extension of the adjacent boundaries and without disturbing these. If, however, the cell has more than three sides, at some stage in its shrinking four edges from surrounding cells must be brought together at a point, a condition of instability which will be immediately followed by readjustment to give two three-ray corners. This is shown in figure 1.13, which depicts the outlines traced from two bubbles whose volumes were varied by withdrawing air. When the bubbles had shrunk to the point where four adjacent edges met, there was a sudden readjustment to give a triangle and a tetragon to replace the initial tetragon and pentagon. Note that an adjustment of this kind, if it occurs anywhere but in a peripheral cell, changes neither the number of sides nor the number of cells, but it has given rise to a three-sided cell that now can shrink to the vanishing point without further discontinuous motion. The behavior of any bubble depends on its neighbors, but in general those with few sides will on the average tend to get smaller, and those with many sides to grow. The instability is perpetual and is not removed as cells disappear. The process will continue indefinitely in a froth, although at a progressively slower rate because pressure differences become less as the average cell becomes larger. In a large array, starting with a random distribution of sizes, there is probably a tendency toward a fixed distribution of shapes and relative cell sizes determined by the topological requirements and by the equation of rate of volume change as a function of curvature. When this equilibrium has been established, the rate of change of area of the average cell is a function of the product of the pressure difference and the length of the cell wall, and hence the area should be a linear function of time.

Diffusion depends on pressure difference, not on the absolute pressure. The change in volume resulting from this diffusion will clearly be much greater if the total pressure is low than if it is high. For this reason, most soap froths at atmospheric pressure are very slow to change (except for mechanical damage to the films), and the mechanism of growth has not previously been noted. However, if a froth is maintained under low pressure it will grow rapidly. Acting on this assumption, the writer has produced models—both two- and three-dimensional in character (figures 1.14 and 1.15)—in which the bubbles grow according to the mechanism outlined above. A fine froth of bubbles 1 or 2 millimeters in diameter, obtained by vigorously shaking, changes to cells the size of the container in about an hour.[19] It is believed that this is geometrically, though not mechanistically, a nearly exact model of grain growth as it occurs in metals. The slow motion until an unstable contact occurs, and the rapid readjustment

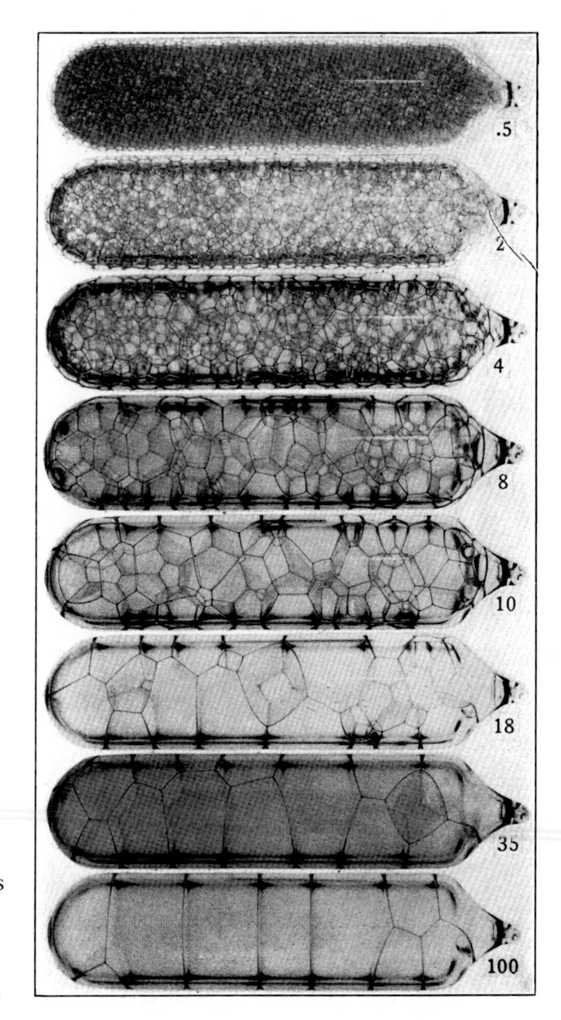

Figure 1.14
Growth of bubbles by diffusion of air under reduced pressure. Numbers indicate time in minutes after cessation of agitation.

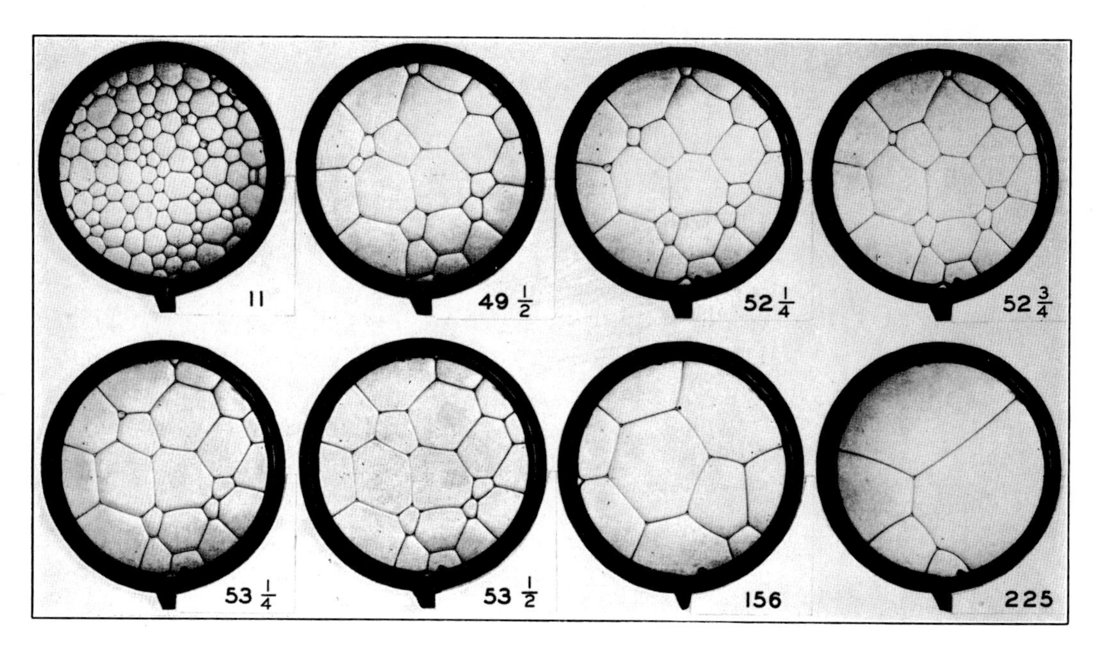

Figure 1.15
Growth and disappearance of bubbles in a flat cell.

immediately thereafter, is striking to watch. This type of movement is exactly duplicated by grain boundaries in metals, and is easily observed, for example, in zinc on a hot stage under the polarizing microscope. This effect was noticed, though not explained, by Carpenter and Elam in their classic work on grain growth.[20]

The photographs of the flat cell are shown in figure 1.15 and illustrate clearly the mechanism whereby grain growth occurs. Particularly notice the sequence of events in the period $52\frac{1}{4}$ to $53\frac{1}{4}$ minutes from the start, during which the four- and five-cornered cells at 9 to 10 o'clock shrink to the point of instability, and readjust to triangles which disappear, all in a short time during which very little change occurs elsewhere in the cell where lower curvatures exist. Figure 1.16 shows the average area of bubble in this cell plotted as a function of time, and it can be seen that during a period where the bubbles are small compared with the cell, they follow quite closely a linear relation of area as a function of time. The distribution of shape does not vary with time or number of bubbles, provided that the latter is large. The frequency of 3-, 4-, 5-, 6-, 7-, and 8-sided bubbles was, respectively, about 1, 8, 27, 35, 18, and 11 percent.

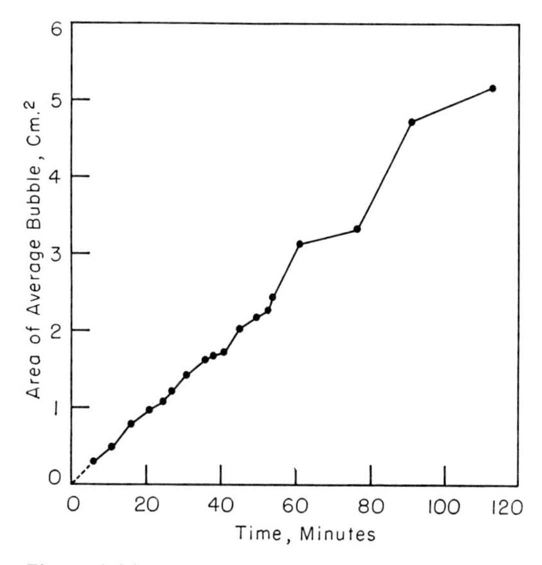

Figure 1.16
Change with time of the average area of bubbles in a flat cell.

Grain growth in a metal does not, of course, depend on diffusion of gas through a membrane. Although pressure differences do exist between grains as a result of the curvature of their boundaries, it seems unlikely that pressure as such can be the driving force for growth, and pressure will certainly not be equalized throughout a given grain. Rather, one can adopt the viewpoint expressed by Harker and Parker, that the mechanism is associated with the probability of an atom at the grain boundary finding itself, as a result of random thermal motion, on one or the other of the adjacent lattices.[7] Clearly the probability of transfer from a boundary to a grain whose surface is concave is greater than to one whose surface is convex, and the migration rate will increase with curvature (probably as $1/R$ in two dimensions and $1/R_1 + 1/R_2$ in three). The rate of adjustment will be greatest at a grain corner where one boundary meets other boundaries nearly at 120 degrees, for any movement there will introduce curvature that decreases away from the end of the boundary, and the center of the boundary will therefore lag behind motion of the corners. This results in a general curved shape superficially not unlike that of a bubble under pressure, though its origin is quite different, and it is not truly an arc of a circle. The curvature is similar to that of a string initially straight being pulled by its ends through treacle, an analogy suggested to the writer by Dr. E. Orowan.

General Topological Relations in Two Dimensions

Equations 2–6 and 8 apply to the surface-tension-limited case of three polygons meeting at a vertex. Certain other relations hold generally and are often of value in considering structural problems. When the number of edges meeting at each vertex is not limited, $E = \Sigma rC_r/2$, where C_r is the number of vertices at which r edges meet, that is, the number of vertices having valence r. From equation 1 it follows that

$$P - \Sigma rC_r/2 + C = 1 \tag{12}$$

or

$$P + \Sigma(1 - r/2)C_r = 1, \tag{12a}$$

which in expanded form becomes

$$P = 1 - C_1/2 - 0C_2 + C_3/2 + C_4 + (3/2)C_5 + \cdots + (r/2 - 1)C_r. \tag{13}$$

This is true for any two-dimensional network of lines and polygons, and gives the total number of polygons as a function of the number of corners of various types. The use of this relation is perhaps most obvious in connection with structural organic chemistry, for it shows from the empirical formula of a compound how many double bonds or benzene rings must be involved in the molecule, i.e., the number of hydrogen atoms that must be added to saturate. Though a simpler relation based on equation 1 has been used to some extent,[21] the above form involving the valences of the individual atoms may be of wider utility. A more general relation results if we express E in terms of any fraction we wish (for example, $2/m$) of the polygon-based expression (equation 3) and the residual fraction $(1 - 2/m)$ of the corner-based equivalent (equation 12). Thus we can write

$$P - (2/m)(\Sigma nP_n + E_b)/2 + (1 - 2/m)\Sigma rC_r/2 + C = 1 \tag{14}$$

or, simplifying,

$$\Sigma(m - n)P_n + \Sigma[m - (m - 2)r/2]C_r - E_b = m.$$

With selected values of m, we get the following:

$$m = 2: \quad \Sigma(2 - n)P_n + 2C - E_b = 2, \tag{15a}$$

$$m = 4: \quad \Sigma(4 - n)P_n + \Sigma(4 - r)C_r - E_b = 4, \tag{15b}$$

$$m = 6: \quad \Sigma(6 - n)P_n + \Sigma(6 - 2r)C_r - E_b = 6, \tag{15c}$$

$$m = 0: \quad \Sigma rC_r - \Sigma nP_n - E_b = 0. \tag{15d}$$

These equations are true under all conditions if r has a single value throughout. Equation 15a is useful as a counterpart of equation 13, for it depends only on the total number of corners in the array and is independent of the number of edges meeting at any of these corners. Equation 15b has the virtue of assigning the same factor to both polygons and corners, while 15c gives an alternate form equivalent to the surface-tension rule of equation 8. Other relations involving combinations of polygons and corners may be derived for use when a particular type of polygon or corner is to be singled out for study.[22]

The Faces and Edges of a Polyhedron

Euler's law relates the faces, edges, and corners of a polyhedron as follows:

$$F - E + C = 2. \tag{16}$$

From this, relationships like the two-dimensional ones of equation 15 can be derived and in the same manner, giving

$$\Sigma(m - n)F_n + \Sigma[m - r(m/2 - 1)]C_r = 2m. \tag{17}$$

This does not require a boundary correction as in the previous two-dimensional treatment, for the boundary therein is now replaced by a face with E_b edges. The various special relations for particular polygons or corners are similar to those in equation 15, except for the absence of E_b, and include the relation for surface-tension corners (see note 13). These will be useful in studying separated solid polyhedral grains, the external shapes of crystal idiomorphs, and the coordination polyhedra in crystal lattices.

If one considers only polyhedra in which three faces and three edges meet at every vertex (as in separate metal grains or other bodies under surface-tension limitations), the following simple but exact and useful relations emerge, wherein $\bar{n}$ is the average number of edges to the average face for a given polyhedron:

$$F - C/2 = 2, \tag{17a}$$

$$F = 12/(6 - \bar{n}), \tag{17b}$$

$$\bar{n} = 6(1 - 2/F). \tag{17c}$$

Three-Dimensional Relations

Relations similar to the two-dimensional ones discussed above can also be developed in the three-dimensional case. The basic characteristic of a three-dimensional network of cells is

$$C - E + P - B = 1, \tag{18}$$

where C, E, and P have the same significance as before (i.e., zero-, one-, and two-dimensional cells) and B is the number of three-dimensional cells or complete polyhedra ("bodies"). This general relation allowing any number of three-, two-, or one-dimensional cells to meet at an edge or corner has too many variables to be valuable. The relations that occur when the junctions are determined by surface tension (which are of the greatest metallurgical interest) are simpler. Surface tension causes all interfaces to seek the lowest interfacial area, which means that all edges are shared by three two-dimensional and three three-dimensional cells, and all corners are formed by the junction of four bodies, four edges, and six polygons. Since each edge is shared by three polygons, a polygon of n sides will contribute $n/3$ edges to the array. We then have $E = \Sigma nP_n/3$. Also, since each edge joins two corners and there are four edges at each corner, $C = E/2$. Substituting these values in equation 18 gives

$$\Sigma nP_n/6 - \Sigma nP_n/3 + P = B + 1 \tag{19}$$

or, simplifying,

$$\sum(6 - n)P_n = 6(B + 1). \quad (20)$$

Note that this equation requires no correction for the number of edges that are shared by less than three bodies and is true even though some of the polygons are internal shared ones and others are on the periphery, belonging to only one three-dimensional cell. This relation is true only if all corners are four-valent, and the minimum number of bodies that can conform to this criterion is three. Thereafter, additional bodies added to the assembly must individually change the relation by an amount so that, for each cell,

$$\Delta\sum(6 - n)P_n = 6. \quad (21)$$

If we limit ourselves to internal cells, in which all polygons are shared by two cells, then for the whole array of B cells,

$$\sum(6 - n)P_n = 6B. \quad (22)$$

Equation 21 is related to equation 5, for if the three-dimensional array were separated into unconnected polyhedra for individual study according to Euler's law, a single polygon would provide identical faces on each of two polyhedra. There are, of course, very many polyhedra that meet this requirement, and there are almost innumerable combinations of them which can conform to the *local* requirements of surface tension as in a froth of soap bubbles. If an additional requirement is imposed, that only one type of polyhedron be used and that this by duplication fill all space, the only possibility is an assemblage of Kelvin's tetrakaidecahedra, each separated polyhedron having 6 four-sided and 8 six-sided faces.[23] Any departure from this decreases the average number of sides and corners per cell, as can be seen empirically by inserting various polyhedra between the edges of such an assembly.

Relation between Space-Filling and Surface-Tension Requirements

When the detailed requirements of local surface-tension equilibrium are superimposed on the topological relations, some interesting facts emerge. In two dimensions both the surface-tension requirements and those of topology can be satisfied with regular straight-sided hexagons, all angles being 120 degrees. In three dimensions the surfaces must meet each other in groups of three, at angles of 120 degrees, along lines which themselves meet in groups of four mutually at the angle whose cosine is −1/3, namely 109° 28′ 16″. This is the angle subtended by straight lines joining the corners of a regular tetrahedron to a point equidistant from them all and is beautifully demonstrated by the films that form on dipping a tetrahedral wire frame in soap solution (figure 1.17). No regular polyhedron with plane sides has exactly this angle between its edges. A pentagonal dodecahedron approaches it closely, for a regular pentagon has a corner angle of 108 degrees; however, it is impossible to fill space with dodecahedra, for the unit lacks the three planes of symmetry necessary for stacking. The only body possessing appropriate symmetry that simultaneously conforms to equation 22, and on stacking meets four at a point, is the truncated octahedron (cuboctahedron, figure 1.18) with 6 four-sided and 8 six-sided faces if it is considered as a separate polyhedron (though contributing three P_4's and four P_6's to the three-dimensional array of equation 22). As Lord Kelvin has shown, if double curvatures are introduced in the hexagonal faces so that the four 120-degree angles and two 90-degree angles of the six polygons that meet at a common point in a stack of plane-sided bodies all become 109.5 degrees, the body will simultaneously meet all surface-tension requirements.[23] In an array of soap films of this geometry, there would be no pressure differences between adjacent bubbles and the froth would be stable. By convention the name α-tetrakaihedron has been often restricted to this body of Kelvin's, distinguishing it from the many 14-faced polyhedra having plane faces.[24]

Kelvin demonstrated the shape of the doubly curved surfaces with zero mean curvature necessary to meet this criterion by dipping into soap solutions a wire frame in the form of a tetragonal prism with two square faces separated by a distance equal to $1/\sqrt{2}$ times their edge. The eight beautiful, curved films that join the square frame to the edges of the central convex-edged but plane quadrilateral represent half of the adjacent hexagonal faces of two minimum-area tetrakaidecahedra in contact (figure 1.19). A few experiments carried out with this tetragonal wire frame and with the simple cube of Plateau from which it is derived serve admirably to fix in the observer's mind the principles of grain shape.

Filling Space with Irregular Cells Conforming to Surface-Tension Rules of Contact

Kelvin's α-tetrakaidecahedra fill space when stacked on a body-centered cubic lattice. Local departures from this ideal arrangement of interfaces decrease the number of four- and six-sided polygons, produce many five-sided faces, and soon approach the less symmetrical, but in a way more uniform shape of the typical bubble, grain, or biological cell. That the shapes are a result of surface tension and mathematical necessity applied to the interface rather than to any innate directional crystallographic or biological growth force is obvious when the typical metal grains of figure 1.20 are compared with the cells in human fat tissue, figure 1.21.

[The writer cannot resist introducing into his picture gallery at this point an illustration purely of historical interest. Figure 1.22 is taken from P. C. Grignon's *Memoires de Physique sur l'Art de Fabriquer le Fer* (Paris, 1775), and shows the grains in a piece of wrought iron, supposedly after breaking in a hot-short range. Grignon believed the unit grain to be a 14-sided figure with 2 hexagonal and 12 quadrilateral faces. This, of course, cannot stack, even on a symmetrical ordered array, but it is at least as near an approximation to the truth as the dodecahedron which in intervening years has often been regarded as the archetype of the grain or biological cell.]

Even more striking is the stereoscopic microradiograph, figure 1.23, prepared by William M. Williams at the writer's suggestion. This shows annealed polycrystalline aluminum containing a little tin. Since the liquid tin-rich phase has a dihedral angle at the annealing temperature a little below 60 degrees, it spreads continuously along the edges of the grains in three dimensions and delineates the true shape of the grain, leaving the faces undisturbed. The slight trouble involved in using a stereoscope on this microradiograph will be amply rewarded by a view, heretofore impossible, of the real shape and interrelation of metal grains. They are virtually indistinguishable from a three-dimensional soap froth such as shown in figure 1.14, a fact which further supports the belief that interfacial tensions are the most important factors in determining microstructures.

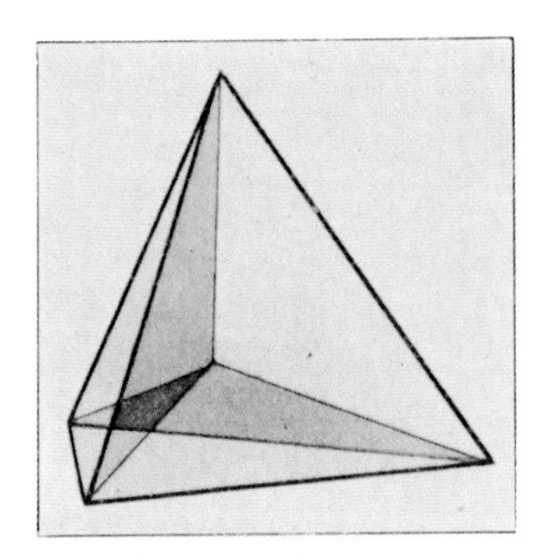

Figure 1.17
Soap bubbles in a tetrahedral wire frame. Drawing by C. S. Barrett.

Figure 1.18
A group of regular truncated octahedra. Drawing by F. T. Lewis.

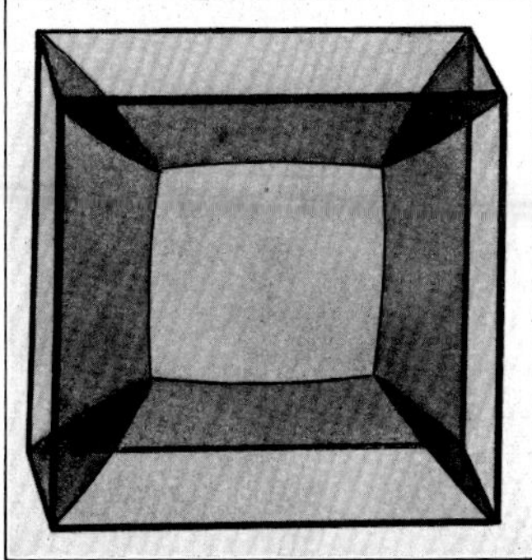

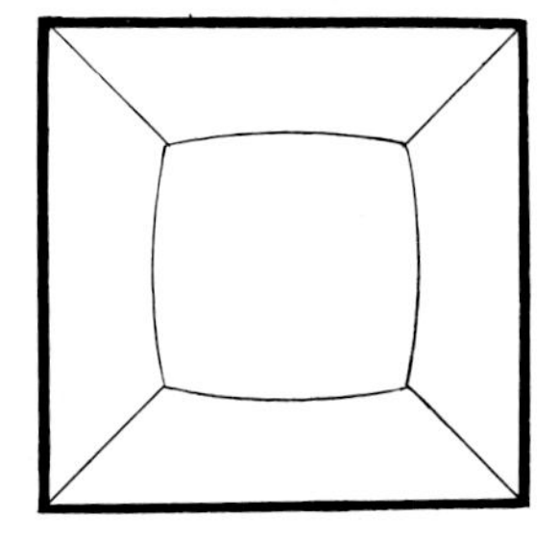

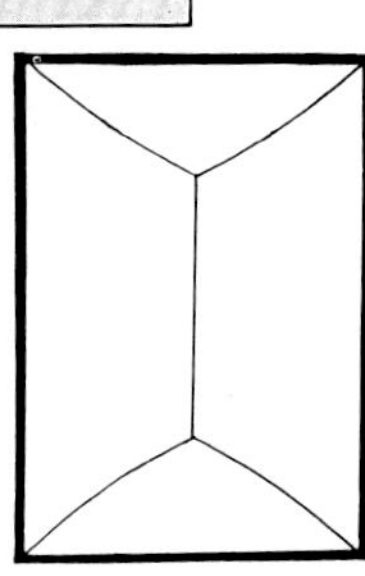

Figure 1.19
(Lower) Projection of soap films in Kelvin's wire frame in the shape of a tetragonal prism having an axial ratio $1/\sqrt{2}$. (Upper) View in perspective. Drawings by C. S. Barrett.

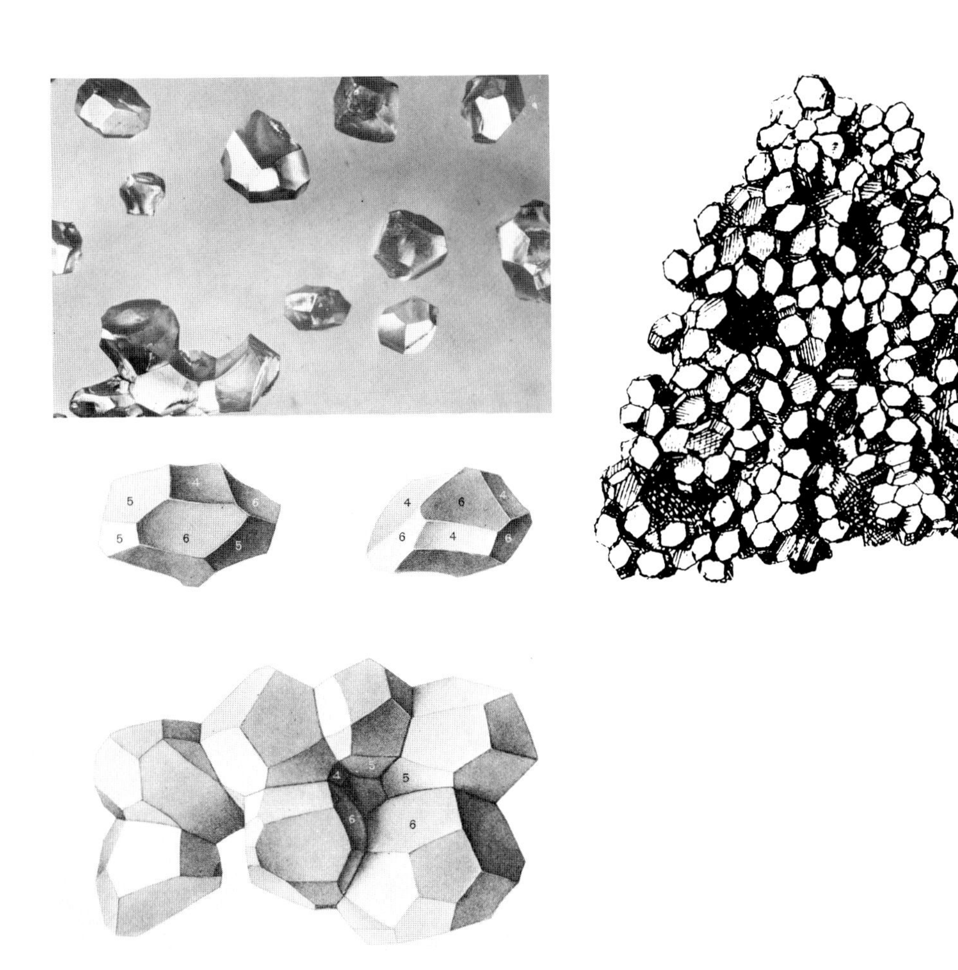

Figure 1.20
Grains of beta brass, separated by crushing a fine-grained casting in the brittle range at about 400°C.

Figure 1.21
Models of cells in human fat (×275). Reproduced by permission of F. T. Lewis (see note 2b) and the American Academy of Arts and Sciences.

Figure 1.22
Aggregate of iron grains as pictured by P. C. Grignon (1775).

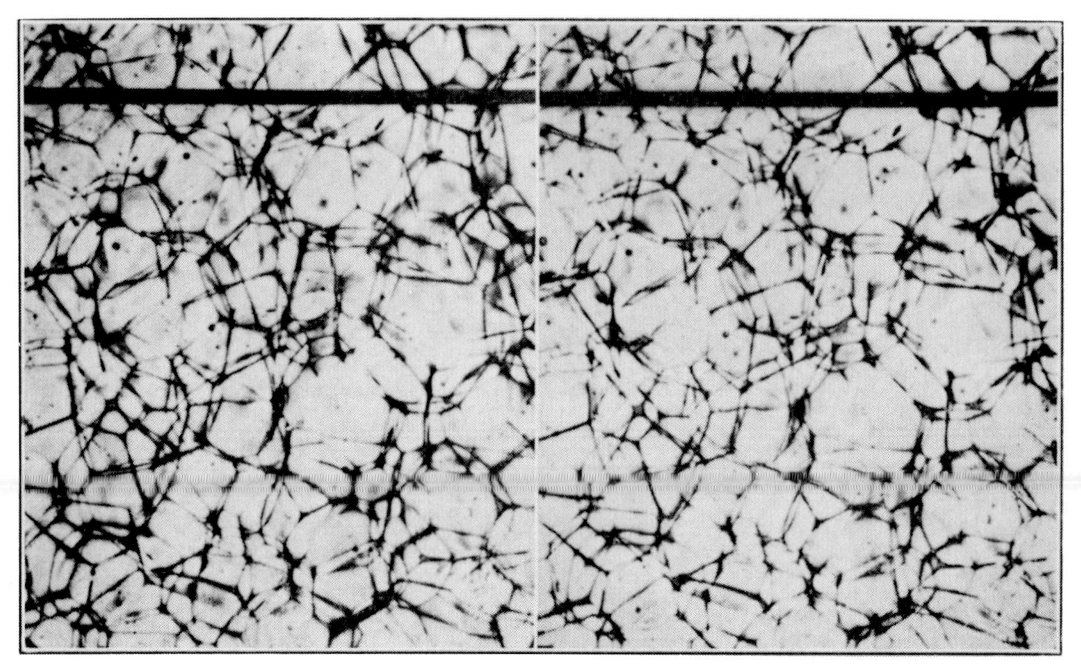

Figure 1.23
Stereoscopic pair of microradiographs showing actual shape in three dimensions of grains in aluminum-tin alloy (2 volume % Sn) annealed 18 hours at 575°C (×25). Photo by W. M. Williams.

It is now obvious that, starting from an ordered array of tetrakaidecahedra, but relaxing the requirement of symmetry, while retaining those for space-filling and surface-tension contacts, neighboring polyhedra may differ from each other, and innumerable choices of cell and neighbors become possible. They are limited only by the requirement that they must individually conform to equation 21, collectively to equation 22 (if peripheral cells are omitted), and the whole array must conform to equation 20. However, any departure from the ideal stacking will decrease the average number of both faces and edges associated with each three-dimensional cell, as well as the number of edges per face.

It can be shown from equation 20 that the average number of corners and faces per three-dimensional cell and the average number of edges per face ($\bar{n}$) are related as follows:

$$C/B = \bar{n}/[(6 - \bar{n}) - 6/P], \quad (23)$$

$$P/B = 6/[(6 - \bar{n}) - \bar{n}/C], \quad (24)$$

$$\bar{n} = 6[1 - (B + 1)/P], \quad (25)$$

$$P - C = B + 1. \quad (26)$$

If a large number of cells is considered, so that the effects of the peripheral cells are unimportant, these relations simplify to the following:

$$C/B = \bar{n}/(6 - \bar{n}), \quad (23a)$$

$$P/B = 6/(6 - \bar{n}), \quad (24a)$$

$$\bar{n} = 6(1 - B/P), \quad (25a)$$

$$P/B - C/B = 1. \quad (26a)$$

The minimum interface surface for a given number of cells is obtained when there is a maximum number of faces shared between cells without breaking contact between previously established ones. This is equivalent to saying that there should be a maximum number of corners (or edges) for a given number of bodies or, quantitatively from equation 23a, that $\bar{n}/(6 - \bar{n})$ should be as large as possible. It is empirically true, and supposedly can be proved mathematically, that the only physically realizable solutions correspond to 6 or less corners per body on the average.[25] When C/B is exactly 6, P/B is 7 (corresponding respectively to 24 corners and to 14 sides on each cell when considered as a separate polyhedron) and $\bar{n}$ is exactly $5\frac{1}{7}$, provided that peripheral cells are avoided, as in the derivation of equations 23a to 26a. If all cells are counted, $\bar{n}$ tends toward $5\frac{1}{7}$ as a larger number of cells is involved and the peripheral cells become less important. This rather unexpected number, $5\frac{1}{7}$, is the three-dimensional equivalent of the hexagon in two dimensions. It is the average number of sides to the polygonal faces of polyhedra that fill space with a minimum number of interfaces. In the Kelvin tetrakaidecahedra, it results from the average of six quadrilateral and eight hexagonal faces. Only one other simple body seems to be known that can by itself fulfill this relation—the polyhedron consisting of twelve pentagonal faces and two hexagonal faces discussed by Goldberg.[26] But this, as well as possible polyhedra with multiples of fourteen sides, lacks the symmetry needed for stacking. A soap froth or an array of metal grains has an infinitesimal probability of making random contacts in the manner dictated by the Kelvin body, and in reality consists of an assemblage of many kinds of polyhedra conforming exactly to the relations of equations 20 to 26 and tending toward the values $C/B = 6$, $P/B = 7$, and $\bar{n} = 36/7$, though never attaining them, even if peripheral cells are not counted. The more nearly they conform to these relations, the greater the number of

Table I Summary of shape determinations on bubbles, vegetable cells, and metal grains

	600 uniform bubbles (0.1 or 0.2 cc)[a]		100 small bubbles (0.05 cc) in mixture[b]		50 large bubbles (0.4 cc) in mixture[b]		Mixture of 50 large and 100 small bubbles[b]		450 vegetable cells[a]		30 beta brass grains[c]		Tetra-hedron	Cube	Pentagonal dodeca-hedron	Rhombic dodeca-hedron	Kelvin tetrakai-decahedron	Goldberg tetrakai-decahedron
	$2P_n$	%	$2P_n$	%	$2P_n$	%	$2P_n$	%	$2P_n$	%	$2P_n$	%	F_n					
cells with $n = 3$	...	...	...	...	...	...	...	...	32	5.1	11	2.5	4	...	...	...	...	...
4	866	10.5	318	32.9	115	11.3	433	22.9	1694	27.3	88	20.2	...	6	...	12	6	12
5	5503	67.0	564	58.1	498	48.1	1062	56.1	2465	39.7	190	43.6	...	...	12	...	...	...
6	1817	22.1	86	8.9	288	28.3	374	19.8	1575	25.4	123	28.7	...	...	...	...	8	2
7	35	0.4	...	...	113	11.2	113	6.0	390	6.3	20	4.6	...	...	...	...	...	...
8	...	...	...	...	6	5.9	6	0.3	50	0.8	3	0.7	...	...	...	...	...	...
9	...	...	...	...	1	1.1	1	0.05	5	0.1	...	...	...	...	...	...	...	...
$\sum 2P_n$	8221		968		1021		1989		6211		435		4	6	12	12	14	14
$\bar{n}$	5.111		4.76		5.412		5.095		5.123		5.142		3	4	5	4	5.143	5.143
P/B	6.851		4.840		10.210		6.630		6.901		7.250		2	3	6	6	7	7
C/B	5.851		3.840		9.210		5.630		5.893		6.213		1	2 (1)	5	4 (3)	6 (6)	6
$6 - C/B$	0.149		2.160		−3.210		0.370		0.107		−0.213		5.0	4.0	1.0	2.0	0.0	0.0
$P/B - C/B$	1.000		1.000		1.000		1.000		1.008		1.037		1.0	1.0	1.0	2.0	1.0	1.0

[a] Matzke (note 3a).
[b] Matzke and Nestler (note 3b). Note that the counts for mixed bubbles are adjusted to conform to the correct volumetric mixture of large and small bubbles.
[c] Desch (note 6).

shared faces and the lower the total interface area will be. The number $6 - C/B$ can therefore serve as a criterion of additional surface and hence grain growth tendency.[27]

Experimental Observations on Shape of Bubbles, Grains, and Biological Cells

Starting with the work of C. H. Desch,[6] D'Arcy Thompson,[5] F. T. Lewis,[2a–f] E. B. Matzke,[3] G. van Iterson,[2g] and others have experimentally examined the shape of various systems in which surface-tension forces were suspected. The overwhelming importance of the five-sided face has been observed as well as the approach to fourteen sides for the average separated polyhedron, but no attempts have been made to apply topological principles in three dimensions, and the significance of the number $5\frac{1}{7}$ seems not to have been realized. Even the shapes of compressed lead shot studied by Marvin (unless symmetrically arrayed in close packing when they give rhombic or trapezoidal dodecahedra) do not show extreme deviation from the same relations—a surprising fact until one realizes that random assemblages of cells not enormously differing in size will only rarely give contacts with more than four cells at a point.[4]

Matzke made a most painstaking study of the shape of soap bubbles, reporting the number of faces of all types on 600 cells in a froth of bubbles of uniform size, and later, with Nestler, he examined 300 bubbles in a nonuniform assembly.[3] They give statistical data on the various polyhedron and polygon shapes, and include drawings of the commonest types of polyhedra. Their data on nonperipheral bubbles are listed in table I, together with their summary of earlier measurements on vegetable cells, Desch's values

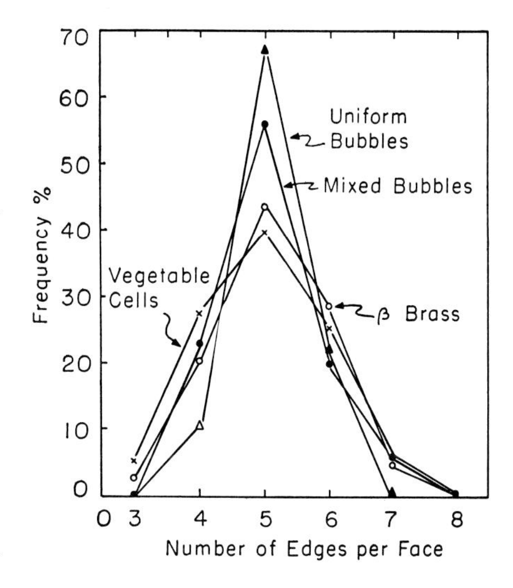

Figure 1.24
Frequency of various polygonal faces in grains, cells, and bubbles.

for beta-brass grains, and equivalent values for some simple polyhedra of both stackable and unstackable kinds. These are summarized in figure 1.24. Together these provide a useful illustration of the correctness and utility of the principles developed in the previous section of the present paper. Table I includes newly computed values of the average number of corners and polygons per cell, the average number of edges per face, and other criteria suggested by equations 23 to 26. The frequencies of different polygons are listed as the actual number of faces observed on bubbles, grains, or cells counted as separate polyhedra, which is equivalent to twice the value of P in the equations. All the computed functions assume three-dimensional arrays wherein each polygon is shared by two three-dimensional cells, and each corner by four.

It should be noted that the value of $P/B - C/B$ is exactly one when the separate polyhedra have only three-ray vertices, regardless of their stacking potentialities or the number of faces that meet at a point in a stack. This function, which is the same as $P/B[1 - (\bar{n}/6)]$, serves only to check conformity with equations 5 and 21. On the other hand, the value of $[6 - (C/B)]$, or $[7 - (P/B)]$ which is identical, is an index of departure from the maximum possible sharing of faces and corners, and is zero with ideal stacking.

The numbers of corners given in parentheses under the stackable regular polyhedra in table I indicate the value of C/B that obtains in an ordered stack of such polyhedra, with the appropriate number of them (for example, 8 with cubes) meeting at a point. The first number, as in all other columns, gives the number of corners that would be contributed by a single polyhedron if it were surrounded by others in such a way that all corners are shared by three other polyhedra.

Though Matzke did not invoke any topological principles, his observations are clearly precise, for every one of his hundreds of polyhedra conforms to the $(6 - n)$ rule and polygons outnumber corners by exactly one as they must under surface-tension conditions. It cannot, however, be proved that his sampling was correct, for he apparently did not study every bubble in a given volume. Desch, on the other hand, clearly did not use a true sample, and must have selected grains with more than the average number of sides, a natural enough result of choosing well-formed grains for examination. Further studies of grain shape in metals are badly needed, for if five-grain corners are present in significant fraction, they would explain the diminution of grain growth rate as the size increases.

It should be apparent from the above discussion that there is considerable freedom in space filling under the given conditions, and that no one shape can possibly be regarded as that of the "typical" grain unless the grains are arranged with the symmetry of a lattice. Pentagons are clearly by far the most common grain face, though they cannot constitute the only polygons. The average number of sides to the separated polyhedra will approach fourteen. There are, however, innumerable combinations that in the total assembly will meet the requirements once the limitation is removed that all units be the same. With grains of nearly uniform size, polygons with other than four to six sides will be rare, as will grains deviating greatly from fourteen sides. If the grains differ in size, clearly the large grains will, in general, have more sides and the smaller ones less, the average deviating negatively from fourteen as the discrepancy in size increases. It is obvious that as far as local conditions at a single corner are concerned, the junction of six pentagons at a point represents a better conformity with surface-tension conditions than does the interaction of squares and hexagons called for by the tetrakaidecahedron. Moreover, random contacts of bodies of approximately uniform size would tend to give faces more nearly of equal area than those of the 14-sided body. This conflict between the local and the overall solutions of the surface-tension requirements renders it highly unlikely that the ideal solution will occur, and it simultaneously makes local departures energetically not very costly. Despite the great difference in appearance between a random froth of equivolume bubbles and that of the ideal array, there is actually not very much difference between the curvature of the interfaces; indeed the lack of doubly curved surfaces even gives the former a seemingly more stable appearance, for it allows the films to be more nearly plane and the cell edges more nearly straight. The curvature cannot, of course, be zero and, as bubbles depart from uniformity in volume, it becomes increasingly high.

It is interesting to consider the applicability of the relations of equations 21 to 24 in another nonmetallic field, that of glass. A true glass is not a supercooled liquid, but an extended network structure in which the local interatomic bonds and geometry are satisfied almost as well as in a crystal, though the assembly lacks long-range symmetry. The analogy between a froth of rather uniform-sized bubbles and a glass is almost perfect. Compare, for example, the two-dimensional A_2O_3 "crystal" (figure 1.25) with the tracing of a soap froth with "atoms" superimposed (figure 1.26) and the latter with the classic drawing of a two-dimensional A_2O_3 glass structure (figure

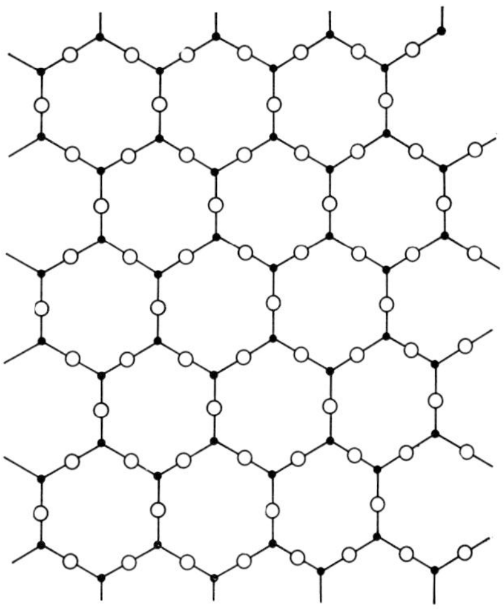

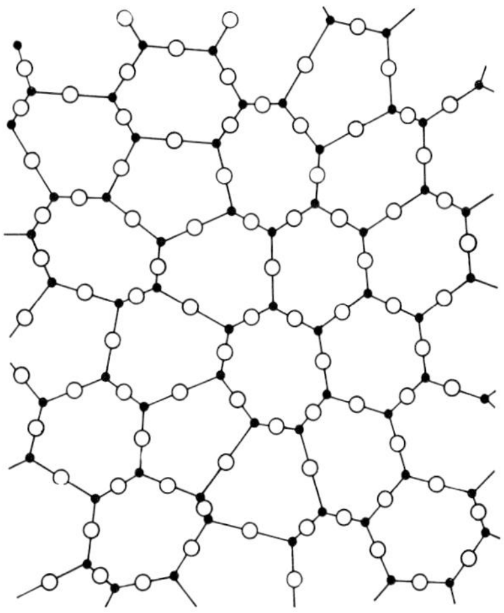

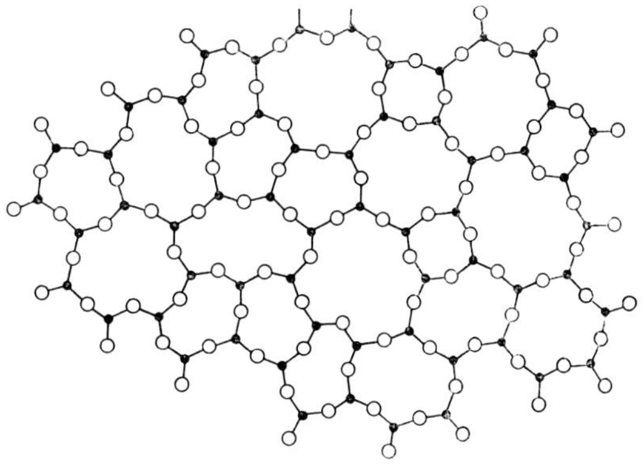

Figure 1.25
Hexagonal lattice of a compound of type A_2O_3.

Figure 1.26
Glasslike network having a composition A_2O_3, drawn by tracing over a two-dimensional froth of soap bubbles.

Figure 1.27
Structure of A_2O_3 glass proposed by Zachariasen.[28]

1.27) by Zachariasen.[28] In three dimensions the angle corresponding to the tetrahedral coordination of oxygen ions around silicon in an SiO_2 glass is almost exactly matched by the junction of four soap bubbles, and the structure as a whole must be a frothlike assemblage, principally of pentagons with fewer hexagons and occasional tetragons, more nearly of equal area and with edges of less curvature than in a crystalline tetrakaidecahedral array. The tolerance for foreign cations would depend upon the number and distribution of polygons with various numbers of sides, and the effect of temperature is undoubtedly to increase the spread in polygon distribution so as to cause a lower density of packing. The soap froth analogy can go even further, for a model such as that shown in figure 1.15 of uniform bubbles but at atmospheric pressure so as to be stable and with arrangements for displacing at least one side of the cell serves also as a model to demonstrate the viscous deformation of glass. The manner in which various bubbles in this model change alignment of neighbors under an external directed stress must be closely akin to the manner in which an SiO_2 tetrahedron transfers its neighbors in a flowing glass. Similar structural principles may also apply to simple organic polymers.

One- and Two-Dimensional Sections of Three-Dimensional Structures

In considering the relation between the area of a grain boundary and the volume of the grains, metallurgists have in the past made various assumptions as to grain shape. Actually this is not necessary, for there exist simple and exact relations between the average random intercepts of a line or a plane and the ratio of line length to area in two dimensions or of surface area to volume in three dimensions.

In a two-dimensional array containing lines of total length l within an area A, with $\overline{N}'$ representing the average number of intercepts with these lines of all possible random lines of average length $\overline{L}'$ which completely traverse this area, then

$$\overline{N}'/\overline{L}' = 2l/\pi A. \qquad (27)$$

Now consider a three-dimensional array of two-dimensional surfaces having a total area S within an isolated volume V, which is to be studied. The surfaces will terminate in lines, and there may be other lines with no relation to any surface. Let the total one-dimensional line length be λ. Now traverse this array with random lines and with random planes in all possible directions. Let L be the length of a given intersecting line and N the number of intercepts that it makes with the surfaces; similarly let A be the area of an intersecting plane, l the total length of its intercepts with the surfaces, and n the number of points where the λ lines intersect it. $\overline{N}$, $\bar{n}$, $\overline{L}$, $\bar{l}$, and $\overline{A}$ represent the average of these quantities for all possible intersections. The following relations then hold:

$$\overline{N}/\overline{L} = 2\bar{l}/\pi\overline{A} = S/2V, \qquad (28)$$

$$\bar{l}/\overline{A} = \pi S/4V, \qquad (29)$$

$$\bar{n}/\overline{A} = \lambda 2V, \qquad (30)$$

$$\bar{n}/\bar{l} = 2\lambda/\pi S, \qquad (31)$$

$$\bar{n}\ \overline{L}/\overline{A}\overline{N} = \lambda/S. \qquad (32)$$

These relations depend on the probability of intersection of surfaces and lines and are exactly true as long as all possible random sections are averaged.[29] No assumptions are necessary as to actual shape or disposition of the surfaces or lines.

To the metallurgist, perhaps the most useful of these relations will be 28, which gives the true ratio of grain boundary surface to volume by simple linear intercept counting on a random two-dimensional section. Relations 31 and 32 use the ratio of number of grain corners to the length of grain boundary trace in a two-dimensional slice to give the ratio of perimeter to area of the faces of the three-dimensional grains. This number usefully supplements the purely topological number of equation 25, which requires actual three-dimensional study. (In both relations care must be taken to see that the sharing of edges by adjacent polygons is properly considered.) By appropriate subdivision of N, n, and l into the intercepts with the boundaries of various phases, the characteristics of each phase in a polyphase or porous structure can be separately studied. If, in addition, the fractional linear intercept method of determining relative volumes of various constituents is used, a virtually complete analysis of the structure is possible with only two-dimensional sections.[30] Anisotropy in shape and size may be studied by measurements on sections taken in various directions.

Note that the average grain intercept, $\bar{L}/\bar{N}$, is the true average diameter of the three-dimensional grains. It is perhaps a better figure for reporting grain sizes than the square root of the average grain area on a two-dimensional section, as commonly used following the Jeffries method.

The application of these principles to the study of cell shape and other biological problems, as well as to similar problems in other fields, will be obvious.[31]

Acknowledgments

The author wishes to acknowledge helpful discussions on topology with N. H. Kuyper in 1948. Saunders MacLane kindly read an early draft of the manuscript and uncovered some mathematically loose statements, though he is in no way responsible for any errors that may remain.

Notes and References

1
See, for example, O. Veblen, *Analysis Situs* (New York, 1931), and R. Courant and H. Robbins, *What Is Mathematics?* (New York, 1941). The latter is an excellent elementary introduction to topological reasoning.

2
(a) F. T. Lewis, "The typical shape of polyhedral cells in vegetable parenchyma and the restoration of that shape following cell division," *Proceedings*, American Academy of Arts and Sciences, 58:537–552 (1923).

(b) F. T. Lewis, "A further study of the polyhedral shapes of cells," *Proceedings*, American Academy of Arts and Sciences, 61:1–34 (1925).

(c) F. T. Lewis, "The significance of cells as revealed by their polyhedral shapes, with special reference to precartilage, and a surmise concerning nerve cells and neuroglia," *Proceedings*, American Academy of Arts and Sciences, 68:251–254 (1933).

(d) F. T. Lewis, "The geometry of growth and cell division in columnar parenchyma," *American Journal of Botany* 31:619–629 (1944).

(e) F. T. Lewis, "The shape of cells as a mathematical problem," *American Scientist* 34:359–369 (1946).

(f) F. T. Lewis, "The analogous shapes of cells and bubbles," *Proceedings*, American Academy of Arts and Sciences, 77:147–186 (1949).

(g) G. van Iterson and A. D. J. Meeuse, "The shape of cells in homogeneous plant tissues," *Proceedings*, Koninklijke Nederlandsche Akademie van Wetenschappen, 44:770–778, 897–906 (1941).

(h) N. Higinbotham, "The three-dimensional shapes of undifferentiated cells in the petiole of *Angiopteris evecta*," *American Journal of Botany* 29:851–858 (1942).

(i) R. L. Hulbary, "The influence of air spaces on the three-dimensional shapes of cells in *Elodea* stems, and a comparison with pith cells of *Ailanthus*," *American Journal of Botany* 31:561–580 (1944).

(j) J. W. Marvin, "Cell shape and cell volume relations in the pith of *Eupatorium perfoliatum* L.," *American Journal of Botany* 31:201–218 (1944).

(k) E. B. Matzke, "Three-dimensional shape changes during cell division in the epidermis of the apical meristem of *Anacharis densa* (*Elodea*)," *American Journal of Botany* 36:584–595 (1949).

3

(a) E. B. Matzke, "The three-dimensional shape of bubbles in foam: An analysis of the role of surface forces in three-dimensional cell shape determination," *American Journal of Botany* 33:58–80 (1946).

(b) E. B. Matzke and J. Nestler, "Volume-shape relationships in variant foams: A further study of the role of surface forces in three-dimensional cell shape determination," *American Journal of Botany* 33:130–144 (1946).

4

J. W. Marvin, "The shape of compressed lead shot and its relation to cell shape," *American Journal of Botany* 26:280–288 (1939). Johannes Kepler had made similar, though qualitative, observations in 1611, and so had Stephen Hales in his *Vegetable Staticks* (1727).

5

D'Arcy W. Thompson, *On Growth and Form,* 2nd edition, Cambridge, 1942.

6

C. H. Desch, "The solidification of metals from the liquid state," *Journal,* Institute of Metals, 22:241–263 (1919).

7

D. Harker and E. Parker, "Grain shape and grain growth," *Transactions,* American Society for Metals, 34:156–195 (1945).

8

C. S. Smith, "Grains, phases and interfaces: An interpretation of microstructure," *Transactions,* American Institute of Mining and Metallurgical Engineers, 175:15–51 (1948).

9

E. Scheil, "Statistical investigations of the structures of alloys," *Zeitschrift für Metallkunde* 27:199–208 (1935), 28:340–343 (1936). W. A. Johnson, "Estimation of spatial grain size," *Metal Progress* 49(1):89–92 (1946). J. B. Rutherford, R. H. Aborn, and E. C. Bain, "Relation of grain area on a plane section and the grain size of a metal," *Metals and Alloys* 8:345–348 (1937); *ASM Metals Handbook* (1948), p. 405.

10

This is, in fact, a definition of dimensionality. Two cells (i.e., lines) in one-dimensional space can be separated by a point, three cells (polygons) in two dimensions, four cells in three dimensions; in general, $n + 1$ cells constitute the minimum number of cells that must meet at a point (a zero-dimensional cell) if the space is to be subdivided into an extensive array of separate cells in n dimensions.

11

This relation becomes identical with the better-known Euler's law applying to the surface of a polyhedron ($P - E + C = 2$) if the area outside the array is counted as a single polygon covering the rest of the polyhedron surface. For present purposes it seems better to consider the isolated two-dimensional array dealt with by the characteristic noted in the text. Note that the relation does not apply to a single endless line: at least one corner is needed, as on an Euler polyhedron. A more general mathematical form is $N_0 - N_1 + N_2 - N_3 + \cdots + (-1)^n N_n = 1$.

12

Thompson (*On Growth and Form,* pp. 596–598) shows the 12 possible assemblages of 8 noninsular cells and quotes M. Brüchner as showing that with 13 cells there are 50,000 possible arrangements, and with 16, nearly 30 million.

13

Another method is to consider the array of grains as a portion of the surface of a polyhedron, each grain constituting one face, with an additional face having E_b sides completing the surface of the polyhedron. Invoking Euler's law and utilizing the relation $2E = 3C$ as in the determination of equation 5, one then gets

$$\sum(6 - n)F_n = 12,$$

where F_n is the number of faces with n sides. This derivation was used by D'Arcy Thompson, but it seems less immediately useful when considering limited two-dimensional networks than the present treatment in which the boundary is clearly treated as such.

14
W. C. Graustein, "On the average number of sides of polygons of a net," *Annals of Mathematics* 32:149–153 (1932).

15
Since the average grain must be a hexagon, it is possible to replace the grains within a given boundary with hexagons alone. Conversely, it is possible to fit the group of grains into a hole cut in an extensive set of hexagons and maintain exact topological conformity with the boundary. Such an arrangement is shown in figure 1.6C.

16
It is well known that the energies of grain boundaries are not independent of the relative orientation of grains or of the orientation of the boundary between two fixed grains. Statistically, however, most boundaries will have large orientation differences, and the energies will be relatively unaffected by small changes of orientation. The equal-energy assumption is therefore acceptable as a first approximation.

17
J. Plateau, *Statique Expérimentale et Théorique des Liquides Soumis aux Seules Forces Moléculaires,* two volumes, Ghent, 1873.

18
Lord Kelvin, "On the division of space with minimum partitional area," *Philosophical Magazine* 24:503–514 (1887).

19
This model was first made in 1947, and though it had been frequently demonstrated, this was its first appearance in print. It is easy to make. Any aqueous soap solution or dilute liquid detergent seems to be satisfactory. The glass containers should not be too large, since soap films are not mechanically stable if their minimum dimension exceeds about 1 inch. The liquid should occupy about a third of the volume. It does not show in the illustrations, as these have been photographed vertically from above. The cells should be evacuated nearly to the vapor pressure of the solution before sealing, or the rate of change of grain size will be very small. Boiling for a few minutes with the container at about 50°C and connected to a rough vacuum line generally suffices. Too good a vacuum makes it impossible to produce a froth. Once a cell is sealed, it will last indefinitely.

20
H. C. H. Carpenter and C. F. Elam, "Crystal growth and recrystallization in metals," *Journal,* Institute of Metals, 24:83–131 (1920).

21
J. K. Senior, private communication.

22
Two sections of the original paper that dealt with metallurgical problems are of limited interest and are here omitted.

23
Lord Kelvin, "On the homogeneous division of space," *Proceedings*, Royal Society of London, 55:1 (1894).

24
(*Note added 1980*) R. E. Williams in his paper "Space-filling polyhedron: Its relationship to aggregates of soap bubbles, plant cells, and metal crystallites" (*Science* 161:276–277, 1968) describes a β-tetrakaihedron having two quadrilateral, four hexagonal, and eight pentagonal faces, replication of which will pack in associated pairs to fill space. Its faces are more nearly plane than the hexagons on the Kelvin body, from which it is obtained by slight distortion. The distribution of polygons among its faces is close to that in random foams (cf. figure 1.24), but it remains a hypothetical archetype. Cells having this particular symmetry are encountered extremely rarely in practice.

25
I must apologize for passing over this essential and basic point so lightly. The empirical truth can be shown by experiments in which additional polyhedra are inserted in a stack of Kelvin bodies; rigorous proof must await the services of a mathematician.

26
M. Goldberg, "The isoperimetric problem for polyhedra," *Tohoku Mathematical Journal* 40:226–236 (1934).

27
A better approach to the optimum value of $\bar{n}$ was given in a later paper (number 129 in the bibliography). What hypothetical plane polygon would have corner angles equal to 109.471. . . degrees, the angle of the three-dimensional surface-tension junction? Since the sum of corner angles in a plane polygon is $n\theta = \pi(n - 2)$, we have $n = 2/(1 - \theta/\pi)$; with the above value of θ, n becomes 5.10430. . . . This is an irrational number, so there can be no plane-faced polyhedron meeting the three-dimensional requirements, though the Kelvin body, with its $5\frac{1}{7}$ sides, comes close. In actual random foams of approximately uniform bubbles, five sided faces with only slight curvature are common. Extending the above argument to entire polyhedra gives an "ideal" cell having 13.394. . . faces, 34.195. . . edges, and 22.789. . . vertices. (See also note 24.) H. S. M. Coxeter arrived at a similar solution by a different route. He believes that the Kelvin minimum-area cell is a misleading ideal, and that in practical foams n should be closer to 5.1043 than to the Kelvin body's 5.1425. All such foams must, however, involve some, though perhaps small, curvature and have a pressure difference between adjacent cells, and the Kelvin body is the only known configuration that could be stable in systems in which diffusion occurs.

28
W. H. Zachariasen, "The atomic arrangement in glass," *Journal,* American Chemical Society, 54:3841–3851 (1932).

29
I would like to acknowledge the help of Lester Guttman and Saunders MacLane in establishing the correct form of these relations (for their derivation see paper number 80 in the bibliography). At the time of writing I was unaware that similar relations had been published by the Russian metallurgist S. A. Saltyakov. They are the basis of the presently burgeoning field of quantitative microscopy, so well adapted to modern computerized equipment, and they gave birth to a new field and a new international organization, the Society for Stereology. The present state of the field is well presented by E. E. Underwood in his book *Quantitative Stereology* (Reading, MA, 1970). The journal *Metallography* continues to publish papers on quantitative aspects of microscopic morphology.

30
R. T. Howard and M. Cohen, "Quantitative metallography by point-counting and lineal analysis," *Transactions,* American Institute of Mining and Metallurgical Engineers, 172:413–426 (1947).

31
(*Note added 1980*) An outmoded section of purely metallurgical interest has been omitted.

In his discussion of this paper John von Neumann showed that:

In a two-dimensional bubble-froth the total gas-gain-rate of any bubble is (positively) proportional to $n - 6$, where n is the number of sides of the bubble (i.e., of its bounding circular-arc-polygon). The (positive) coefficient of proportionality depends only on the general properties of the froth and of its containing "Flat Cell." Thus every hexagonal bubble (irrespective of further details of shape!) has a constant-gas-content; every pentagonal bubble loses gas at the same rate; every heptagonal bubble gains gas at the same rate as the pentagonal ones lose it; every tetragonal (octagonal) bubble loses (gains) gas at twice the rate at which pentagonal (heptagonal) bubbles lose (gain) it; etc. Note that these results apply only to the continuous changes of gas-content due to diffusion.

The bubble-shape problem is of considerable interest to architects and designers. See, for example, R. E. Williams, *Natural Structure* (Morpark, CA, 1972); P. Pearce, *Structure in Nature Is a Strategy for Design* (Cambridge, MA, 1978); and C. Thywissen, *Wachsende und sich teilende Pneus: Growing and Dividing Pneus* (sic) (number 19 of the *Mitteilungen des Instituts für Leichte Flachentragwerke,* University of Stuttgart, 1979).

2
The Discovery of Carbon in Steel

Early Concepts of the Nature of Steel

No metallic material has had more influence upon man's history than iron and its simple alloy with carbon, steel. As more archaeological sites are explored in eastern Anatolia and contiguous areas of Iraq and Iran, man's first knowledge of iron is carried farther and farther back. The earliest man-made iron we now know of dates from about 2800 B.C. Iron did not, however, become important until methods of relatively large-scale production had been developed, supposedly by the Hittites, and it did not challenge the superiority of bronze (save in the important economic aspect of its availability) until smiths had found how to convert soft iron into that magnificent material, steel, which can be relatively easily worked to shape and metamorphosized by a final heat treatment into a material of strength and hardness almost unsurpassed. Steel's combination of availability, fabricability, and final extreme or adjustable hardness sets it apart from all other materials.

Though a few details of the hardening mechanism are still a little uncertain, it is well established that steel differs in composition from pure iron essentially only by the presence of a small amount of carbon (between about 0.2 and 1.5 percent)[1] and that the hardness derives from a transformation in the arrangement of atoms in the iron crystal which occurs in the pure metal at a definite temperature, but both the temperature and the residual structure are modified by the presence of the carbon. The structure and the resulting properties depend greatly upon cooling rate, and great hardness results only after quenching or other very rapid cooling which suppresses a softening reaction that occurs at high temperatures but does not prevent a structural change to the hard form which occurs about 20°C.

Through most of history steel meant a material capable of being given a considerable hardness by quenching.[2] It was the material for tools and swords, not for general structural application.

Once man had discovered the startling effects of fire on materials, he was prompt and ingenious in discovering a wide range of alloys—even before the beginning of the first millennium B.C., he had exploited practically every useful metal or alloy that can be made by carbon reduction of readily identifiable ore minerals. The relation between properties and composition (i.e., mixed selected raw materials) was fairly clear in the case of the bronzes, the precious metal alloys, the hard and soft solders, and even brass. The fact that steel was also an alloy was not so clear; indeed, it was not definitely accepted until the very end of the eighteenth century A.D., 3000 years after the practical discovery. This knowledge arose out of and contributed to the Chemical Revolution in an intimate way.[3]

The source of the difficulty in recognizing steel as an alloy is not hard to find: it is simply that the alloying element is carbon, and carbon was the inevitable accompaniment of any metallurgical operation, first as a fuel to provide heat by its combustion, second, though more obscurely, as a reducing agent to remove the oxygen from the ore. A third role, which it plays *only* in the case of iron among all the ancient metals, is to dissolve in the reduced metal, profoundly modifying its properties.

Oxygen's role in combustion and calcination remained unsuspected until discovered independently by Scheele and by Priestley in 1772–1774. Equally important was the realization that charcoal was not just the source of some vague principle but that it consisted mostly of a material chemical element (soon to be named carbon), which in the age-old reduction of metal ores combined with the previously unsuspected oxygen to produce "fixed air." It was during the excitement over the discovery of the new gases and of their reaction and of the new antiphlogistic explanation of reduction that the role of carbon in distinguishing steel from iron was realized. When "minus oxygen" placed "phlogiston" in calcination-reduction reactions, a comparable material explanation had to be invoked in connection with the diverse properties of the different siderurgic materials. From then on, the extension of this knowledge to the improvement of steel becomes applied science. In its early stages, however, the science derived far more from technology than did technology from science.

With such profusion of charcoal in the smith's hearth it was unlikely that the minor amount dissolving in the metal itself would be at first appreciated. Some people regarded steel as a different metal, having its own ore and manner of working.[4] Aristotle, and many who followed him, believed that steel was actually a purer form of iron, iron carried more nearly to the state of metallic perfection,[5] though its brittleness belied this and, when it was still further "purified," as cast iron, it had lost all malleability though it had gained the essential metallic property of fusability. Biringuccio summed this up in 1540 by saying that

> steel is nothing other than iron, well purified by means of art and given a more perfect elemental mixture and quality by the great decoction of the fire. By the attraction of some suitable substances in the things that are added to it, its natural dryness is mollified by a certain amount of moisture and it becomes whiter and more dense so that it seems almost to have been removed from its original nature.[6]

This was not entirely illogical, since steel was produced by prolonged heating in a fire, and purification by fire is a common experience and provides a popular figure of speech. The earliest steels were probably made simply by holding iron for prolonged periods beneath charcoal in a hearth. The famed Indian steel wootz, the basis of the Damascus sword, was made by heating iron to fusion in crucibles with pieces of wood, but this suggestive process did not penetrate Europe until after 1800. When cast iron from the blast furnace became available, steel was made either by its partial refining in the hearth or by immersing wrought iron in molten cast iron (a process which may even have encouraged the development of the blast furnace),[7] but cast iron is just another mystery coming from the furnace when operated in a particular way. In the cementation process (which as a production process for massive steel is probably a sixteenth-century development, though case-hardening is earlier), iron is converted to steel by prolonged heating in boxes packed with charcoal or other carbonaceous material. This, too, seemed unmistakably to be a purification process analogous to the analytically proven purification of gold by a similar process bearing the same name.

Iron is distinguished markedly from the precious metals by the lack of early literature. Unlike the educated assayers, the unhappy workmen who knew about iron were, in Biringuccio's words, "never able to enjoy any quiet except in the evening when they are exhausted by the long and laborious day that began for them with the first crowing of the cock. Sometimes they even fall asleep without bothering about supper." Small wonder that there is little first-hand literature on the subject! If we omit Theophilus's account of the superficial hardening of tools in the early twelfth century, Biringuccio himself gives the first good description of steelmaking. He achieves the "purification" by making a bath of molten cast iron into which lumps of wrought iron of thirty to forty pounds are immersed for four to six hours,

> so that all the solid iron may take into its pores those subtle substances that are found in the melted [cast] iron by whose virtue the coarse substances that are in the bloom are consumed and expanded and all of them become soft and pasty. When the masters observe this, they judge that the subtle virtue has penetrated fully within; and they make sure of it by testing, taking out one of the masses and bringing it under a forge hammer to beat it out, and then throwing it into the water while it is as hot as possible they harden it; and when it has been hardened they break it and look to see whether every little part has changed its nature and become entirely free inside from every layer of iron.

It should be noted that the subtle substances entering into the iron do not contaminate, they purify.

The cementation of iron to give massive steel was first described in 1589 by that Italian of limitless curiosity, G. B. della Porta;[8] as an industrial process it appeared very early in the seventeenth century and flourished particularly in England. It was the best steel for fine tools, but expensive. Agricultural implements were most commonly made from "natural steel," resulting from direct reduction and carburisation in the hearth with an inevitably large variation of carbon content only partly eliminated by selection on the basis of fracture.[9]

The furnaces used in the slow cementation process for making steel are described in Robert Plot's *Natural History of Staffordshire* (1686). After referring to Aristotle, Plot went on to state that the ancient conversion of iron to steel was performed by frequent ignition, "as it is now by a *long* one; whereby the *Vitriolic Salt* of the Iron being thus strongly press't by the Violence of the fire for so long a time, is forced out of the *pores* of it wherein it was lodg'd, . . . and flyes quite away, leaving the *Iron* wholly void of all *Salt* to the center. . . ."

At the end of the seventeenth century, then, we have the practical man (guided as he always will be by the knowledge in his fingers and his eyes) unconsciously putting carbon into iron by his steelmaking processes, while the philosopher thought that some deleterious principle was being removed. This was soon to be changed, however.

The Phlogiston Theory of Steel

In 1703 G. E. Stahl reformulated J. J. Becher's chemical philosophy into the phlogiston theory, which provided the focus for a heightened attack on the chemistry of metals throughout the eighteenth century. Its first effect was to entrench the old idea of steel being more thoroughly metallic, for it was made by methods that would add phlogiston to already-reduced iron.[10]

The first serious scientific investigation on iron and steel—indeed, the first book devoted entirely to iron—was R. A. F. de Réaumur's *L'Art de Convertir le Fer Forgé en Acier* . . . published in Paris in 1722. Réaumur, whose work has been discussed elsewhere,[11] first approached steel through a study of practice rather than from chemical theory. Contrasting the four principal ways of making steel he decided that, contrary to the general opinion, something was added, not subtracted, in going from iron to steel. He was enough of the times that he called that substance "sulphurs and salts," the sulphurs, of course, not being today's elemental sulphur but rather something undefined but related to the reducing combustible principle of the chemists. It had slightly more reality than Stahl's phlogiston and, though a material plentifully contained in the charcoal, was definitely not the principal component of the charcoal itself. Nevertheless, Réaumur recognized clearly that steel was intermediate in nature between wrought iron and cast iron and could be made starting with either, and he was practically modern in his insistence on the manner in which the added material modified the structure of the iron which, being changed by heat treatment, was the basis of hardening. His work, however, could not bear fruit until oxygen had been discovered and phlogiston demolished.

The phlogiston theory inspired an immense amount of experimental work on the chemistry of metals, but this work had little connection with technology. It was centered in France, Germany, and Sweden, while in Great Britain, where iron and steel technology was burgeoning with the Industrial Revolution, the imaginative contributions were more pneumatic than metallic. Neither Darby's use of coke for smelting iron nor Cort's improvement in wrought iron manufacture, which was necessitated both by the increased scale of operation and by the inferior quality of the coke-made pig, had any scientific basis. Perhaps, however, it was England's economic success which forced other iron-making countries, particularly Sweden and France, to seek the science behind the practice. This is the next stage of our story.

The German chemist, J. A. Cramer, whose *Elements of the Art of Assaying* published in Leyden in 1739 was the first book to attempt to supply a framework of theory to the old assayer's practices, was a staunch phlogistonist. Cramer gives very clear and succinct descriptions of the preparation of steel in the laboratory, and he regards it as being nothing but pure iron impregnated with a great quantity of phlogiston:

> . . . All you do in this Operation [of making steel] is to apply oily Vapours to pure Iron, the rigid body of which being mollifyed by the Heat, and made quite red hot, is penetrated by the said Vapours, which then strictly unite to it. . . . For this reason the essential difference between pure Iron and Steel, consists in the greater Proportion of Phlogiston more intimately joined to one than to the other. . . . That every oily substance free from the acid of sulphur is fit for changing iron into steel is plain from the several experiments of Workmen.[12]

C. E. Gellert, in a later work on metallurgical chemistry (which leaned heavily upon Cramer though it was somewhat more methodical and complete), wrote that the general opinion "that calces of metal should recover their metallic form only by restoring to them that phlogiston which they are supposed to have lost, does not seem to be sufficiently grounded."[13] He appears to be the first to suspect that the reduction of metallic calces results from the action of the fire removing a material part of the calx. However, he is completely satisfied with phlogiston in steel. He gives details of a laboratory experiment involving the cementation of malleable iron with charcoal dust and wood ashes (or pounded animal substances such as claws, or horns, which convey "not only a greater quantity but in the same time a much tenderer phlogiston, which acts far quicker and better upon the iron than that of the vegetable kingdom") and clearly thinks of the process as simply adding phlogiston by slow penetration into the iron to convert it to steel.

Perhaps the best summary of French chemical thinking in the middle of the eighteenth century is the two-volume *Dictionnaire de Chymie* by P. J. Macquer, published anonymously in 1776. In reporting Cramer's conclusion that iron acquires the quality of steel solely by the acquisition of a new quantity of phlogiston, Macquer comments on the peculiar property by which the earth of iron is capable of combining with the inflammable principle and of becoming metallized without undergoing the fusion which is necessary for the reduction of all other metallic calces. Then,

> From what we have said, we may judge that steel is much better purified iron than any other iron, impregnated with a larger quantity of inflammable principle, and hardened by the temper. Some celebrated natural philosophers, but who were not chemists, have advanced, that steel was only iron which still retained something of its mineral matter and that its state was intermediate betwixt that of cast iron and soft forged iron. But this opinion is manifestly erroneous.

The translator of the English edition of the dictionary, James Keir, added a note on the English process of cast steel which he believed to be efficacious because of the separation of earthly matters, acids, and sulphur. He believed, however, that steel did contain some substances [sic] derived from the cementing compound, and "with great probability at least believed to be phlogiston."

Oriental Textures Inspire European Science

The phlogistonic explanation of steel fitted squarely into and even lent its weight to the support of the chemical theory of the day. The idea that iron was but a stage in the reduction to the purest state of steel could easily have persisted long after the demise of phlogiston in chemical theory, and the truth could have been found only slowly as a result of improved chemical analysis. Metallurgy, however, is a complex art, and the users of metals, if not primary producers, have been concerned with problems of texture even more than with composition. In Europe this was usually an exploitation of the fairly obvious relationship between the grain or fiber of a broken surface of a metal and its serviceability. Oriental metallurgists, however, had long used metals of a composite nature, with a texture visible to the naked eye. The best known was Damascus steel, consisting of layers of high- and low-carbon steels interwoven either by crystallization or by forging (see figures 4.1 and 12.7). It owed its superiority to this texture, which was developed by chemical etching on a finished sword or gun barrel and thus gave a visible guarantee of correct processing and an aesthetic quality which European steels lacked.[14]

European attempts to duplicate Chinese porcelains in the early eighteenth century had inspired the development of high-temperature techniques and analytical procedures, together with a great deal of work on clays and other minerals, both silicates and otherwise. At the end of the eighteenth century metallurgists were exposed to the same Oriental sensitivity to the texture of materials, and this provided the stimulus that led directly to the discovery of carbon in steel.

There are two quite distinct kinds of textured Oriental steel: (1) the true Damascus steel, which is a high-carbon steel in which the texture originates as intracrystalline segregation during very slow solidification of the molten alloy, subsequent forging being carried out in such a way as to preserve the identity of high- and low-carbon areas in an irregular lamellar structure; and (2) a composite structure made by welding a mixed faggot of iron and steel, forging and contorting this in a manner to give irregular exposure of the two metals as a visible pattern on the finished surface. The former was used mainly for swords and was not duplicated in Europe until 1821. Both steels were called by the name "Damascus," as were other Oriental textures in linen and inlay. The forged structure was used primarily for gun barrels made in India and Turkey, and these were duplicated in Europe in the eighteenth century, for the origin of their structure was fairly obvious.

A factory to make Damascus gun barrels from forged interlaminated iron and steel was established in Sweden, and in 1773 the manager of the plant, Peter Wäsström, described their manufacture in a detailed paper to the Swedish Academy of Sciences.[15] His description of the final etching of the barrels to reveal the damask caught the attention of the well-known metallurgist, Sven Rinman. In discussing Wäsström's paper, Rinman remarked that the surface with elevations and excavations in strong relief arose from the different rates of corrosion of the two different materials distributed in the texture. Rinman reports that he had been able to produce a more subtle damask "in flames and waves of darker and lighter colors" by using metal composed of five different kinds of iron and steel laid together and welded.[16] A year later Rinman was back with a comprehensive paper, "Experiments of etching upon iron and steel."[17] This paper was of great importance for it showed clearly that the different colors were due to a material difference between the different kinds of iron. He found that there was a material residue insoluble even in nitric acid which "has a strong attraction for inflammable matter," and that grey cast iron leaves in solution a black sediment, sometimes of equal bulk and shape with the bit of cast iron itself, consisting of a matter like blacklead, which is found to be iron earth overcharged with phlogiston: Take away a part of this grey iron's redundant phlogiston, either by melting it into white iron, or by cementation with absorbents in Réaumur's way into malleable iron or steel, and it becomes soluble in aqua fortis completely without residuum.

Rinman was puzzled by the fact that steel, supposedly containing more phlogiston, was less slowly acted upon by nitric acid, until he saw that this was a result of

> a sediment, consisting partly of inflammable matter, and partly of iron calx, which covers the steel and hinders the action of the etching water, and likewise gives the steel a more or less black surface, according as it contains more or less of the inflammable matters, or as it is more or less hard, insomuch that we are in condition to judge in some degree of the steels hardness from the degrees of lighter or darker grey colour which it takes in the etching.

This paper of Rinman gave the phlogiston responsible for the differentiating between steel, cast iron, and wrought iron a palpability that it had previously lacked. It became a separable material substance, which (in the extremely phlogisticated case of grey cast iron) was tentatively identified as a known mineral, black lead (also called plumbago and graphite), which was under active study by Swedish chemists at the time.

It should be noted that the above contribution arose entirely from technology. The desire to make something led to a puzzling observation which was elucidated by empirical experiment. It was not a prediction based on science. This, I believe, has been a very common pattern in past advances, though it will perhaps be less true in the future because of the discouragement of empirical experiment unrelated to theory.

Bergman's Essay on the Analysis of Iron

Rinman's intriguing observations caught the eye of the great chemist, Tobern Bergman (1735–1784), who was particularly interested in the composition of mineral matter and whose paper was published in Upsala in 1781 with the title, *Dissertatio Chemica de Analysi Ferri.*[18]

This was the exciting period when pneumatic chemistry was well founded. The relationship of inflammable air (today's hydrogen) to the reducing principle in metals was known. The discovery of dephlogisticated air (oxygen) was less than a decade old, and the formation of fixed air (carbon dioxide) by the combustion of charcoal had been shown. A thorough phlogistonist, Bergman believed that the properties of substances arose from phlogiston and caloric entering into their composition in varying manners and amounts.

In 1775 Bergman had published an edition of Sheffer's lectures of 1749–1751 with important original additions, in which he remarked that the properties of steel appeared to depend principally on a certain quantity of phlogiston. By the time of his research on iron, Bergman believed that calces (oxides) were acid radicals that had been congealed by a certain amount of phlogiston which was tightly bound. A further addition of phlogiston, which he called reducing phlogiston, would convert the calx to a truly metallic state. The latter portion of phlogiston was easily separated and restored, and the ease of its separation marked the division of metals into noble and base. He proceeded to measure quantitatively the amounts of both kinds of phlogiston in several types of iron and steel.

This he did first by measuring the volume of inflammable air evolved during the solution of various types of iron in acid and also by comparing the amounts of different irons needed to precipitate a known weight of silver from a silver sulphate solution. He concluded that the same volume of inflammable air came from equal weights of the same iron with either vitriolic or muriatic acids (H_2SO_4 and HCl, respectively), though the rate of solution was quite different. He concludes that

> In one cubic inch of inflammable air there is present about the same amount of phlogiston as in 2.17 assay pounds of Österby steel, 2.08 pounds of Husaby cast iron, 2.08 of Österby wrought iron and 1.96 pounds of Grängens wrought iron. . . . It appears that *the least amount of phlogiston is present in cast iron, the median amount of steel, and the greatest amount in wrought iron.*[19]

The above is precisely the opposite of the earlier opinion. Bergman notes, however, that his experiments would measure only that part of phlogiston which is proper to the iron molecules, and a part of the inflammable principle is contained in the plumbago and other foreign materials which would not be attacked by the acid and remained as the residue which Rinman had observed. Later in the paper Bergman remarks more specifically that grey cast iron *is* saturated with phlogiston, and the small evolution of gas is due to the fact that most of the phlogiston is associated with the graphite and not available for combination with the iron which alone gives rise to inflammable gas. Certain principles were present in any of the states of iron, but in different amounts:

> That is to say, ductile iron contains almost no plumbago, but it enjoys the smallest amount of combined caloric matter and an abundance of inflammable air; steel has a lesser amount of phlogiston, but is richer in caloric matter; and cast iron saturated with plumbago possesses little phlogiston and a modicum of caloric matter.[20]

From Scheele's experiments, Bergman knew plumbago to be composed of aerial acid and phlogiston: the refining of cast iron which took place in fire and blast of the refining hearth was a result of the decomposition of the plumbago by air, and is marked by both the boiling resulting from the departure of the aerial acid and the bright incandescence that occurs as the copious phlogiston seizes the vital air and generates light.

After some studies of caloric and an examination of possible effects of arsenic, zinc, manganese, and other foreign materials, Bergman returns to report quantitatively the amount of acid-insoluble residue of the kind discovered by Rinman.[21] This he found varied between almost nothing in wrought iron to as much as 6.7 percent, though more usually 3–4 percent, in cast irons. He confirmed Rinman's identification of the material as plumbago and proved that the most phlogisticated, i.e., grey cast iron, contained a residue that matched common plumbago in every way except for a minor siliceous part which remained as a white residue after calcination. Both forms of plumbago consisted of aeric acid and phlogiston, as he proved by combustion experiments (following Cronstadt), by using them for quantitatively reducing iron oxide, and by their reactions with niter, though he cautioned that the amount of material was insufficient for quite conclusive tests. Bergman summarized his analytical results in a table (see table I, slightly rearranged from Bergman).

Included in the paper are studies of sulphur content in hot-short iron (which inexplicably failed to precipitate from solution as barium sulphate) and of phosphorous in cold-short iron, which in acid solution threw down a white precipitate (of ferric phosphate [?]) on long standing.[22] Altogether there are 273 separate experiments, most of them quantitative, reported in this remarkable work. The summary table includes the elements—carbon, silicon, and manganese—which today we regard as being most important, but it should be noted that Bergman considered that the various materials found by analysis were actually incidental to the true nature of the distinction between the different forms of iron, which arose from the form in which the phlogiston occurred. He is, however, quite specific that the differences are due to heterogeneous substances which are present in greater amounts in cast iron than the others, that the siliceous matter is an admixture, and that plumbago is necessary to produce steel. He also specifically states that the hardness depends in some manner on a moderate plumbago content which combines with phlogiston so that the latter is not available for promoting the

Table I The composition of various forms of iron (Bergman, 1781)

		Cast iron	Steel	Ductile wrought iron	Hot-short wrought iron	Cold-short wrought iron
Siliceous matter, percent	Min.	1.0	0.3	0.05		0.05
	Max.	3.4	0.9	0.3	0.8	0.3
Plumbago, percent	Min.	1.0	0.2	0.05		0.05
	Max.	3.3	0.8	0.2	0.7	0.3
Manganese, percent	Min.	0.5	0.5	0.5		0.5
	Max.	30.0	30.0	30.0	0.5	4.0
Iron, percent	Min.	63.3	68.3	99.4		95.4
	Max.	97.5	99.0	99.5	98	99.4
Inflammable air, cu. in. from 100 assay lbs.	Min.	38	44	48		50
	Max.	48	48	51	48	52
Combined caloric, degrees*	Min.	20	74	122		122
	Max.	52	114	136	130	134
Specific gravity, typical		7.662	7.002	7.798		7.792
		7.759	7.643	7.754		7.751
			7.775	7.827	7.753	7.791
			7.727			
			7.784**			

*The caloric figures are meaningless because they record only the maximum change in temperature of the solution (in degrees Réaumur), and though they are potentially significant their value is vitiated by innumerable factors that affect the rate of solution and hence, in nonadiabatic measurements, the maximum temperature rise. Bergman later saw that the evolved heat was not the combined heat, and in the second edition gives the latter as proportional to the reciprocals of the numbers listed here.
**7.693 if hardened.

metallicity of the iron. There is no suggestion that an iron carbide is formed and no true distinction between the residue obtained from a white cast iron or a steel and a grey cast iron.

Bergman's paper contains a correct, even quantitative, description of the distinctions between the forms of iron and steel. He believed that the behavior of phlogiston was the true explanation rather than the composition per se. Both the phlogiston and the caloric matter of the eighteenth-century chemists reflect a concern with physical properties which the chemist abandoned in the Chemical Revolution. The complete demolition of phlogiston in steel naturally occurred at the hands of the French chemists, and it promptly followed the translation of Bergman into their language.

The active Swedish period of discovery in this field was now temporarily over. It will be noted how much of the sequence of events depended upon the local circumstances. The initial discoveries followed upon the installation of a factory for making Damascus gun barrels, bringing to Europe a glimpse of the age-old Oriental sensitivity to the texture of materials; this was developed by Rinman, one of the many educated Swedish engineers concerned with industrial processes, typified earlier in the century by Triewald and Polhem, fully immersed in industry, and having a practical man's awareness of the difference between the many forms of iron but with an insatiable thirst for facts dealing with its chemical behavior. The next step required a more logically scientific environment. It was fortunate (though not fortuitous, since both were an outgrowth of cultural factors) that the observations of the action of acid on iron happened to coincide with a vast interest in gases, for this was the time of the birth of pneumatic chemistry.

C. W. Scheele (1742–1786) had studied the gas resulting from the combustion of graphite. In 1770 he had been studying the evolution of hydrogen by the action of acids on iron filings, following Cavendish's 1766 paper on his three kinds of factitious air, and in 1779 reported himself satisfied that "plumbago is a kind of mineral sulphur or charcoal of which the constituents are aerial acid united with a large quantity of phlogiston. The small quantity of iron can scarcely come into any consideration." Then he studied the black residue left when pig iron was dissolved in sulphuric acid, remarking that "it was black, shining and felt between the fingers like plumbago"; he concluded, after combustion tests, that it must certainly be plumbago, though containing less phlogiston than plumbago and harboring some material that left a white residue.

Joseph Priestley in England, probably inspired by the Swedish research, worked on the subject with inconclusive results. He measured gas evolution and weighed the black residue from cast iron. Priestley concluded that a quarter of an ounce of cast iron gave about 90 ounce measures of inflammable air, while after "white-heart" malleablizing it gave about 106; steel gave between 90.5 and 103 cubic inches per quarter ounce in three experiments of different size samples. He noted that the differences in the quantities of air was always in reciprocal proportion to the quantity of residue. The latter, he concluded, contained no sulphur but, by combustion tests, contained much plumbago.[23] Priestley's independent discovery of oxygen provided the keystone for Lavoisier's theory of combustion, and chemists everywhere were exploring the role of carbon as a fuel and reducing agent with the excitement of pioneers in a new land.

History and geology had combined to give Sweden by far the strongest school of mineralogical chemistry in the eighteenth century.[24] Besides Scheele, J. G. Wallerius, A. F. Cronstedt, and H. T. Scheffer all did particularly important work, but it was in Bergman that the whole mass of new information was concentrated into a more or less systematic analytical scheme which gave emphasis upon composition even if, for a moment, the phlogiston theory dominated explanation. His concepts on structure and the unit-step origin of crystal faces, published in 1775, were the qualitative basis of the mathematical relations in the crystallography of Häuy. The very title of Bergman's paper, "The analysis of iron," suggests his approach in general. But the center of chemical interest was to move to France. Phlogiston lost its importance with the recognition of oxygen. The Chemical Revolution under the leadership of Lavoisier inevitably brought with it a simplification of the understanding of the various forms of iron and steel.

The Research Frontier Moves to France

There was close personal contact and correspondence between chemists in Scandinavia and those in France. Bergman's *De Analysi Ferri* was translated into French by Pierre C. Grignon and published in 1783 with extensive but not very helpful notes. Grignon had first heard of Bergman's work from a Swedish visitor to France in 1779 who reported the role of graphite (*molybdena*) and also Bergman's discovery of the white precipitate from phosphorous in cold-short iron. In a letter from the same man in November 1781, Grignon heard of Bergman's published paper and he eventually got a copy from Berthollet at the Academy of Sciences, for no bookseller could find one.

Bergman had personal contact with Guyton de Morveau (1737–1816), a very able chemist at Dijon, who was responsible for the translation of Bergman's collected works into French. Secretary of the Dijon Academy, de Morveau published extensively, was editor of the first volume of the chemical section of the *Encyclopédie Méthodique,* and later achieved merited fame in connection with the new nomenclature of chemistry. He began this alone, but eventually collaborated with Lavoisier, Berthollet, and Fourcroy in the famous *Méthode de Nomenclature Chimique* (Paris, 1787), which, by embedding the new chemistry in a sensible system of nomenclature, did much to propagate it.[25]

Writing in 1772, de Morveau had followed the usual belief that steel was more heavily phlogisticated iron.[26] In 1786 he recorded some doubt about the phlogiston theory in general, and a year later, after he had established close contact with Lavoisier and his companions in Paris, he had become a full convert to the new theories. In his 32-page article on steel in the *Encyclopédie Méthodique* (written in 1786, but not published until 1789), de Morveau limited himself to the chemical aspects of his subject.[27]

On the nature of steel, de Morveau concludes that cast iron, steel, and wrought iron can all come from the same terrestrial material and can easily be passed one into the other by appropriate treatment. After critical analysis of all of the evidence, he decides that steel differs from iron neither in its content of heat, nor phlogiston, nor vital air (i.e., oxygen, which had been introduced into the discussion by a misguided paper of Lavoisier's in 1782). The analyses of

Bergman pointed directly to plumbago. Then he shows that charcoal and plumbago are the same substances (he had reported this in 1783) not only chemically, but also by showing that charcoal fired at the highest temperature in Tennant's blowpipe approached more and more the physical condition of plumbago. De Morveau summarizes as follows:

> We thus conclude that steel, regardless of the way in which it had been formed, is nothing but iron which in its nature approaches that of ductile iron, because the martial earth in it is more freed from heterogeneous parts and, if not more perfectly metallized than in cast iron, at least more completely so; that it is distinguished from that [ductile iron] because it admits into its composition an appreciable amount of plumbago; that steel approaches cast iron more closely than ductile iron because of the presence of this mephitic sulphur;[28] that it differs little from grey iron except for the greater abundance of the "sulphur" in that; that it is farther removed from white iron because that harbors earthy parts, not metallized and even extraneous, which can be separated by a second quiet fusion in sealed vessels without any addition; that, therefore, the passage from cast iron to the state of steel is accomplished always by the purging of the iron and the removal of the excessive plumbago; that the conversion of iron to steel operates principally because either there is formed in it, or it receives [from outside] an appreciable amount of plumbago; that heat only influences the changes by producing and maintaining fluidity without which there could be no combination at all; that the composition which constitutes steel can very well, by its own affinity, bind a great quantity of the matter of heat; and, in a word, that the properties of steel in general depend on a proper proportion of these principles, just as the different qualities of steel depend on the accidents which vary their proportions.[29]

Scientists and Revolutionaries

In May 1786, at about the same time that de Morveau was writing, three distinguished authors presented to the Academy of Sciences a paper which is a landmark in siderurgical literature. This is the *Mémoire sur le Fer Considéré dans Ses Différens États Métalliques* ("On the different metallic states of iron") by C. A. Vandermonde (1735–1796), C. L. Berthollet (1748–1822), and Gaspard Monge (1746–1818).[30] All three were scientists of the highest eminence, and the factors which led to their association on the subject of iron would merit extensive study. Like de Morveau, both Berthollet and Monge became important public figures in the Revolutionary and Napoleonic periods. They were favorites of Bonaparte and accompanied him on the Egyptian campaign.

Vandermonde is perhaps best known as a mathematician. His interest in steel undoubtedly arose in connection with his duties as director of the Conservatoire pour les Arts et Métiers, to which office he had been appointed in 1782. Berthollet, though trained as a doctor, made many contributions to industrial chemistry but is perhaps best known for his powerful championing of the antiphlogiston theory, his many researches on chlorine and dyeing, and his book on chemical statics in which he attacked the law of constant combining proportions. Monge, perhaps most famous for his development of descriptive geometry, had synthesized water from its elements independently of Cavendish and Lavoisier in 1783. After the overthrow of the French monarchy, all three men played a prominent role in promoting arms production under the aegis of the Committee of Public Safety. Vandermonde reported on the Klingenthal factory's methods for producing bayonets and swords. Monge wrote a treatise on the casting of cannon, in which he introduced several labor-saving devices.

At the specific request of the French Minister of Commerce, Berthollet wrote a pamphlet, *Précis d'Une Théorie sur la Nature de l'Acier*, summarizing the conclusions of the Academy paper. This appeared in 1789, on the eve of the Revolution, but it was republished three years later under the auspices of the Committee of Public Safety and the imprint of the War Department. This was considerably expanded and more practical in style as befitted its title, *Avis aux Ouvriers en Fer sur la Fabrication de l'Acier*.[31] It carried the names of all three authors, with that of Berthollet last, perhaps because he was a less ardent revolutionary. It was finished on 13 Brumaire, Ann. II (November 3, 1793), and L. M. N. Carnot announced that 15,000 copies were to be printed and distributed *avec profusion*.[32]

There is, however, nothing to indicate that it was concern with national policy that produced the first collaboration of the three men. The continuing activity of the Academy of Sciences in matters of industrial concern, which had produced Réaumur's classic work of 1722, had prepared the way, and Berthollet, at least, would have looked at Bergman's essay, if not when Guyton de Morveau importuned him for the Academy's copy, at least when the translation appeared. Though perhaps not central to the main chemical questions of the day, it at least touched closely upon them, and its author's name alone would suffice to gain attention.

In the introduction to their paper, Vandermonde, Berthollet, and Monge say that they were led by analogy to conclude that the variation of the properties of iron comes only from foreign substances alloyed with the iron, and they support this by extensive quotations from Bergman and other writers. The report of their own work begins with an unfortunate statement of disagreement with Réaumur's and Bergman's correct belief that steel was intermediate between wrought and cast irons. De Morveau also had thought that white cast iron had "nonmetallized earthy parts" in it, and this error was to plague the literature for some time to come.[33] It began with Lavoisier's singularly poorly reasoned paper four years earlier in which he had discussed the composition of cast iron.[34] It had slight substantiation in that a blast furnace operated with a small amount of charcoal would give white cast iron while more would give grey; hence the white might be thought to have unreduced calx still in it. Vandermonde, Berthollet, and Monge even believed that the quantity of charcoal contained in white cast iron was less than in steel.[35]

They proceed to reinterpret Bergman's experiments and theory without using his untenable phlogistonic overtones. Following Lavoisier's new concepts, the authors say that when metals are dissolved in dilute acids they begin by decomposing the water, taking from the acid the dephlogisticated air (oxygen) which they need in order to be soluble in acids without evolving gas. The inflammable air of the water, having lost its oxygen to the metal, escapes as elastic gas with effervescence and heat. This gas comes entirely from the decomposition of water and is not related to phlogiston in the metal. The fact that a white cast iron evolved less inflammable air supported their erroneous view of its constitution, for it was reasonable to assume that it contained already oxidized metal. They repeated some of Bergman's experiments with more care, finding that the conversion of iron into steel by cementation involved an increase in weight of about 1/180 (0.56 percent), though they saw that there may have been some compensating loss due to the removal of oxygen in blister formation. Their new hydrogen-evolution measurements were corrected for both pressure and temperature, the latter empirically, for Charles's law had not yet been enunciated. Though cast iron gives less inflammable air by virtue of the fact it is not completely reduced, steel gives less, they think, because (modernizing the words but not the thought) the carbon itself combines with hydrogen to give hydrocarbons of less volume, in support of which view they cite Rinman's observation that the combustion of inflammable gas from the solution of steel gave rise to fixed air (carbon dioxide).

The French chemists used their dephlogisticated theory to explain the processes which were used practically to smelt and refine iron and to make steel. They naturally came to regard the refining process as purging cast iron from two foreign substances, dephlogisticated air and charcoal. Two distinct functions of charcoal in the blast furnace are clearly distinguished: the creation of a high temperature to melt the slag and separate the gangue, and the reduction of the ore. Furnace operation logically gives grey or white iron, they explain, depending on the amount of blast, tuyère angle, and ratio of charcoal to ore. They wiggle out of the paradox of the antagonistic substances dephlogisticated air and charcoal being together in cast iron by first supposing that only the surfaces of the melting drops become carburized and then that the charcoal is in the form of graphite, which is more difficult to burn.

They saw that the solubility of carbonaceous matter in the iron changes with temperature, and on cooling the solution becomes turbid by a rejection of the excess. If cooling is done slowly, the rejected material will form plumbago—the kish appearing at the surface of very grey cast iron and coating the ladles—but quick cooling will surprise the graphite before it can float out and it is found disseminated in the metal as in grey cast iron. "The affinities of two substances always being reciprocal, the charcoal must in turn be regarded as retaining iron." Perhaps a confusion arose from an examination of residues from white cast iron in which the carbon really is present as a carbide, not graphite, and the distinction between the different residues was not at first noticed. Although Scheele had already shown that iron in natural graphite was an impurity which could be extracted by acid, the French savants believed iron to be an integral part of the substance.[36] The *New System of Chemical Nomenclature* perpetuated the confusion by identifying graphite as *carbure de fer*. They did, however, prove conclusively that plumbago from cast iron was primarily identical with charcoal by using it to reduce litharge, arsenic oxide, sulphuric and phosphoric acids, and did other tests, including burning it in a Priestley apparatus in oxygen. They conclude their lengthy paper with a summary, quoted in full below:

> Cast iron may be regarded as a regulus the reduction of which is incomplete, that is to say, a regulus which still conserves a portion of the base of dephlogisticated air [oxygen]. This follows (1) because this metallic substance, in order to dissolve in vitriolic and muriatic acids, evolves less inflammable air [hydrogen], decomposes less water and absorbs less dephlogisticated air than does soft iron for this purpose, which proves that it already contains a portion of the dephlogisticated air that is necessary for its solution; and (2) because as a result of temperature alone,

cast iron, especially when it is grey, refines and whitens without addition and without the contact with air,[37] which could not occur if it did not contain dephlogisticated air to bring about the combustion of the charcoal which renders it grey.

In addition, cast iron, especially when grey or black, contains charcoal which it has absorbed naturally. This is proved (1) by the faculty which it has of cementing soft iron and of transmitting to it enough charcoal to turn it into true steel; (2) by the black residue which is always found at the bottom of its solutions in vitriolic acid, when the solution is done cold—a residue which, like charcoal, dissolves when hot in inflammable air,[38] and gives fixed air [carbon dioxide] on combustion. It is to the greater or smaller quantity of carbonaceous matter that cast iron owes the different colors displayed on its fracture, which may be controlled by varying the proportion of charcoal in the furnace charge.

Cement steel is nothing other than iron reduced as well as possible, combined in addition with a certain proportion of natural charcoal. The existence of charcoal in steel appears to us to be proved (1) by the increase in weight of the iron when it is cemented in pure and degassed charcoal; (2) by the carbonaceous residue which the steel resulting from such cementation leaves at the bottom after solution in acids, and which, like that of cast iron, dissolves when hot in inflammable air, and then gives fixed air by its combustion. As to the metallic reduction, that it is pushed further in cement steel than in soft iron is proved by the fact the presence of the bubbles which are seen in blister steel, and which can only come from the fixed air formed by the combination of the charcoal with the dephlogisticated air which was still in the iron.[39] Overcemented steel differs from the preceding only by a great quantity of absorbed carbonaceous matter, which is proved by a greater increase of weight during cementation, by a great black residue in the solutions, and principally because iron is only given that quality by forcing the conditions which favor cementation, such as temperature and duration.

Perfectly soft iron would be a regulus in the state of greatest purity; but the softest iron of commerce always contains (1) a little charcoal, which is proved by a light black residue after solution; (2) a little dephlogisticated air, which, being evolved during cementation, produces fixed air, and forms the bubbles which are always encountered in blister steel, even that made from ther softest iron. Moreover, the variations which are observed in the volumes of inflammable gas produced by the solutions of different wrought irons prove that the metallic reduction is not always carried to the same point.

Finally, charcoal, after having been held in solution by cast iron or steel in the state of fusion, and finding itself rejected by the metal at the moment of the cooling, leaves the combination while retaining all the iron that can remain united with it. That charcoal saturated with iron is then plumbago, which separates from the metal, and which, when the cooling is slow, comes to swim at the surface, where it can be gathered in its natural state. But when the cooling is rapid, and the pasty state of the metal opposes that purging, the rejected plumbago remains disseminated in the mass, and communicates to it qualities of steel. So steel, when in the cold state, may be considered as the result of a disturbed solution; and the charcoal which it contains having at first been held in solution, then rejected because of the cooling, is nothing other than finely divided plumbago, scattered, and not combined.[40]

As mentioned above, the new theory was expounded in popular form in a special pamphlet issued under the aegis of the revolutionary Committee of Public Safety in 1793. This was reprinted promptly in the *Observations sur la Physique* and, much later, in the first issue of the *Annales des Chimie* to appear after a four-year hiatus due to the

Revolution. Curiously, no English translation of the full Academy paper was published. The theory was quoted, however, by Thomas Beddoes, who gave a reinterpretation of the puddling process in its terms,[41] and by Joseph Black, who saw its obvious inconsistencies as mentioned above.

Carbon Is Accepted

The charcoal (French, *charbone*) of this paper soon became the carbon (*carbone*, Latin *carbonum*) of the elementary substance in the new chemical nomenclature.[42] Coming as part of the Chemical Revolution, the recognition of carbon as an element was second in importance only to the identification of oxygen.[43] These metallurgical experiments played some role in unraveling the allotropy of carbon, and quite early Clouet, with Guyton de Morveau, showed that steel could be made by heating iron in contact with diamond. Despite good analyses and measurements of the amount of CO_2 formed, it was not unnatural to assume that the big difference in properties depended upon minor impurities, for the structural basis for properties had not yet been accepted.[44]

Although some metallurgists were very slow to accept the dominant role accorded to carbon by the French investigators, subsequent steps in siderurgical science were largely based upon the successive refinement of the understanding of the state of carbon in the different forms of iron and steel, and the unraveling of the true significance of the various impurities. Silicon, which was well identified by Bergman, was at times thought to be the principal hardening element. Its rather mysterious role in cast iron, primarily that of promoting graphitization during cooling, was not fully elucidated until the end of the nineteenth century.

With the chemical role of carbon elucidated, the production of steel was much more easily controlled. Cast steel was made in France by Clouet in 1798. He believed that the only difference betweeen the various forms of iron was in the amount of carbon, a small quantity giving steel a larger amount white cast iron, and a still larger amount grey cast iron.[45] He was also a leader in the production of elaborate patterns of forged Damascus steel. The producers of cast crucible steel showed a curious conservatism, for cemented blister steel remained the chief raw material for cast steel until well into the twentieth century, despite Clouet and Mushet's successful experiments with the far quicker process using wrought iron with pig iron or with charcoal as a charge. David Mushet in 1804 had the first glimpse as to the reason for the superiority of wootz, the very-high-carbon Indian steel from which the true Damascus blade was made.[46] Then followed Faraday's studies, which indicated that wootz owed its properties to its crystalline nature, and the work of Bréant in 1821, which beautifully tied together both the compositional and the structural aspects of this particular steel.[47] With carbon understood, Bessemer found control of his process easy, though its invention was not a deduction from theory, as the Martins's probably was.

Though analysis proved immensely helpful in explaining failures and in controlling the smelting and refining of metals, further basic theoretical advance was slow until it was possible to follow the relationship between composition and structure on a microscopic level (beginning with the work of Sorby), on an atomic level (beginning with x-ray diffraction after 1912), and on the intermediate scale of the electron microscope in the 1950s. For over a century following the contributions of the great Swedish and French chemists whose work we have discussed, metallurgical problems rarely impinged upon the interests of the foremost pure scientists. Partly this was because most materials of metallic interest were too complex for the physicist and, being solid solutions, refused to conform to the simple multiple proportions that alone were encompassed by the developing molecular theory of the chemist.

In the twentieth century, however, the metallurgists' accumulated empirical alloy lore had a highly stimulating effect on the development of the quantum theory of solids, and the dislocations and other imperfections that arose from a consideration of metallic plasticity and diffusion gave a richness to solid-state physics that it was in danger of losing with the picture of the rigid ordered lattice that arose from x-ray diffraction studies. In the future it seems probable that the pure and applied sciences will grow even closer together and both will merge continually with technology. The empirical complexities of the practical man will undoubtedly continue to suggest more and more fertile fields for study by the theoretically inclined.

Summary

We have seen the end of the eighteenth century as a particularly crucial period in the understanding of iron and steel. For many centuries steel had been made with well-tried practical processes which had eluded scientific explanation. Since iron is the only common metal which will dissolve carbon, the triple role of charcoal as a fuel, a reductant, and an alloying element was unique and hard to unravel. The first stage beyond Aristotle's belief that steel was a purer form of iron because it had been longer heated in the fire came with the phlogiston theory, which saw steel as a more refined iron containing an additional quantity of phlogiston. Réaumur made this substantial with his sulphurs and salts, but most eighteenth century chemists followed the phlogiston theory.

The next stage of scientific understanding arose not from science but from a practical observation of an Oriental technique, in which a surface pattern left by acid attack revealed to the naked eye a material difference between iron and steel. This came about in Sweden at a time when analytical methods were being rapidly developed and where the new gases and combustion were under active investigation. The amounts of gas evolved by solution of steels in acids were measured by Bergman, who related the properties of steel to the presence of carbonaceous residues, though he continued to attribute the essential function to phlogiston, associated partly with the iron and partly with the "plumbago."

In France, however, where Lavoisier's clear thinking had just explained combustion in purely compositional terms, with oxygen, carbon, and hydrogen and metallic oxides essentially as understood today, it soon became apparent that the phlogistic overtones were unnecessary. An inability to distinguish between carbon and carbides caused a misinterpretation of the difference between white and grey cast iron (which was attributed to oxygen). This was only slowly removed, but an enormously important step had been taken. Chemical composition had replaced mysterious principles.

If the practical iron and steel man was rather slow to benefit from the new knowledge, at least he can feel that his practical knowledge, by helping the identification of carbon as a chemical element, had contributed in a far from trivial way to the Chemical Revolution itself.

Notes and References

1
This ignores the effect of impurities and intentional alloying elements which modify the behavior but which are not essential to the property of hardening on quenching. By extension, the word steel is today used very loosely, not only to include "mild" (low-carbon) steels but practically any iron-rich alloys, even some stainless steels in which carbon is an undesirable impurity. The change in meaning occurred at the end of the nineteenth century under the same commercial debasement of terminology that spawned also many "bronzes" devoid of tin.

2
All early steel hardening involved the direct quenching of the steel in a manner to give the desired final hardness as it came from the quenching bath. It was a difficult and precarious operation. This is what used to be meant by "tempered" steel. The modern process of a full quench and a tempering (reheating) operation to soften to the desired degree is not mentioned in the literature prior to the sixteenth century A.D., and the date of its introduction is unknown.

3
This connection was suggested by T. A. Wertime, "The discovery of the element carbon," *Osiris* 11:211–220 (1954). See also Wertime, *The Coming of the Age of Steel* (Chicago, 1962), ch. 5.

4
In 1962 a skilled blacksmith in Mehris, Iran, in answer to a question by the author stated that steel and iron were of quite different species, "as different as a willow tree from an oak." Though he rejected the "true" explanation, this had no effect on his work. He was using steel to tip first-class agricultural tools, and he knew that iron and each of his principal steels (pieces cut from crankshafts and springs from worn-out U.S. trucks) had to be handled quite differently. But an earlier smith, who would have been also a smelter, would have been more aware of the fact that a change in furnace regimen would produce vastly different products.

5
Even as late as 1783 Buffon, who should have known better for he was both an ironmaster and a naturalist, was to write that "steel should be regarded as iron even more pure than the best iron; the one and the other are only the same metal in two different states, and steel is, so to say, an iron more metallic than simple iron; it is certainly denser, more magnetic, less dark in color, and of a much finer and more compact texture." *Histoîre Naturelle des Mineraux*, vol. 1 (Paris, 1783), p. 477.

6
Vannocio Biringuccio, *De la Pirotechnia* (Venice, 1540; for the English translation see item 46 in the bibliography).

7
See item 151 in the bibliography.

8
G. B. della Porta, *Magiae Naturalis Libri XX* (Naples, 1589; English translation, London, 1658; reprinted, New York, 1957). This and many of the other original texts cited in this article have been gathered in my *Sources for the History of the Science of Steel, 1532–1786* (item 150 in the bibliography).

9
Mathurin Jousse, *La Fidelle Ouverture de l'Art de Serrurier* (Paris, 1627; for a partial English translation see item 113 in the bibliography).

10
The Iatrochemists' sulphur passed to the *terra pinguis* of Becher and the phlogiston of Stahl. Again, logic was misleading. Physically, indeed, the phlogiston theory was in some degree right, for phlogiston is simply physically real electrons in the state defined by the conduction band of the quantum theory. The joining of one ounce of phlogiston to aluminum ions in an electrolytic cell produces about one ton of metal: the old chemists had no such neat way of performing the transfer, for their reactions were confused by the gases oxygen and oxides of carbon, which were not identified until late. Chemically—and the argument in the eighteenth century *was* mostly

chemical—phlogiston was wrong for it overlooked the presence of oxygen. Even the Iatrochemists had good physical (but again, poor chemical) basis for their principles, for salt, sulphur, and mercury lacked only diamond to represent the four chief mechanisms of atomic binding in modern solid-state theory. And, around 1800, it was chemists who laid the foundation for the modern physical theory of heat and thermodynamics.

11
See item 120 in the bibliography and also the introduction to the English translation of Réaumur (item 92 in the bibliography).

12
J. A. Cramer, *Elementa Artis Docimasticae* (Leyden, 1739). Quotation is from the English translation by C. Mortimer (London, 1741), pp. 346–347.

13
C. E. Gellert, *Anfangsgrunde zur metallurgischen Chymie* (Leipzig, 1751); French translation, *Chimie Métallurgique*, 2 vols. (Paris, 1758); English translation, *Metallurgic Chemistry* (London, 1776). The section on steel is Process 74.

14
This is discussed in some detail in C. S. Smith, *A History of Metallography* (Chicago, 1960).

15
Peter Wässtrôm, "Beskrifning på damascherade skjut-gevär af jårn och stål," *Kongl. Vetenskaps Akademiens Handlingar* 34:311–318 (1773).

16
Sven Rinman, addition to the paper by Wässtrôm, loc. cit., pp. 318–321.

17
Sven Rinman, "Rön om etsning på jårn och stål," *Kongl. Vetenskaps Akademiens Handlingar* 35:3–14 (1774). For an English translation see Smith, *History of Metallography*, pp. 249–255.

18
This work was reprinted, with some changes, in Bergman's Collected Works (*Opuscula Physica et Chemica . . .*, vol. III [Upsala, 1783], pp. 1–107), and from this source was translated into German (Tobern Bergman, "Von der Grundtheilen des Eisens," in *Kleine Physische und Chemische Werke . . . übersetzt von Heinrich Tabor,* vol. II [Frankfurt, 1785]). However, this essay was not included in the English translation (Tobern Bergman, *Physical and Chemical Essays*, vols. I and II translated by E. Cullen [London, 1784], vol. III [anonymous translator, Edinburgh, 1791]). The scientifically inclined ironmaster, P. C. Grignon, published a French edition (Bergman, *Analyse de Fer . . .* [Paris, 1783]) based on the 1781 pamphlet. The quotations that follow are from an English translation by Patrick Hickey and the present author (see item 150 in the bibliography).

19
Bergman, *Dissertatio*, pp. 17–18.

20
Ibid., p. 34.

21
Bergman's thermal measurements (which showed the amount of both bound and free "caloric matter" in steels and other substances) are important in connection with the history of specific heats and heats of reaction. Bergman had worked on latent heat in 1772, and in the same year as this paper on iron, he had published a famous paper on specific heat. Johannes Gadolin, who supposedly was involved in the experimental thermometric measurements on iron, later contributed significantly to the theory of specific heat.

22
The causes of brittleness in metals was the subject of a special essay by Bergman. See his *Opuscula Physica et Chemica*, vol. III (1783), p. 109. This was not included in the English edition of the essays.

23
Joseph Priestley, "Experiments and observations relating to iron," in *Experiments and Observations on Different Kinds of Air, and Other Branches of Natural Philosophy* (Birmingham, 1790), vol. III, pp. 480–507. This edition is a rearranged abridgment of papers published between 1774 and 1786.

24
An excellent discussion of the chemical environment in the last half of the seventeenth century is given in vol. III of J. R. Partington, *A History of Chemistry* (London, 1962), pp. 158–236. It is particularly good on the Scandinavian chemists.

25
For the background and significance of the revolution in nomenclature, see Morris P. Crosland, *Historical Studies in the Language of Chemistry* (London, 1962).

26
L. B. Guyton de Morveau, "Digressions sur la phlogistique," in *Digressions Academiques, ou essais sur quelques sujets de physique, de chemie, et d'histoire naturelle* (Dijon, 1772).
27
De Morveau, "Acier," in *Encyclopédie Méthodique. Chymie, Pharmacie et Métallurgie*, vol. I, part ii (Paris, 1789), pp. 420–451. The article was in type before September 1786 (see de Morveau's letter to Berthollet in *Observations sur la Physique* 29:308–312 [1786]). There is a good biography of de Morveau by W. A. Smeaton, *Ambix* 6:18–34 (1957/1958). See also Partington, *History of Chemistry*, pp. 516–534.
28
De Morveau is following Scheele in regarding plumbago as a carbonaceous reducing material.
29
De Morveau, *Encyclopédie Méthodique*, vol. I, p. 450.
30
Mém. Acad. Sci., 1786 (published in 1788), pp. 132–200; *Observations sur la Physique* 29:210–287 (1786). For an English translation see item 150 in the bibliography.
31
This booklet was reprinted, minus the political harangue and the four plates with their description, in *Observations sur la Physique* 43:373–386 (1793), and again in both the *Journal des Arts and Manufactures* 2:572–619 (Ann. III [1795]) and the *Annales de Chimie* 19:13–43 (1797). There is an English translation in *Nicholson's Journal of Natural Philosophy* 2:64–70, 102–105 (1798).
32
An excellent study of the organization of arms production and the role played by many savants is that by Camille Richard, *Le Comité de Salut Public et les Fabrications de Guerre sous la Terreur* (Paris, 1921). The activities of scientists in promoting war production during this period and their later involvement in affairs of state is strongly reminiscent of the recent scene in the United States.
33
To understand this error, it must be remembered that cast iron is an extremely complicated material. It becomes grey in fracture only when it has a high carbon content *and* enough silicon (about 2 percent) to promote the growth of numerous discrete particles of graphite during normal cooling of the cast metal in a mold. In the absence of silicon the carbon will form iron carbide, hard and white on the fracture, and remaining as a residue on solution in acids under certain conditions. Silicon, of course, is an earthy material, and its presence was recognized by Bergman. To complicate matters, the physical state of the carbon and the color of the fracture depend on the cooling rate, and a thin casting made from a grey pig will be white. Though Réaumur in 1726 recognized the truth of this, later scientists ignored it.
The truth regarding oxygen in white iron was clear enough to Joseph Black, who, in the last decade of the eighteenth century, objected to the idea of oxide being in the metal for "it is inconceivable that there should be present in fusible iron . . . white hot . . . both an abundance of carbon and an oxyd of iron without acting on each other originally, especially in the grey iron which most abounds in carbon." Steel is simply "refined iron, intimately combined with a small quantity of pure carbonaceous matter, or the carbon of the French chemists," which must be introduced into the iron. (Joseph Black, *Lectures on the Elements of Chemistry* . . . , edited by J. Robinson, 3 vols. [Edinburgh, 1803]. Iron and steel appear in vol. II, pp. 402–506.)
34
A. L. Lavoisier, "Mémoire sur l'union des principe oxigine avec le fer," presented Dec. 20, 1783, *Mém. Acad. Sci. 1782* (published 1785), pp. 541–559.
35
Throughout the paper the authors refer to charcoal (*charbon*), for the element *carbone* had not yet been christened: they do treat it, however, as a definite substance, and "carbon" would not be a misleading translation.
36
Some of the confusion arises from the fact that the residue on acid solution, which they correctly assumed to be carbonaceous matter, is only graphite in the case of grey cast iron. In annealed steel and in white cast iron it is actually iron carbide, Fe_3C, which is partly decomposed on subsequent digestion in acid. The graphite remains unaffected. Vandermonde and his co-authors did know that the character of the deposit changed

during prolonged heating with sulphuric acid and also that the gases evolved do not consist entirely of hydrogen but contain some carbon (hydrocarbon), which they found by combustion and a limewater test.

37
An erroneous observation.

38
See note 36.

39
Though this explanation of the blisters in cement steel is correct, the authors were misled by the fact that the oxide is an extraneous inclusion of slag and is not essential to the nature of the iron itself.

40
Vandermonde, Berthollet, and Monge, *Mémoire sur le Fer . . .*, pp. 198–200.

41
T. Beddoes, "An account of some appearances attending the conversion of cast into malleable iron," *Philosophical Transactions* 81:173–181 (1791).

42
G. de Morveau, A. L. Lavoisier, C. L. Berthollet, and A. F. de Fourcroy, *Méthode de Nomenclature Chimique* (Paris, 1787).

43
For further discussion of this point, see the paper by Wertime cited in note 2.

44
This point is developed by L. P. Williams, "Faraday and the alloys of steel," in C. S. Smith, ed., *The Sorby Centennial Symposium on the History of Metallurgy* (New York, 1965).

45
L. Clouet, "Resultats d'experiences sur les differens états du fer," *Journal des Mines* no. 49:3–12. (Ann. VII [1798]).

46
David Mushet, "Experiments on Wootz or Indian Steel," *Philosophical Transactions* 95:175 (1804). Reprinted with additions in Mushet's *Papers on Iron and Steel* (London, 1840), pp. 650–678.

47
See item 120 in the bibliography, pp. 17–26.

3
Structure, Substructure, and Superstructure

Anyone who works with the microscope for an intellectual or a practical purpose will frequently pause for a moment of sheer enjoyment of the patterns that he sees, for they have much in common with formal art. What follows is an attempt to extend into a more general field some views on the nature of organization and relationships that arose during many years of study of the microstructures of metals and alloys.[1] In a landscape painting of the Far East, a rock in the foreground with cracks and crystalline texture is often echoed in a distant mountain with cliffs, chasms, wrinkles, and valleys; a tree may be related to a distant forest, or a turbulent and eddied stream to a distant tranquil pond. Each part with its own structure merges into a structure on a larger scale. Underlying structures are imagined as a necessary basis for the visible features. The connectivity of all is suggested by the branching treelike element of the design. Both separateness and continuity are interwoven, each necessary to the other and demonstrating the relationship between different features on a single scale and between units and aggregates on differing scales. There is an analogy between a work of art which suggests an interplay of dimensions and the real internal structure of a piece of metal or rock which results from physical interactions between the atoms and electrons composing it.

The study of microstructure on the scale within the range of the optical microscope (dimensions between a micron and a millimeter) is a somewhat old-fashioned branch of science, and it still involves a high degree of empirical observation and deduction. Far more "highbrow" is the rigorous science and simple elegant mathematics of the ideal crystal lattice considered as point groups in space. The whole field of crystal structure, mathematically developed in the nineteenth century by Bravais, Federov, and Schoenflies, was experimentally opened up by Von Laue and especially the Braggs in 1912–13, using the diffraction of x-rays to reveal and to measure the periodicities and symmetries in the arrangement of planes of atoms in crystals. But the mathematical physicist must simplify in order to get a manageable model, and although his concepts are of great beauty, they are austere in the extreme, and the more complicated crystal patterns observed by the metallurgist or geologist, being based on partly imperfect reality, often have a richer aesthetic content. Those who are concerned with structure on a superatomic scale find that there is more significance and interest in the imperfections in crystals than in the monotonous perfection of the crystal lattice itself. Like the biologist, the metallurgist is concerned with aggregates and assemblies in which repeated or extended irregularities in the arrangement of atoms become the basis of major structural features on a larger scale, eventually bridging the gap between the atom and things perceptible to human senses.

The symmetry of crystals in relation to decorative ornament has been treated by many writers, none better than by Weyl in his *Symmetry* (Princeton, 1952). The patterns of crystal imperfection are less commonly known, despite their prevalence and despite their relationship to so many aesthetically satisfying forms in which regularity and irregularity are intricately intertwined.

Crystalline Aggregates and Foam Structures

Aggregates of crystals have structures which are defined by the atomically thin layer of disordered material between the crystals. Many characteristics of their shape are shared with simple undifferentiated biological cells and the simplest common soap froth. In all these, the pertinent features are the two-dimensional surfaces that separate volumes of matter which, on this scale, is featureless. Two-dimensional interfaces are necessary to define the separate identity of things in three dimensions. Junctions of the interfaces themselves produce linear (one-dimensional) features, and these, in turn, meet at points of zero dimension. This interaction between dimensions, the very essence of form, is expressed in mathematical beauty as Euler's law. This states that, in a simply connected array, the number of points minus the number of line segments plus the number of surfaces and minus the number of polyhedral cells is equal to one, i.e.,

$$n_0 - n_1 + n_2 - n_3 = 1,$$

where n_0, n_1, n_2, and n_3 are the numbers of zero-, one-, two-, and three-dimensional features. There are no limitations to this, beyond the requirement of simple connectivity. Even more than Euclid, hath Euler gazed on beauty bare.

A pure metal, when cast (or, better, after a little working and heating), has a structure like that of Carrara marble—hosts of little crystals packed together irregularly. The units do not look like crystals, for they lack the symmetrical vertices and plane faces of a regular polyhedron, but internal order is there nevertheless. Although for centuries man has been fascinated by the geometrical shape and glitter of natural crystals, he has only recently come to see that the essence of crystallinity lies not in external shape but in the uniformity of the relationship of atoms to their neighbors within the crystal. A single isolated crystal growing from a solution or melt can grow uninterruptedly in accordance with the dictates of the atomic steps on its surface. Usually this will result in a simple polyhedron (as in figure 3.1), reflecting the internal order because of its effect on the rate of growth in different directions. If many crystals start to grow in the same region, sooner or later they will interfere with each other. Neighboring crystals differing in no way whatever save in the direction of their atom rows in space cannot join without some imperfection. Figure 3.2 illustrates this. It is a magnified photograph of an array of tiny uniform bubbles floating on soapy water. The lines of disorder that form between the differently oriented areas of regularly arranged bubbles in this two-dimensional model are believed to be closely analogous to the planes of disorder constituting the boundaries between the three-dimensional crystal grains in metals, rocks, and other polycrystalline materials. The boundaries are a source of both strength and weakness and they provide sites for the beginning of any crystalline change. Though themselves invisible except at the extreme limit of resolution of modern electron and ion microscopy, they differ so much in energy from the body of the crystals that they are easily revealed as lines of enhanced chemical attack (figure 3.3), early melting (figure 3.4), or they can be inferred from the sudden change of crystal direction revealed by some kinds of chemical attack on the surface (figure 3.5). Patterns like these can often be seen with the naked eye on the weathered surface of a cast brass doorknob or hand rail, or internally in clear ice which has been kept just at its melting point for several hours.

Now these boundaries, which on an atomic scale are just imperfections in a uniform stacking array, on a larger scale them-

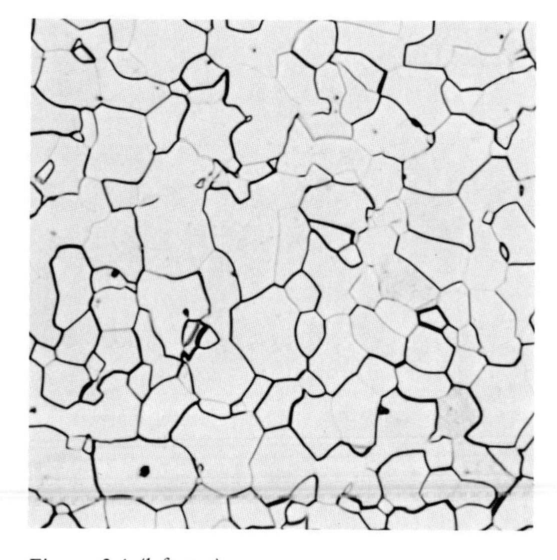

Figure 3.1 (*left, top*)
A group of polyhedral salt crystals growing individually from solution. × 80. Photo by C. W. Mason.

Figure 3.2 (*left, bottom*)
Raft of tiny uniform soap bubbles showing "grain boundaries" where zones of differing orientation meet. × 7.

Figure 3.3
Deeply etched section of a piece of niobium metal, showing a network of grain boundaries revealed by selective attack. × 200. Photo courtesy of R. J. Gray, Oak Ridge National Laboratory.

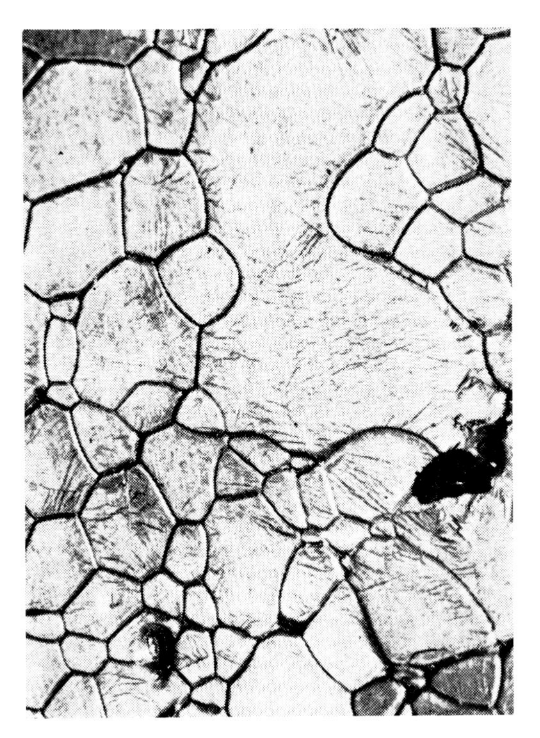

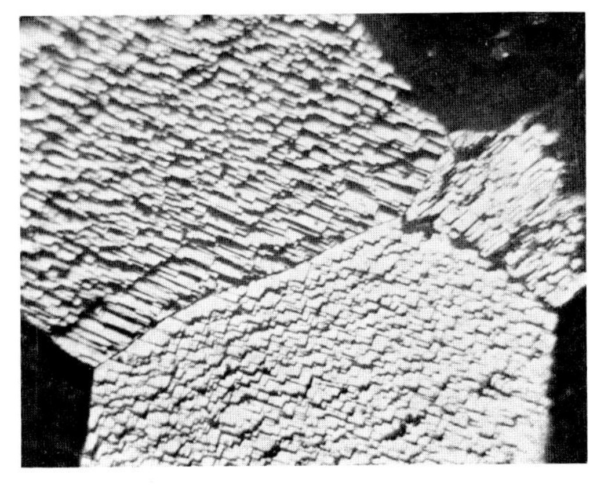

Figure 3.4
Surface of overheated aluminum sheet showing the beginning of melting at the grain boundaries. × 5. Photo courtesy of the British Non-Ferrous Metals Research Association.

Figure 3.5
Etched section of silicon-iron alloy showing the junction of three crystals. This historic photograph was taken in 1898 by J. E. Stead.

selves become the basis of structure. They are, in fact, films of matter, distinguished by structure rather than composition. They must completely surround every crystal in a mass and extend in foamlike fashion continuously throughout the entire mass. Having high energy and mobility, they tend to adjust to a configuration of small area, which makes them join each other always in groups of three at an angle of 120 degrees, just as do the films in a froth of soap bubbles. In a mass of large bubbles of irregular size (figure 3.6), there will be differences in pressure between adjacent bubbles to match the surface tension in the curved films and to reconcile the 120-degree angle with the necessity to fill space. Since three bubbles meet at each junction, Euler's law requires the average bubble in an extended array in two dimensions to have exactly six sides, but there is no requirement that each one be a hexagon, only that if there are some with more than six sides, there must be a matching number with less. The froth therefore, though lacking long-range symmetry, nevertheless has very definite rules as to its composition. It is pleasing in appearance because the eye senses this interplay between regularity and irregularity. The topological requirements of space-filling rigidly determine the relationships of the whole, but allow any one cell to be of pretty much any shape, while surface-tension equilibrium requires only that the films be at 120 degrees to each other at the point of meeting, always three together, and it produces the pressure differences that are needed to balance the resulting curvatures. Beyond this, all depends on the accidents which brought a bubble of a particular size to a given place and surrounded it with its particular neighbors, each also with its private history.

It is interesting to compare a two-dimensional soap froth with the topologically similar but geometrically different pattern of craze marks in a ceramic glaze (figure 3.7). Though the cracks divide the surface into cells meeting three of each junction, the geometry is different from the froth because the cracks must follow the direction of stress in the glaze and a new crack joins an old one perpendicularly.

A foam in three dimensions is a bit more complicated but depends on the same principles. To divide space into three-dimensional cells, at least six two-dimensional interfaces must meet at each point; and if surface tension dominates they will join in groups of three at 120 degrees to each other along lines, forming cell edges, which meet symmetrically at the tetrahedral angle of 109.47 degrees (the angle whose cosine is $-1/3$). This configuration of three-, two-, and one-dimensional junctions is repeated at every vertex. Curvature is necessary to connect adjacent vertices and to reconcile the short- and long-range needs. Because the polygons (cell faces) must be in groups which close around each three-dimensional cell, the average polygon will have a smaller number of sides than the hexagon which connectedly fills space in two dimensions. No single plane polygon can meet the requirements, for it would have to have 5.1043 sides in order to have corner angles of 109.47 degrees. The best solution that has been proposed corresponds to a fourteen-sided body with six plane four-sided faces and eight doubly curved hexagonal faces, the mixture of polygons having on the average $5\frac{1}{7}$ sides. This curious irrational number is of the utmost importance, though it is little appreciated. Certainly it is responsible for the prevalence of pentagons in nature, and it probably lies behind the fivefold symmetry

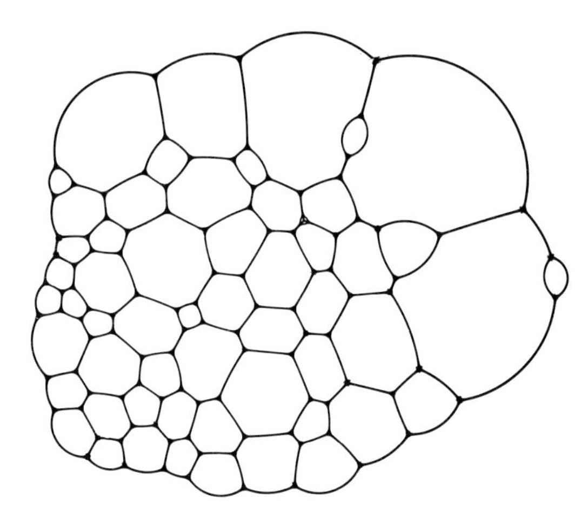

Figure 3.6
Froth of irregular soap bubbles showing a cellular structure analogous to that of metals. These bubbles were blown between parallel glass plates and are essentially two-dimensional.

Figure 3.7
Pattern of craze lines on a glazed ceramic surface. × 0.75.

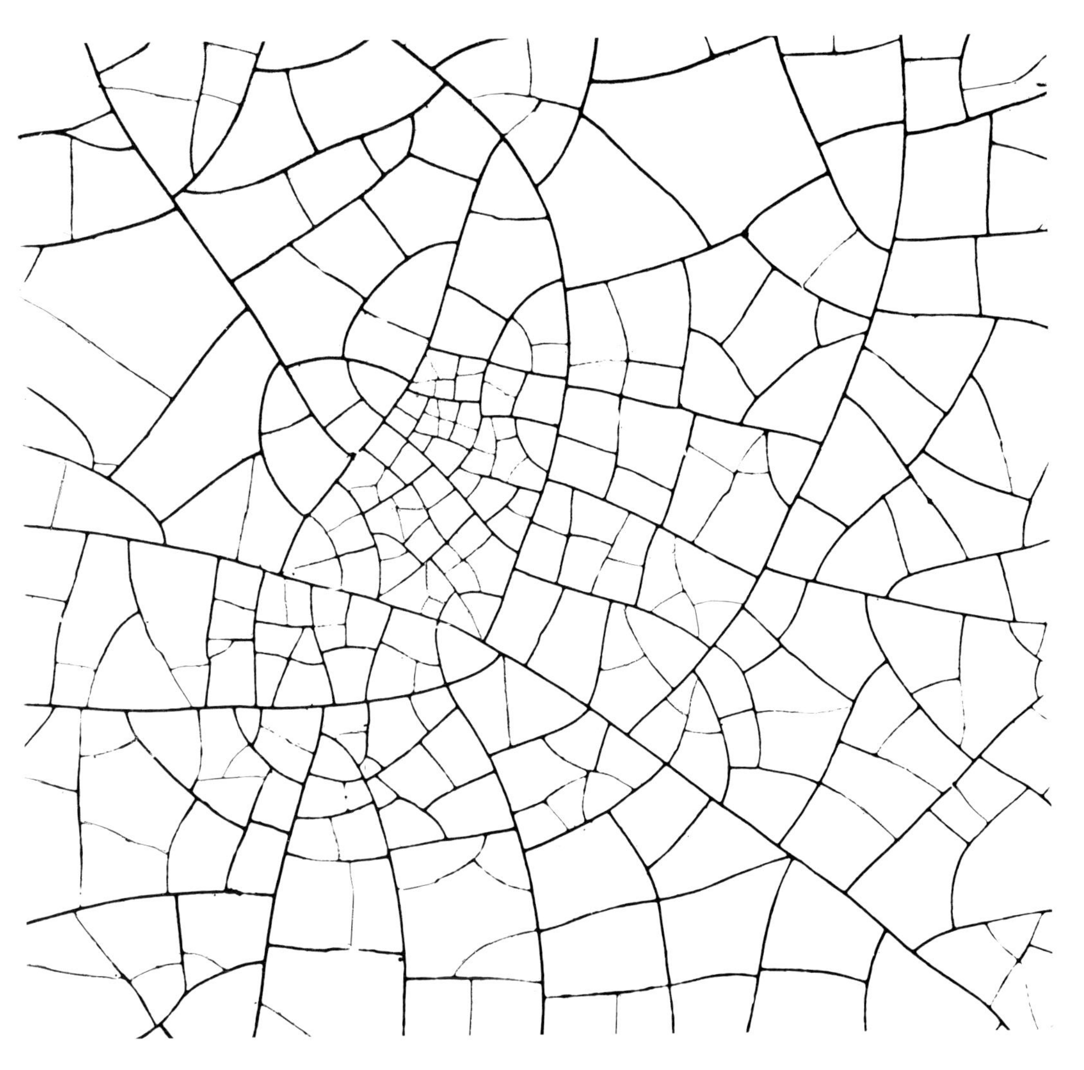

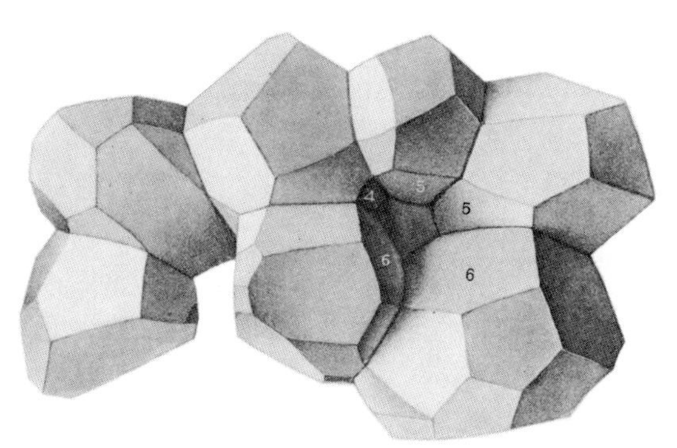

Figure 3.8
The shape of cells in human fat tissue. × 275. Photo by F. T. Lewis, courtesy of the American Academy of Arts and Sciences.

Figure 3.9
Crystal grains of a metal (brass) separated from an aggregate, showing the natural shape of crystals when packed randomly into contact with each other. Note the frequency of pentagons and curved surfaces.

of plants and the five fingers and toes of animals. Pentagonal faces are readily seen within a three-dimensional froth of bubbles on a glass of beer, and they occur also in such disparate bodies as human fat cells or metal grains (figures 3.8 and 3.9). Pentagons are frequent but not universal, for the ideal number is an irrational one and pentagonally faced polyhedra alone cannot fill space.

It should be noted that the external shape of the crystals in figure 3.9 reveals nothing of their inner order, for the shape depends on the property of the disordered boundary, not the ordered crystals that are separated. A more obviously crystalline geometric structure occurs when a second crystalline phase originates in direct contact with a preexisting one, for it automatically forms in whatever definite orientation gives the lowest energy of the interface. Not infrequently two different kinds of crystal will grow in close symbiotic relationship with each other, forming a duplex but well oriented unit, in a larger, irregular, foamlike aggregate. Examples of such oriented duplex structure are shown in figures 3.10 to 3.12.

Branched Structures

The soap froth is the archetype for all cellular systems which for any reason are constrained toward a minimum area of interface. A quite different type of structure, though a common one, is the one that results from the growth of isolated individuals in the branched form best illustrated by a common tree. This occurs whenever a protuberance has an advantage over adjacent areas in getting more matter, heat, light, or other prerequisites for growth. Such structures occur in electric discharge (figure 3.13), in corrosion (figure 3.14), and even in crystal

Figure 3.10 (*left, top*)
Duplex crystals with bands of different composition in exact orientation relationship within one grain, but forming an overall structure like a pure metal, here seen in a copper-silicon alloy, worked and annealed. × 75. (This is an etched section.)

Figure 3.11 (*left, bottom*)
The same alloy as in figure 3.10, deformed by cold rolling. Note the heterogeneity of the distortion. × 10. Photo by B. Nielsen, University of Chicago.

Figure 3.12
Transformation structure in a hardened nickel steel, showing interference between differently oriented crystals growing within the same crystalline matrix. × 250. Photo by Daniel Hoffman.

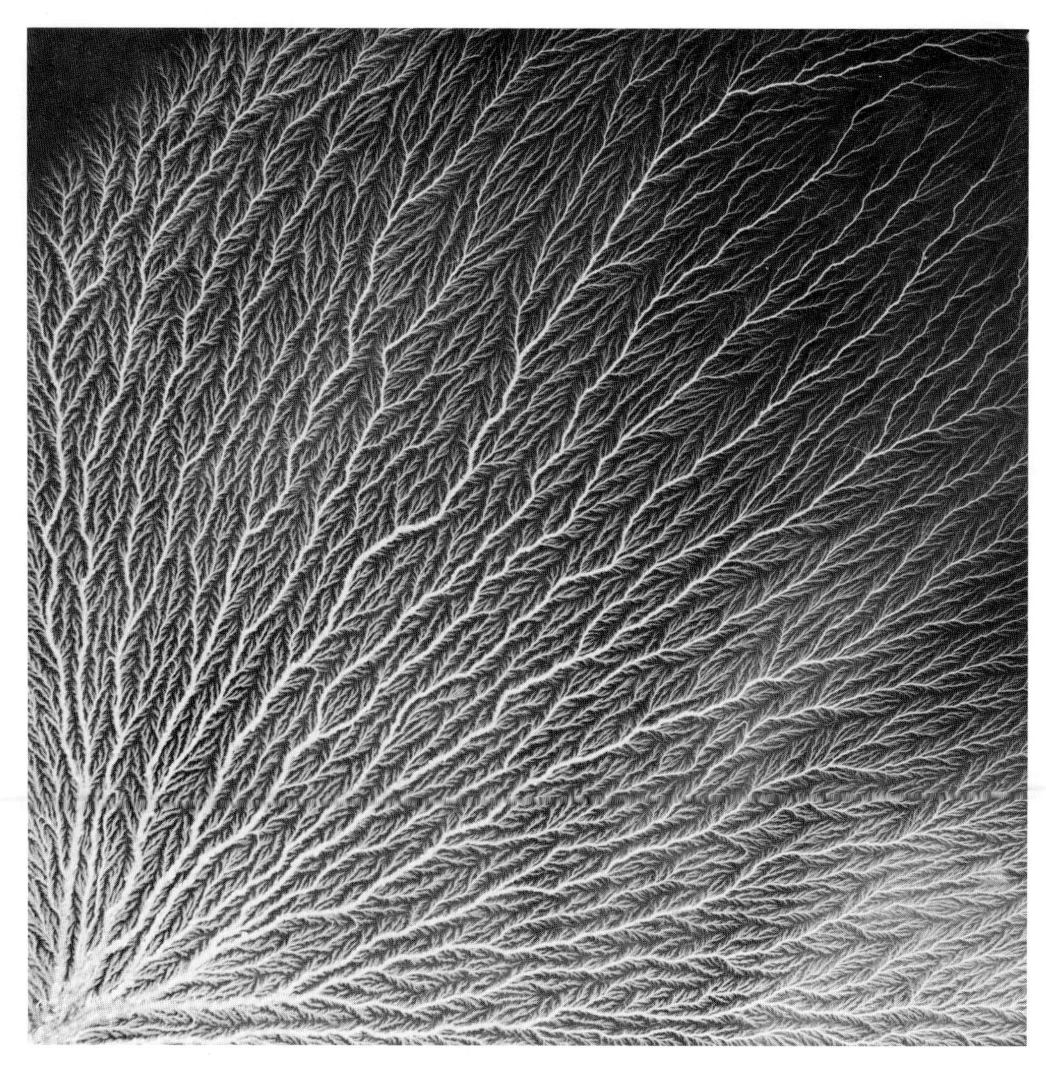

Figure 3.13
Pattern of electrical breakdown in a plate of transparent lucite charged by ion implantation and discharged through a single point. Photo courtesy of Bernard Vonnegut.

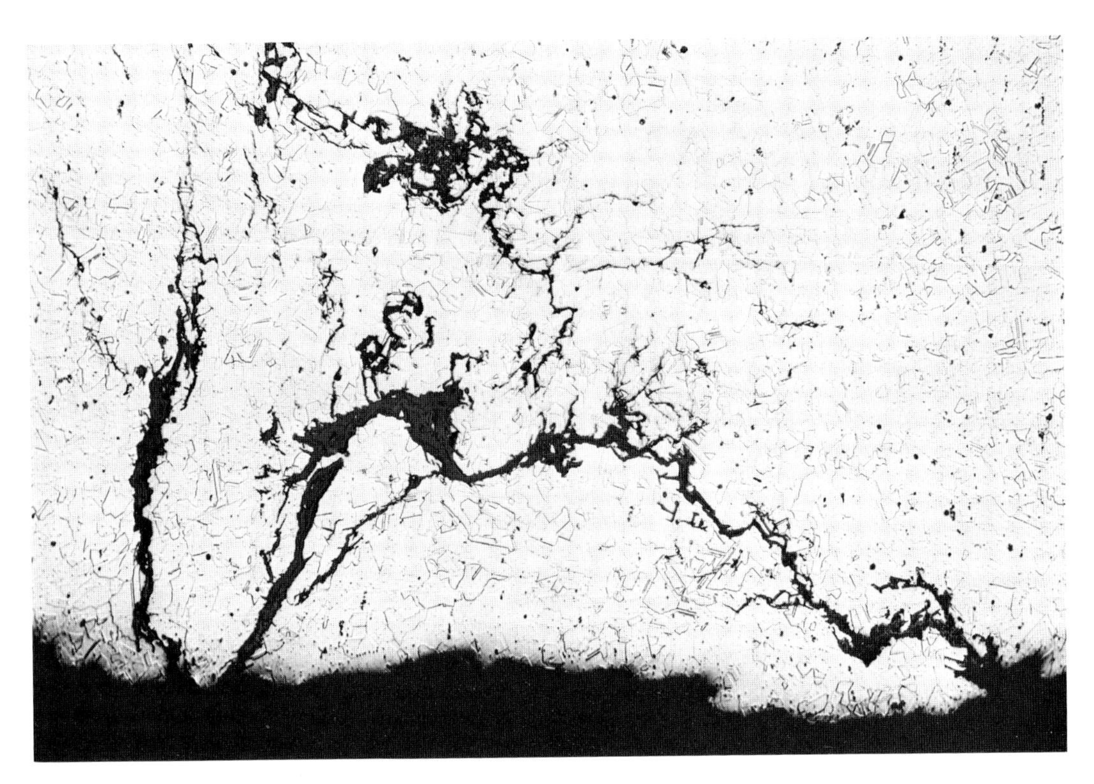

Figure 3.14
Branching pattern in corrosion of stainless steel in uranyl sulphate solution. × 180. Photo courtesy of R. J. Gray, Oak Ridge National Laboratory.

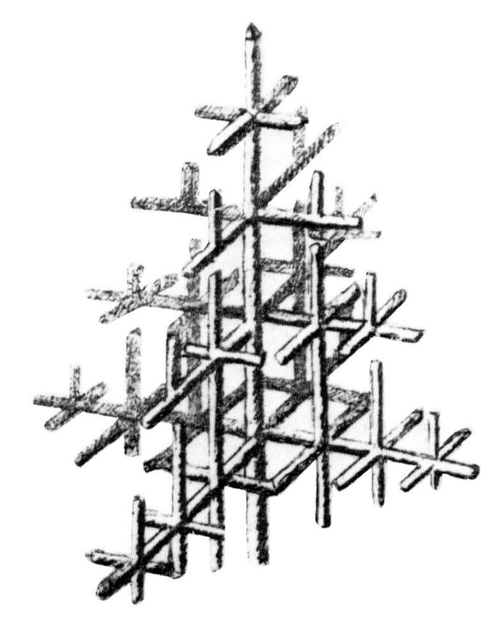

Figure 3.15
Dendritic growth of an iron crystal. × 300. In large steel ingots such crystals are sometimes several inches long. This historic drawing was made in 1876 by the Russian metallurgist D. K. Tschernoff.

growth (figure 3.15), although in the last case the basic mechanism is given an overall symmetry. All these branching structures start from a point and grow lineally, but they eventually stop as the branches interfere with others already present. Until the structure encounters some extraneous obstacle (figure 3.16), the shapes are quite different from the interface-determined shapes discussed above. The structure is that of an individual, not of an aggregate.

There are other structures in which a branched treelike structure arises from an inverse mechanism. The best-known example is the successive joining of many small streams to form a single large river. In brittle solids the merging together of many small cracks to form a single surface gives rise to a similar form (figure 3.17).

The Role of History in Structures

In the typical process amenable to study by physics, the small number of units involved and the simplicity of their interactions gives a definiteness and reproducibility that is either invariant or is dependent in a simple way upon time. In other sciences such as biology or metallurgy, the structure of complex matter must be dealt with, involving myriads of units and interacting interactions, with associations of perfection and imperfection which can be combined in an almost infinite variety of ways. The structures which merit particular study because they happen to exist depend almost completely on their history—quite as much as does, though with more diversity, the present human condition. Although other structures might have been formed with equal a priori probability from the same units and unit processes, the whole unique sequence of atomic-scale events that actually did occur, each adding a little to a preexisting structure, was necessary to give rise to the particular array of molecules, crystals, or cells that form the final structure.[2] Although the ideal crystal lattice of a substance at equilibrium depends only on its composition and temperature, all other aspects of the structure of a given bit of polycrystalline matter depends upon history—the details of the nucleation of individual crystals, usually at sites where imperfections or heterogeneities preexist in the matrix; the locally varying rates at which the individual crystals grow into their environment, incorporating or rejecting matter as a result of the microprocesses of atom transfer; and the manner in which the crystals impinge to produce the grain boundary as a new element of structure which itself changes shape in accordance with its properties and the particular local geometry resulting from historical accidents. Far more complex, but in principle similar, things occur in biological and social organizations.

In the space-filling aggregate, the individuals limit each other. They may be arrayed randomly or regularly, but however undetermined the shape of an individual, the conditions of joining at the points where three or more meet are defined. Structure on one level, by its imperfections or variations, always gives rise to a new kind of structure on a larger scale. (Inversely, it may even be that there is no detectable structure without some underlying structure on a smaller scale. The validity of atomism depends on the tool used to find it.) A local configuration will always have some connection to neighboring ones. In ever-decreasing degree every part is dependent on the whole and vice versa.

Figure 3.16
The surface of an ingot of antimony, showing dendritic crystals which have grown to interference with each other. (This pattern was the mystic Star of Antimony of the alchemists.) About half natural size. Photo courtesy of the Science Museum, London, Crown copyright.

Figure 3.17
Pattern of ridges formed by a crack moving in a crystal of a brittle compound of copper and magnesium, Cu_2Mg. The crack proceeded from the top to the bottom of the figure. × 100. Photo courtesy of Duane Mizer, Dow Metal Products Company.

On Sections and Surfaces

The structures usually observed on metals and rocks are those of plane sections cut through a three-dimensional structure, slicing through the crystal planes and boundaries at various angles, and thus introducing distortions of shape and hiding connections that may exist in the third dimension. We have become very adept at interpreting things from two-dimensional representation, indeed most of our thinking is in such terms. The two-dimensional surface of a painting can represent a straight or distorted projection or a point-perspective view of either real or imaginary things. In sculpture, the surface can be the natural surface of an object, but it is usually a cut through a body of material which has a three-dimensional structure and it reveals a surface texture with its own aesthetic qualities. Sections are subtly different from the same structures when formed against a preexisting surface. (Compare figure 3.4 of the real surface of a polycrystalline metal with figure 3.3 of a section: in the former the angles are nearly all at 120 degrees, while the latter has a pleasing diversity.) Sectioning is simpler than three-dimensional representation because there is no superposition as in projection and no change of scale as in perspective; it gives a single-elevation contour map, with volumes reduced to areas, surfaces to lines, and lines to points. If the structure is cellular and randomly oriented, representations of all possible views will be seen at various places in the section. Depending on the orientation, certain features will be magnified in one direction. Convexity or concavity of a surface in relation to the sectioning plane produces closed isolation or extended connectivity of the linear traces on it. If the structure is not random but irregularly lamellar as the grain of wood, the variations in the third dimension can be seen as a distribution of texture in the two-dimensional slice. Some examples in which such structures are exploited are wood-veneer textures, marbled ceramics, the Damascus sword (figure 3.18), and Japanese swords and *tsuba* (figure 3.19). These all owe much of their charm to the combined aspects of both design and texture that they display, with effects not unlike those of woven textiles but more natural in origin and with three-dimensional overtones.

Conclusion

Do not these simple structures of crystals and the simpler ones of bubbles graphically illustrate some important features of the world and our appreciation of it, aesthetically as well as intellectually? It is the Chinese principle of *yang* and *yin*, balanced positive and negative deviations from uniformity, which, if occurring at many places must form a foam structure of cells no matter what material-space or idea-space is involved. The freedom of a structural unit inflicts and suffers constraints whenever its closer interaction with some neighbors makes cooperation with others less easy. Social order intensifies the interfacial tension against a differently ordered group. Everything that we can see, everything that we can understand, is related to structure, and, as the gestalt psychologists have so beautifully shown, perception itself is in patterns, not fragments. All awareness or mental activity seems to involve the comparison of a sensed or thought pattern with a preexisting one, a pattern formed in the brain's physical structure by biological inheritance and the imprint of experience. Could it be that aesthetic enjoyment is the result of the formation of a

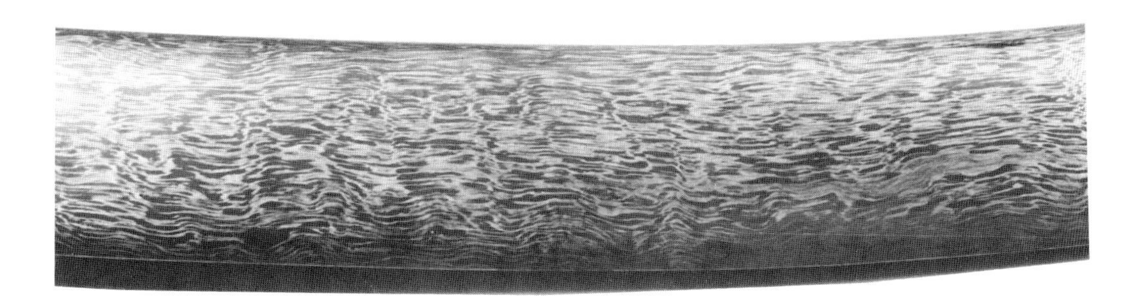

Figure 3.18
Detail of late-seventeenth-century Damascus sword blade (Persian). The surface of the blade had been formed by cutting through the irregular laminar structure which originated in the crystallinization of the high-carbon steel and had maintained its identity during forging. Natural size. Photo courtesy of the Trustees, The Wallace Collection, London.

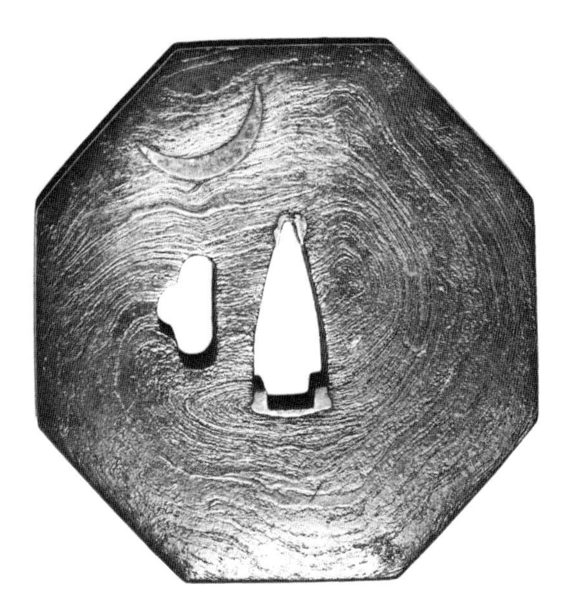

Figure 3.19
A Japanese *mokumé* swordguard. The texture arises from the intentional incorporation of innumerable layers of slightly different steels into a single mass by repeated welding and forging, and then chemically etching the final surface which was cut through the forged lamellae. The moon is inlaid in silver. Natural size. Collection of G. E. Hearn.

kind of moiré pattern between a newly sensed experience and the old; between the different parts of a sensed pattern transposed in space and in orientation and with variations in scale and time by the marvelous properties of the brain? The parts of a sensed whole form many patterns suggesting each other in varying scale and aspect, with patterns of imperfection and disorder of one kind forming the partially ordered framework of another with an almost magical diversity depending on the degree to which local deviations from the ideal pattern are averaged out. Somehow the brain perceives the relationship and actively enjoys the rich interplay possible in patterns composed of the simplest parts, an interplay between local and long-range, between branching extension and consolidation, between substance and surface, between order and disorder.

The very nature of life is pattern-matching, whether in the simple acceptance or rejection of "food" units to fit the RNA molecules within a cell or the joining together of conforming and differentiating cells in the overall pattern of the organism which the parts themselves both dictate and conform to. The growth of ordered but lifeless matter typically occurs by the addition of atoms or molecules to the very surface of a crystal. A not dissimilar process of structural matching is involved in the duplication of protein within a living cell, but a complete organism grows by *internal* multiplication, and the consequent burgeoning of outward movement produces the differing environments for cells which are an essential characteristic of a living organism.

There is a kind of indeterminacy, quite different in essence from the famous principle of Heisenberg but just as effective in limiting our knowledge of nature, which lies in the fact that we can neither consciously sense nor think of very much at any one moment. Understanding can only come from a roving viewpoint and sequential changes of scale of attention. The current precision in science will limit its advance unless a way can be found for relating different but interwoven scales and dimensions.

The elimination of the extraneous, in both experiment and theory, has been the veritable basis of all scientific advance since the seventeenth century, and has led us to a point where practically everything above the atom is understood "in principle." Sooner or later, however, science in its advance will have exhausted the supply of problems that involve only those aspects of nature that can be freshly studied in simple isolation. The great need now is for concern with systems of greater complexity, for methods of dealing with complicated nature as it exists. The artist has long been making meaningful and communicable statements, if not always precise ones, about complex things. If new methods, which will surely owe something to aesthetics, should enable the scientist to move into more complex fields, his area of interest will approach that of the humanist, and science may even once more blend smoothly into the whole range of human activity.

Notes

1
The converse relationship between aesthetics and metalworking—the influence of the techniques disovered by craftsmen making works of art upon the development of the science of metals—was discussed at some length in my *History of Metallography* (Chicago, 1960).

2
I am indebted to John R. Platt for pointing out to me that biology is essentially a historical science.

4
The Interpretation of Microstructures of Metallic Artifacts

Introduction

The conservation of an object can only be done properly when the techniques that were used for making it originally are understood. The conservationist has, in addition to his aesthetic role, an extremely important role to play in developing a better basis for writing the history of technology. As records of the contemporary scene, the objects that pass through a conservation laboratory are very often of greater value to the historian than the writings of the literati. But the meaning of objects is harder to grasp than that of words.

My own attitude toward the microstructure of metals is not unlike that of an art historian regarding a painting or sculpture. There is something akin to style even in a photomicrograph. There are aspects of structure that are not immediately apparent to the untrained eye, and quite minor features may be clues to a deep meaning. Moreover, many structures are really quite beautiful to look at, and part of the pleasure in being a metallographer lies in the moment of aesthetic enjoyment when the microscope first comes into focus to reveal the pattern, before intellectual analysis begins. I take pleasure in the fact that a few of my photomicrographs have appeared in modern art exhibitions as well as several in scientific journals. The connection between visual art and the patterns of science is, I think, far from trivial.

In virtually everything that exists or that can be thought about, there are recognizable units on a small scale that become integrated into a larger organism or structure, and this is accompanied by a reverse influence of the whole on the parts—if not physically, at least in the perceiving mind.[1] An essential aspect of a good work of art (at least as I, a rank amateur, see it) lies in the presence of this reciprocal relationship of the parts to each other at many different levels of organization and to the whole, as well as to widening circles of external things. Such relationships are certainly the essence of the structure that is visible in a piece of metal when it is examined under the microscope, as well as of the understanding of it.

Structure and History

Metallic microstructures are complex, yet there is a certain order to them, with a nice balance between order and disorder, and between local order (which is easily attained) and long-range order (which is energetically favorable but hard to achieve because of local energy barriers). For reasons of history virtually all aggregates containing many atoms are in a configuration that departs from the ideal structure of lowest energy. All structures (whether of atoms, cells, philosophies, or societies) began from something that (at the scale of consideration proper to it) was without form and void; a nucleus of a definite structure somehow formed somewhere, and if it was a structure more desirable than chaos, it then proceeded to grow at the expense of chaos, by rearranging the units at the interface. Later nuclei of a new structure may have formed and grown at the expense of the old one, or there may have been local conflicts, adjustments, and changes. Anything complex *must* have had a history, a sequence of changes in its parts. A complex structure is a result of, and to a large extent a record of, its past. Though a proton and an electron may, as a pair, be able to spring full-panoplied from the head of Jove, more complex things certainly cannot, or at least do not.

A historical view is therefore essential in understanding complexity. There is a microhistory recorded in the internal structure of a metal or ceramic artifact that vies in interest with the history deducible from its external form. And it is *human* history, reflecting man's skills and knowledge. The structural details of a sequence of objects—if properly read—reflect man's growing knowledge and control of the world about him, as definitely as do the words of past philosophers.

But even on a smaller scale, the microstructure is understandable only as a frozen slice of history. Metals are crystalline, that is, most of the atoms that compose them are arranged in regular order, and their history, for our purposes, can be said to begin with a few atoms collecting together in a geometric array to form a crystalline nucleus which somehow forms and then grows in the disordered liquid by a successive aggregation of atoms at the surface. At a temperature below the melting point, the new crystal will continue to grow until it impinges upon some other crystal, or until all the liquid has disappeared. If two crystals impinge, there remains between them a thin film of matter, not more than two or three atoms thick, that belongs to neither crystal but is disordered. A good model of this is made by floating large numbers of identical very tiny bubbles on the surface of soapy water, for surface tension causes them to aggregate together in much the same way, geometrically, as do atoms (see the frontispiece to this book).

The atoms, of course, are far too small to be seen with the optical microscope, but the effects of their arrangement are visible. The crystals are easily revealed if a surface or section of the metal is first carefully polished and then subjected to chemical attack, which can be adjusted to develop geometric steps on the surface or to eat selectively into the boundaries to form a network of lines very reminiscent of the films separating soap bubbles in a froth (see figure 3.3).

It is a characteristic of metal crystals that they can be deformed, which distorts their outline and introduces internal imperfection but does not destroy crystallinity. All effects of deformation can be removed by annealing, i.e., heating to a suitable high temperature below the melting point to cause new unstrained crystals to grow. Many alloys within certain ranges of composition form homogeneous crystals at high temperatures, but become duplex at lower temperatures if time is allowed for the change to occur. All these events require nucleation and growth, and the actual shape and size of the structure that appears is a record of this history.

Understanding these structures takes some experience, or rather a combination of experience and scientific knowledge. It is not, however, a pure matter of intellect, but rather of the mind, the hand and the eye working together.

It is helpful for the student of structures to have an atlas of photomicrographs available. A nice little book has been published by the Institute of Metals—*The Microstructure of Metals* by J. Nutting and R. G. Baker (London, 1965). For steel a compendious three-volume *De Ferri Metallographia* has been issued under the sponsorship of the High Authority of the European Coal and Iron Community. For general technique see G. L. Kehl, *The Principles of Metallographic Laboratory Practice* (2d ed.; New York, 1943). My own article, "Some elementary principles of polycrystalline microstructure" (item 129 in the bibliography), contains a classification of all known types of structure and discusses their origins.

Chemical, spectrographic, x-ray fluorescence or other compositional analysis should usually accompany microscopic study of a museum object. Though to some extent one can infer composition from structure, it is always highly desirable to have an approximate idea of the composition of the metal that is being examined microscopically. Moreover, a detailed analysis of impurities, especially the rarest ones only detectable by the most sensitive methods such as mass spectrometry or neutron activation analysis, can sometimes be of real value in determining provenance.

The written record is very little help in determining the techniques used by ancient metalworkers, though the objects themselves speak loudly to an educated ear. Though there are hints of techniques in the writings of learned men, notably that encyclopedic gossip, Pliny, and there are surviving recipes on clay or parchment from early periods, the information in them has usually become so garbled by transmission or innate verbal obscurity as to be meaningless unless interpreted in the light of an examination of contemporary objects. Not until the *Treatise on Divers Arts* by the German monk Theophilus (ca. 1125 A.D.) is there any surviving record written by a craftsman himself with the obvious intent to instruct. His writings, which have been translated thrice into English,[2] are delightful and essential reading for anyone concerned with the working of metals. In eighteenth-century France a great attempt was made to record all of the arts and crafts. The famed *Encyclopédie* of Denis Diderot and the *Descriptions des Arts et Métiers* of the Academy of Sciences depict contemporary techniques in words and pictures more closely than at any other time in history. Since then techniques have become too diverse and too complicated to be accurately recorded, and the evidence of the objects themselves again must be sought.

The History of Structure

Until a century ago workers in metal had some sense of structure, but no knowledge of microstructure. The texture of broken pieces reveals to an experienced eye something of the nature of a metal and its suitability for various types of service, but it rarely looks crystalline. The essence of crystalhood was originally thought to lie in external polyhedral form, not internal order, and though "grains" were recognized in metal as were cells in wood, their crystalline nature was unsuspected. The modern science of metallography dates only from 1864, when the first structures were clearly seen by an Englishman named Henry Clifton Sorby. Microscopes good enough to have shown the structure existed long before that. There is a drawing of the surface of a razor blade in Robert Hooke's famous *Micrographia* published in 1665. It shows grinding marks on the flat of the blade, fine honing scratches on the edge, a corrosion pit and a streak of slag. Except for the slag, all of these are external to the metal. To show its structure, it would have been necessary to polish the steel carefully without distorting the surface and then to etch it chemically to develop visible differences between adjacent crystals.

Etching was used extensively in Europe for decorative purposes, especially on armor, from early in the fifteenth century, and to produce the intaglio lines of the graphic artist a century later, but these were brutal etchants that revealed no structure. In the Orient, however, macroscopically heterogeneous steels were used, most notably the famed sword of Damascus (figure 4.1), in

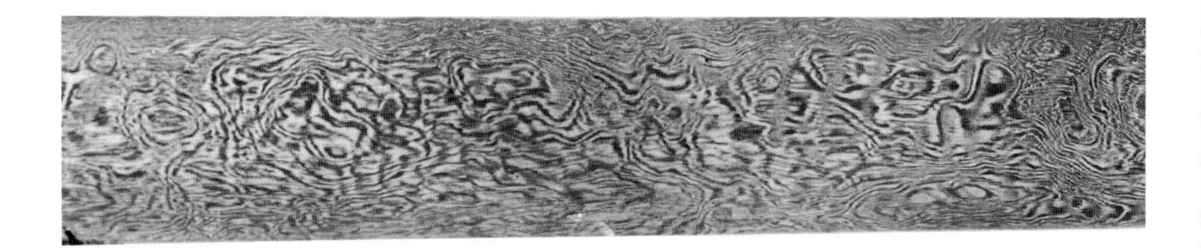

Figure 4.1
Persian straight sword, Isfahan, eighteenth century. Damask shows distorted dendritic structure. × 1.5. Photo courtesy of the Trustees, The Wallace Collection, London.

which layers of metal of different carbon content formed a visible pattern that was revealed by etching and was regarded as a mark of quality as well as an indication by which the smith could control his work. The Japanese swordsmith developed the most marvellous of all metallurgical textures by careful polishing to reveal the pattern of hard and soft metal in his blades, as well as the delicate and beautiful surface texture resulting from the distribution of minute slag particles (figure 4.2).

Europeans succeeded in duplicating the true Damascus sword only in 1821 after centuries of frustration. Turkish and Indian gunbarrels also had a texture, but this was simpler in origin, being produced simply by welding iron and steel bars together. The duplication of this in Europe shortly after the middle of the eighteenth century had interesting scientific consequences, for an observation of the black and white etching of the steel and iron parts of a gunbarrel eventually suggested to a Swedish metallurgist that there was carbon in steel, so upsetting the age-old belief that steel was a more pure form of iron because it had been longer heated in fire. Then in 1804 crystalline texture was developed by etching, not in man-made iron but in a meteorite in which the crystals are huge and the geometric pattern of the two forms of iron that coexist in it on account of its nickel content is easily seen by the naked eye (figure 4.3). Finally came the work of Sorby, and by 1890 enough was known to explain most of the age-old mysteries of metal behavior in terms of the shape and distribution of microconstituents.[3]

The Structure of Castings[4]

A crystal growing into a liquid can take almost any form it wants to: this will usually be a relatively simple geometric shape. If liquid is withdrawn from a partly solidified ingot, the crystals can be seen in their natural shape as in figure 4.4, which is a particularly spectacular crystal about 15 inches high found in the shrinkage cavity in an ingot of gun steel. It looks like a tree, and indeed such a crystal is called a dendrite. In an alloy the material between the branches, which will be the last part to solidify, will generally be of a different composition from the cores, quite easily seen in the structure afterwards. Figure 4.5 is the microstructure of a Chou dynasty Chinese bronze casting from the Freer Gallery. In the rather irregular grain boundaries where adjacent crystals have met can be seen little pools of tin-rich material which mark the last parts to solidify.

Although crystals are geometric, the boundaries along which they meet are rarely so, and in an aggregate of crystals that has had some chance to adjust, the boundaries form a network which is quite similar to a froth of soap bubbles. The boundaries are smoothly curved, and they meet at 120 degrees because only at this angle are the surface-tension forces in local equilibrium. Though the metallurgist usually looks only at plane sections through his structures, it is possible to separate three-dimensional grains that have equilibrated this way, and they have a pleasant irregular shape. Figure 4.6 shows crystal grains separated from a lump of beta brass by dipping it in mercury, which attacks the grain boundaries. The nice curved surfaces and the sharp edges where they meet will be noticed. The faces of these grains, like soap bubbles and the cells in undifferentiated biological tissue, are most frequently pentagonal. The average number of sides is very nearly $5\frac{1}{7}$, the ideal number that is very basic in nature for it marks the interplay of two- and three-dimensional features under conditions of minimum area.

It is characteristic of metal crystals that, unlike most mineral crystals, they can be distorted without losing coherence. Unless this is done hot, however, the distortion introduces innumerable internal imperfections into the crystals and they become hard. If a cold-worked metal is heated to an appropriate temperature, innumerable tiny new unstrained crystals will form and will grow at the expense of the strained material—the

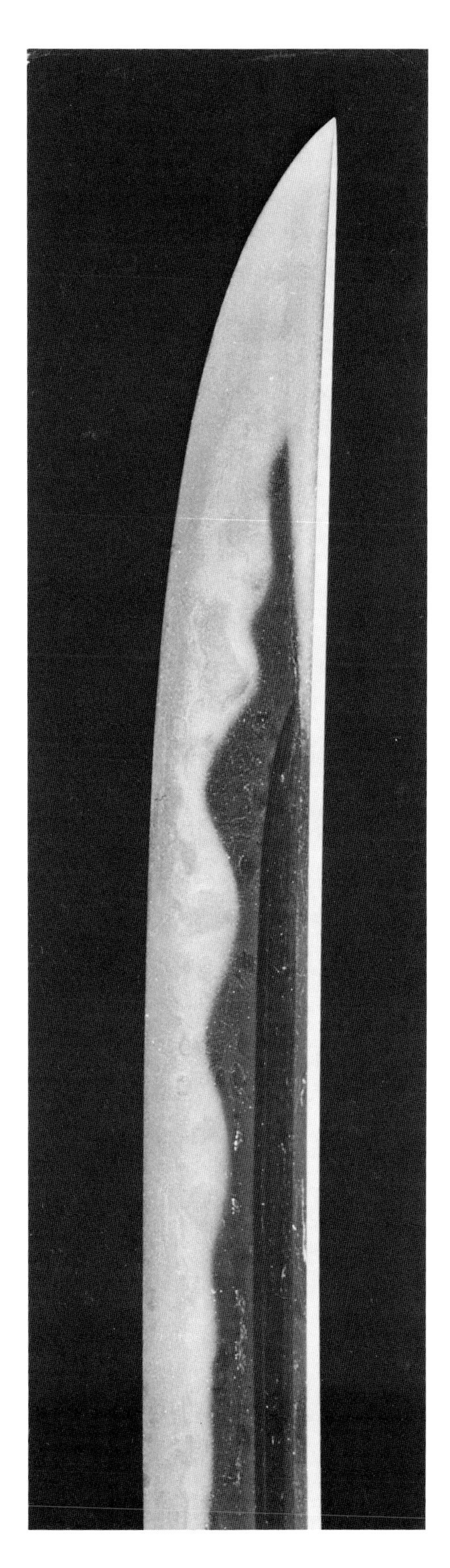

Figure 4.2
Detail of a Japanese sword blade made by Hiromitsu Sagami, dated 1362 A.D. Note the fine texture in the polished surface and the rich structure in the hardened zone near the cutting edge. See also figures 4.49–4.51. Photo courtesy of the Walter A. Compton Collection.

Figure 4.3
The Elbogen Iron Meteorite. This is a direct typographical imprint from the etched surface made by Schreibers and Widmanstätten in 1813. × 0.4.

Figure 4.4
Two views of a 15-inch dendritic crystal found in the shrink head of an ingot of gun steel (Tschernoff, 1899). The sketch shows its original position in the shrink head.

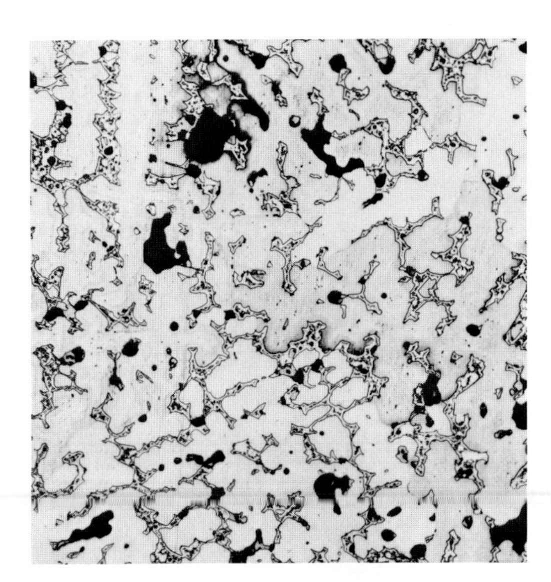

Figure 4.5
Microstructure of cast bronze vessel, Chou dynasty. × 200.

Figure 4.6
Crystal grains of beta brass separated from an aggregate. × 8.5. These show the natural shape of space-filling crystals in random contact with each other. Note that the surfaces are frequently five cornered and are rarely plane.

process known as recrystallization, by which the metal becomes soft again. These crystals will grow into the strained matrix until they meet each other, then forming a new soap-bubble-like network of boundaries. The boundaries are usually curved, for the simple reason that they must join at 120 degrees, but not all crystals (seen on a two-dimensional section) have the six neighbors that would reconcile this angle with straight sides. The crystals with less than six sides are therefore convex, while those with more than six are concave. In a soap froth, this means that the smaller ones are under higher pressure than average, while the larger ones have lower pressure. Gas then diffuses from the former to the latter; the big bubbles eat up the small, but equilibrium is never reached. Exactly the same thing happens with metal grains, leading to bigger and bigger crystals, and softer and softer metal. This whole sequence is controlled by the interfaces, not the crystal bodies which are frozen and uninteresting.

A crystal usually has a composition different from that of the liquid from which it has grown and it may progressively change composition during solidification. These composition changes are conveniently studied in the constitution diagrams of alloy systems, in which the ranges of existence and coexistence of chemically or structurally distinguishable forms of matter are indicated as mapped against temperature and composition. Such a diagram from the copper-tin system is shown in figure 4.7. Solidification of a bronze cooling from the liquid state occurs, of course, in the region marked solid + liquid, and it is easy to see that as the temperature drops and more solid is formed, its composition becomes progressively richer in tin, while the liquid gets even more concentrated in tin. As a result of this, there is a

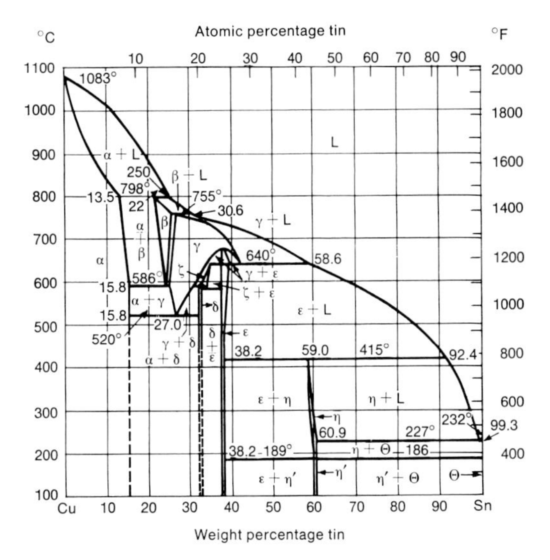

Figure 4.7
Constitution diagram for copper-tin alloys. This is a kind of map showing the regions of temperature and composition wherein the alloys are liquid and where the different crystalline phases, marked alpha, beta, gamma, etc., exist. (The dotted lines below 450°C show the conditions to be expected under normal metallurgical treatment: under true equilibrium conditions additional changes occur.)

gradient of composition from the first to the last parts to solidify, sometimes quite easy to see under the microscope as a gradient of etching within the crystals.

There can be changes in the solid just as in the liquid, for one crystalline phase may, as the temperature changes, form and grow at the expense of another one just as in the first place the crystal formed and grew at the expense of the liquid.

The structures resulting from these solid-state transformations are usually highly oriented since the new crystal forms within or in contact with an old one, and the nucleus forms at an orientation such that the crystal planes match as well as possible so as to minimize the energy of the interface between the two. This was the origin of the meteorite structure in figure 4.3.

In addition to the above changes in structure which occur during fabrication, there can be changes occurring at room temperatures. Of great importance to the archaeologist is corrosion which advances irregularly and eventually obliterates the whole metallic structure. The presence of deep-seated intergranular corrosion is a fair certificate of antiquity. In this area lies a most important field for research, and it will be as informative to the scientist as to the archaeologist for the structures are not those of accelerated laboratory tests, but have more in common with the natural formation of minerals.

Let us now look at some artifacts.

Structure of Copper Artifacts

The earliest known man-made metal objects are of copper. Copper occurs in the metallic state in nature, and it is reasonable to expect that the earliest objects would have been

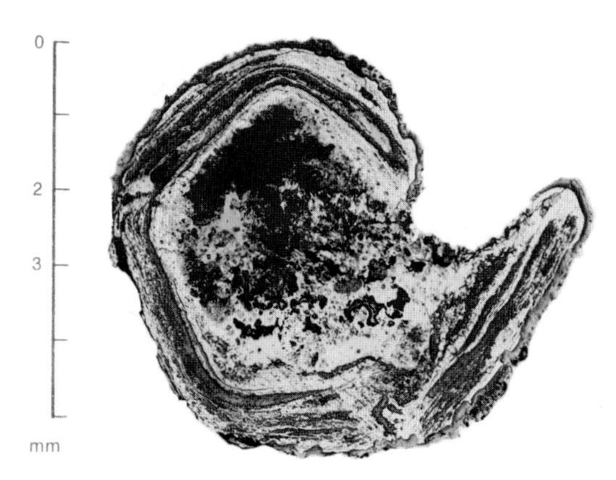

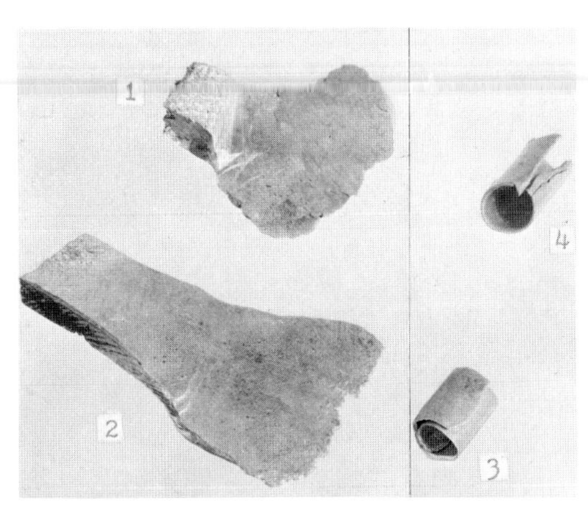

Figure 4.8
Polished cross section of a copper bead from Ali Kosh (Khuzistan province, Iran). The metal is completely corroded, but the resulting cuprite and malachite have preserved the original shape.

Figure 4.9
Worked specimens of native copper. (1) Copper from Talmessi and (2) copper from North Michigan, hammered without annealing to show malleability; (3) and (4) are the results of attempts to duplicate the Ali Kosh bead by coiling hammered sheet: 4 was rolled up after annealing; 3, which was coiled in the work-hardened condition, shows abrupt bends associated with a mechanical yield point. About natural size.

simply hammered to shape with no other treatment than beating between smooth stones. The standard books on metallurgical archaeology and archaeological metallurgy say it is impossible to work native copper very much without annealing it, but this is erroneous, and a selected piece of native copper can be worked almost infinitely without cracking. The earliest copper artifacts known at present date from about 8500 B.C. in Iraq and shortly afterwards in Anatolia. These are unmistakably hammered copper made without annealing. The earliest piece that I have examined is a rolled necklace bead dated about 6500 B.C., found at Ali Kosh in southwestern Iran. It was given me by Professor Hole of Rice University. A cross section is shown in figure 4.8. No metallic copper remains, though the original shape is still clearly preserved in the corrosion-product minerals. Note the rather sudden bend, indicative of what the metallurgist calls yield-point behavior, which is duplicated in modern heavily cold-worked copper but not in soft annealed copper (figure 4.9). There is little doubt that the bead had been made of heavily hammered copper that had never seen an annealing fire. This lack of annealing is quite evident in a slightly later piece from the famous site at Sialk, in which there is still some metal left inside the corrosion product (figure 4.10). This structure is exactly duplicated in a modern piece that I myself hammered from a lump of native copper picked up in the mine at Talmessi near Anarak, about 200 kilometers away from Sialk (figure 4.11). The markings are crossed deformation bands, not grains, for the grain size is enormous. A piece of the original native copper (figure 4.12) shows a curious shading near some of the grain boundaries which is probably a result of their having migrated in geological time.[5]

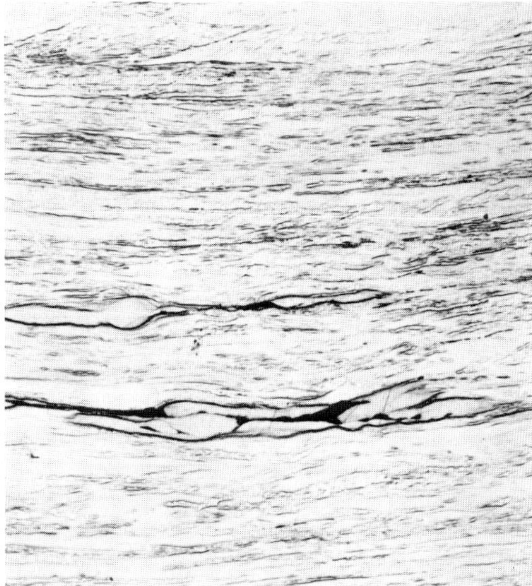

Figure 4.10
Microstructure of copper needle from Sialk, second half fifth millennium B.C. (Ghirshman No. 7281). × 200. Longitudinal section, etched with potassium dichromate. Shows a heavily cold-worked structure with intersecting strain markings, but no grain boundaries are visible and there is no recrystallization. (Hardness VHN 109.)

Figure 4.11 (*right, top*)
Microstructure of native copper from Talmessi mine, Iran, hammered cold to an 82 percent reduction in thickness. (Hardness VHN 130.) The inclusions are a malleable copper-arsenic mineral approximating the compound Cu_8As.

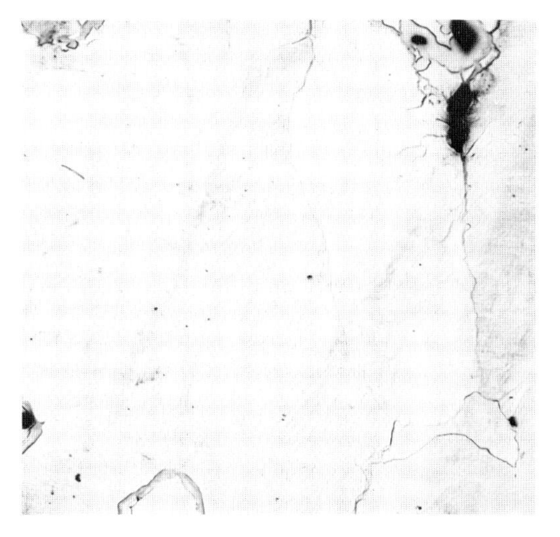

Figure 4.12 (*right, bottom*)
Microstructure of native copper from Talmessi, Iran. Etched with potassium dichromate. × 50. (Hardness VHN 82, except in dark areas which are about 100.)

Compare the structure of the unannealed Middle Eastern objects with that of a spearhead made in North America from native copper at the time of the Old Copper Culture (3000 to 1000 B.C.) (figure 4.13). The microstructure proves unmistakably that this had been well annealed—at a temperature of over 700°C—for the grains are large and have well-formed twins, quite distinct from the structure of the native metal either before or after working and quite unlike that of cast metal.

The Near Eastern objects were small decorative pieces: North American Indians made large serviceable weapons and tools that were often beautifully shaped. It is curious and significant that although metallurgy in the Near East passed relatively soon from the use of native copper to smelting, melting and the whole range of metallurgical techniques, in North America the development stopped with the annealing of native metal. (Incidentally, the Indians rarely made use of cold work to strengthen their copper tools. The structures show that they were usually finished by annealing—or perhaps they had been shaped entirely by hot working; it is impossible to tell which.)

Figure 4.14 shows an untypical American Indian axe that had been heavily cold worked. Note that it is beginning to recrystallize in one area. Extremely pure copper, when heavily worked, will recrystallize in less than a year at room temperature, but the small amount of silver (> 0.02 percent) that is present in most native copper prevents any change. This axe has an unusually low silver content and some recrystallization has occurred during the 3000 years or more since it was made.

Figure 4.13
Microstructure of a copper spearhead, Old Copper Culture, American Great Lakes region. Etched. × 50.

Figure 4.14
Structure near tip of copper spear. Etched. × 50. The metal, heavily cold-worked, has begun to recrystallize for it is unusually low in silver. The microhardness of the cold-worked metal is VHN 86 to 125; of the recrystallized grain, only 56.

At the native copper stage, the American metalworkers were far ahead of those in the Middle East. That they did not proceed to the discovery of melting and casting and the most important step of smelting ores must be associated with the whole of their culture, for they did not know of high-fired or glazed ceramics or show other awareness of the utility of fire in modifying the very nature of matter.[6]

A somewhat later piece of copper from Sialk was also examined. It was rectangular, about 2.1 by 4.4 mm in section, and was shown by analysis to be very pure copper such as would result from melting native copper, rather than from smelting an ore. The structure (figure 4.15) shows a clearly distinguishable network of tiny particles of copper oxide that could only have formed as a eutectic during the solidification of copper containing a little oxygen—perhaps 0.03 percent. This copper had been melted and cast into a rectangular mold before being hammered to its present section. By measuring the approximate size of the eutectic cells as they appear in the three dimensions of the piece, it can be shown that the original casting had been about 8 mm square in section before it was hammered. This is all clearly reflected in the distortion of the eutectic, though had the deformation gone much farther the oxide would have been redistributed irregularly and it could not have been measured. This piece had clearly been annealed, for it has fully recrystallized equiaxed twinned grains growing through and around the copper oxide particles.

Figure 4.15
Transverse section of tool from Sialk (Ghirshman No. 1425). Shows recrystallized structure and distorted copper oxide eutectic. × 200. VHN 54.

Figure 4.16 shows an Iranian chisel which has some impurity in it, probably arsenic and silver, in addition to oxygen. It too has been cast, worked, and annealed. There are two levels of structure. There are rather hazily defined patches that etch more darkly than the background and are a residue of the original interdendritic segregation that occurred during solidification. These cast crystals have been completely broken up by the working and annealing treatment, but the impurity gradients have not been greatly changed. The much smaller twinned crystals result from recrystallization during annealing. Note that their boundaries ignore this composition variation, for there is not enough arsenic to give rise to a new crystal form. Here, too, the difference in structure of a longitudinal cross section (figure 4.17) makes the directionality of the forging obvious. The composition gradients, know as "coring," would have been eliminated by the diffusion that occurs during a long annealing treatment, for example 900°C for two or more hours.

Copper Alloys: Bronze

Copper with natural impurities from selected ores was historically an important introduction to the advantages of alloys. Copper with a few percent of antimony and/or arsenic is easier to melt, gives sounder castings and is mechanically stronger and harder than pure copper. The alloy with arsenic was purposefully made and used many centuries before bronze (copper and tin) appears. However, by 3000 B.C. the outstanding properties of bronze (alloys of copper with tin) had been discovered, probably at first as a result of smelting mixed ores, later by alloying separately smelted metals. The constitution diagram for the copper-rich part of the copper-tin system is shown in figure 4.7. (This diagram shows what occurs in relatively short times of heat treatment. There

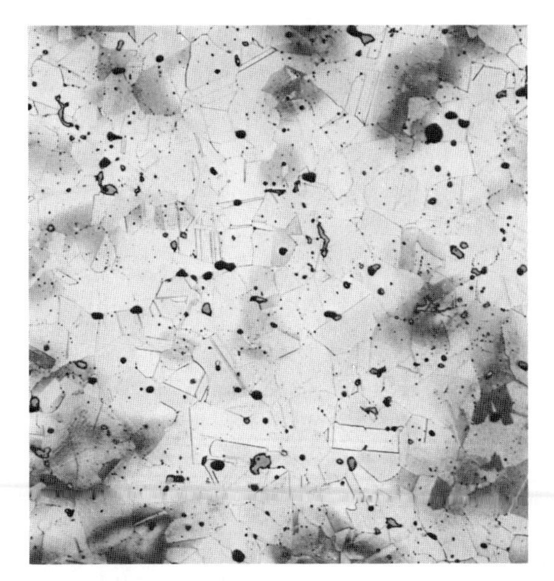

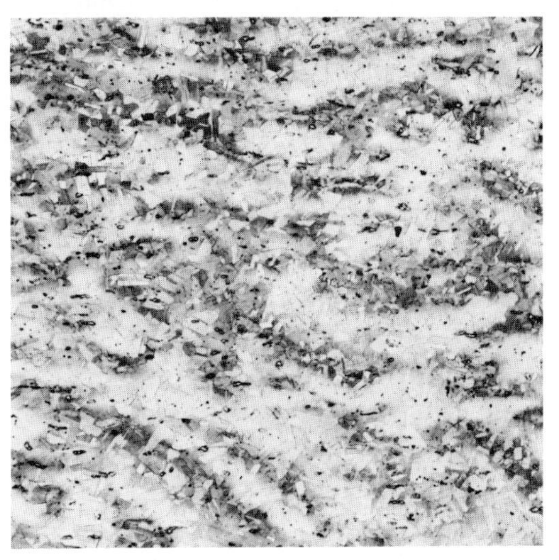

Figure 4.16
Cross section of a copper chisel from Tall-i-Nakhodi, Iran (ca. 3000 B.C.). Specimen from David Stronach. Etched with potassium dichromate. × 200.

Figure 4.17
Same specimen as figure 4.16, longitudinal section. Etched with ammonia and hydrogen peroxide. × 50.

are additional very slow changes below about 350°C that are not shown here.) It will be obvious from the great separation between the compositions of the solid and the liquid that there is plenty of opportunity for segregation during cooling. This results in the presence of some nonequilibrium high-tin phase even in alloys with 10 percent or less tin which in equilibrium are homogeneous alpha phase. The structure of the cast bronze (figure 4.5) consists of dendrites of the main phase, alpha, having a continuous change of composition within their substance; and duplex areas which look like eutectic but actually are of a similar structure, known as eutectoid, that is formed in the solid state from the decomposition of the transitory beta phase, originally the final part that solidified. This casting solidified slowly. If the mold had been colder, or a better conductor of heat, the grain size would have been smaller, and the spacing of the dendritic branches would have been less. Turbulence in the mold may prevent the growth of fully geometric dendrites and give smaller grain size.

In most metals there is a substantial decrease in volume on passing from the liquid to the solid state. Since it is difficult for a liquid to run into the fine crystal interstices and since gases are often evolved by reaction between traces of hydrogen, carbon, and oxygen, which become concentrated in the diminishing liquid, holes are frequently formed in place of metal in the last parts to solidify. These are the black areas in figure 4.18. In addition, most cast bronzes contain lead, which is added to improve the castability and the color, and for economy's sake. The copper-lead constitution diagram is shown in figure 4.19. There are no interme-

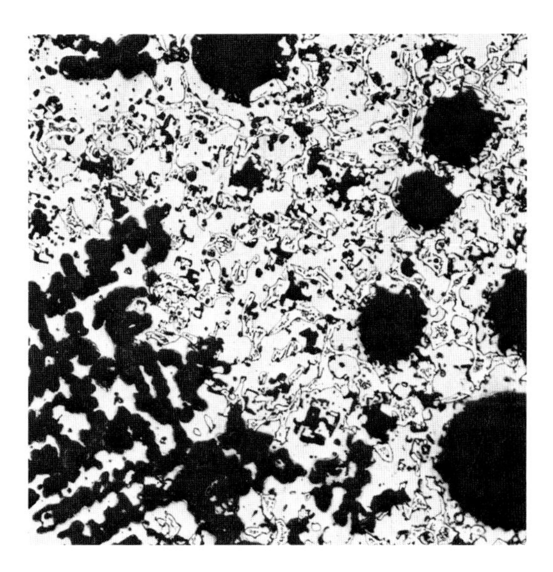

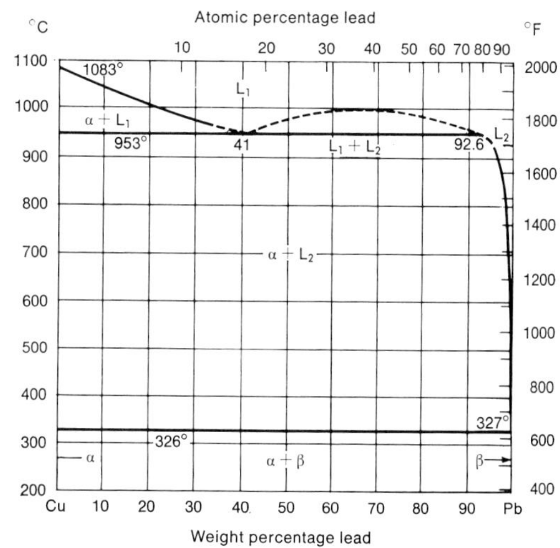

Figure 4.18
Cast bronze arrowhead. Chinese, probably Chou dynasty. Shows interdendritic eutectoid, shrinkage cavities, and lead in both globules and interdendritic form. Etched with potassium dichromate. × 200.

Figure 4.19
Constitution diagram for copper-lead alloys. Note the region in which two immiscible liquids exist. From *Metals Handbook*, A.S.M., 1949 edition.

diate solid solutions or compounds formed, but there is a range of composition and temperature in which two liquids form and can separate, much like oil and water. The bronzes with lead act similarly. At high temperatures the metals mix completely; as the alloy cools, spherical droplets of lead-rich liquid may form in the molten bronze and if there is not time to sink out they will remain to become entangled in the growing crystals. Even if there is not enough lead to form separate drops in the all-liquid field, it is rejected by the solid and concentrated in the liquid during solidification, eventually to occupy irregular interstitial interdendritic spaces. Both this form and spheres can be seen as dark grey areas in figure 4.20. The fact that lead can segregate in the liquid state often results in great differences in composition in different parts of a casting, and a chemical analysis on a small sample may be very misleading.

The microstructure of a bronze casting is quite characteristic and is changed by working or by heat treatment. If the casting is annealed for a sufficiently long time, composition gradients will be erased by diffusion, and the beta eutectoid will disappear unless the alloy contains more than about 13 percent tin. This equalization of composition will proceed much faster if the alloy is cold-worked before annealing. In addition the alloy will completely recrystallize to form more or less equiaxed grains with twin boundaries in them much like the worked-and-annealed crystals of native copper (figure 4.13), but somewhat simpler in form.

Figure 4.20 is a bronze fishhook of the sixth century B.C. from Corinth. In the course of shaping, it had been repeatedly hammered and annealed and the grains are fine and, in most places, well formed. The fine markings across the grains are slip bands

Figure 4.20
Microstructure of bronze fishhook, Corinth, sixth century B.C. Specimen from J. Hawthorne, University of Chicago. Etched with potassium dichromate. × 500. The metal contains 11 percent tin.

from some residual cold work in the final stage of shaping, but the grains in the area photographed are not much distorted. Figure 4.21 is a complete longitudinal section through the hook at low magnification, and it shows clearly the residual structure of the casting, revealed by segregation that was not entirely obliterated by the working operation. The hardness in various parts of this object is plotted in figure 4.22. The point and barb are harder than the thickest part of the body. Evidently a slender cylindrical rod of metal had first been formed and annealed, then the barb was raised by something like a partial chisel cut and shaped by cold hammering. The barb is quite hard—reaching a maximum of 290 VHN—while the less worked parts of the body from which it had been bent are only 145. Both barb and point were clearly cold hammered at the last stage, and no further annealing was given. The shank of this hook is also cold-worked (about 260 VHN). Either it was worked at the same time as the barb, or possibly only that end had been annealed prior to its final shaping. Altogether it is a well-fashioned object.

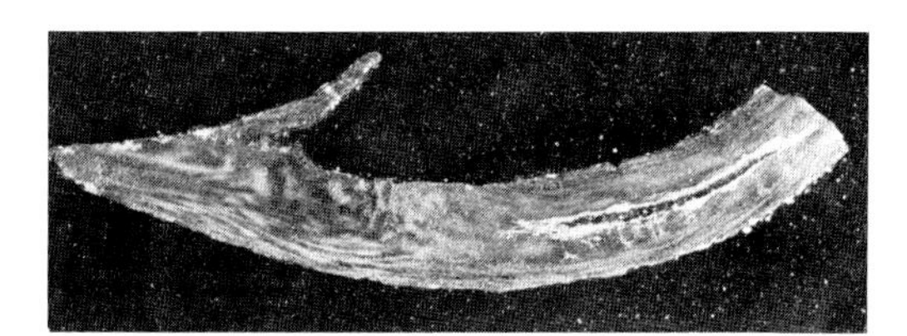

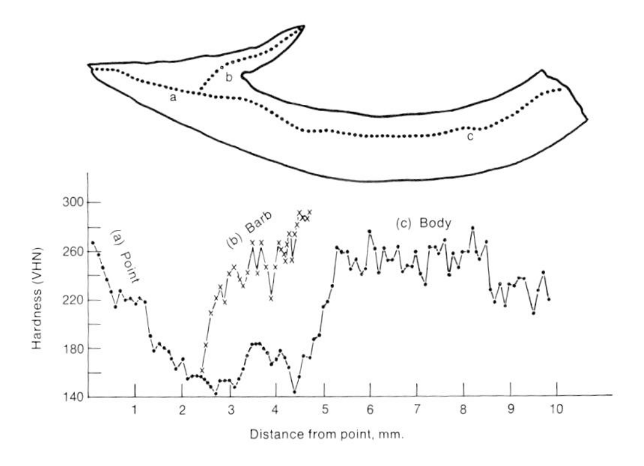

Figure 4.21
Longitudinal section through lower portion of Corinth fishhook. Etched with potassium dichromate, followed by ferric chloride. × 7.

Figure 4.22
Variation of microhardness in different parts of the hook shown in figure 4.21.

Note on Corrosion

In ancient metals, some or all of the microconstituents have usually been replaced by corrosion products. Though conditions vary greatly, the eutectoid in bronzes is usually more easily chemically attacked, and hence corroded to much greater depth below the surface of the metal than is the alpha phase. At an early stage of corrosion in some soils the tin-rich phase has been converted by local electrolytic action into spongy copper with interstices filled with tin oxide, while later the copper turns to cuprite

(Cu_2O) and eventually to green malachite ($Cu_2CO_3(OH)_2$). It should be noted that the corrosion products occupy much more volume than the metal they replace, and the metal ions for which there is no space diffuse through capillary channels to the outside of the piece, where they form an irregular and rather loose mineral layer, usually of malachite—the external coating which is commonly removed by electrolytic or mechanical cleaning in the conservation laboratory. In the early stages of corrosion, the copper formed by the continuing electrolytic action between it and the alloy (which serves as anode) will be deposited in any available space, such as cracks or in spaces left by the still earlier corrosion of lead (figure 4.23). In the course of time this copper too will corrode to cuprite, which is then a pseudomorph of the lead drop that originally existed.

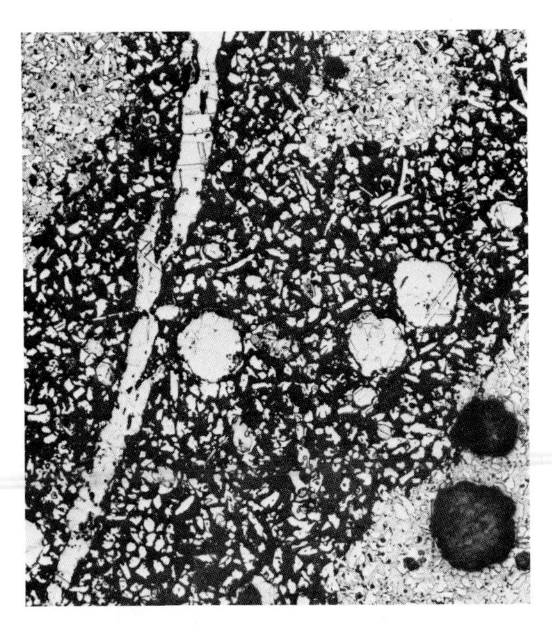

Figure 4.23
Microstructure of a high-tin bronze used as a solder on a Chinese ceremonial vessel, late Chou dynasty. Heavily corroded. Etched with potassium dichromate. × 200. Note the redeposition of copper by electrolytic action in cracks and in the spherical spaces left by corrosion of the lead droplets.

Another type of corrosion in bronzes (which seems to be a form of stress-corrosion, for it occurs only in metal that has been slightly cold-worked) forms a geometric pattern of corrosion along the octahedral planes of the crystal (figure 4.24). For crystallographic reasons, it resembles the structure of the meteorite shown in figure 4.4, although the scale and the chemistry are totally different. The corrosion product is duplex, having a stripe of different mineral in the middle, probably because of differing amounts of tin oxide in the corrosion products formed at different times.

Diffusion in Solid Metals

A solid metal is not changeless even if it is preserved in a noncorrosive environment, although the changes are negligibly slow at room temperatures for all but the low-melting-point metals. They are, however, significant during fabrication. We have already seen the ironing-out of composition gradients in a casting by diffusion occurring during annealing. An early important use of diffusion was in the surface enrichment of alloys containing the precious metals. When an alloy of copper with a small amount of gold in it is heated in air, the copper will diffuse to the surface to form a copper oxide layer, while the underlying metal becomes enriched with gold. This technique was widely used by pre-Columbian metalworkers in South America on alloys containing 25 percent of gold or even less. After the *mise-en-couleur* treatment, these shone like rich solid gold—to the disillusionment of the Conquistadors.[7] In the Old World, similar techniques were used mainly on richer gold alloys for superficially improving the color. The blanching of silver was a related technique, widely used for legal deceit in producing a silver-rich surface on billon coins and for whitening those of sterling composition. Deep diffusion and reaction was the basis of the ancient process of cementation in which granulated gold was freed from silver, copper, and other impurities below its melting point, as well as similar processes for putting carbon into iron in making steel, or zinc into copper in the old calamine process of brass making.

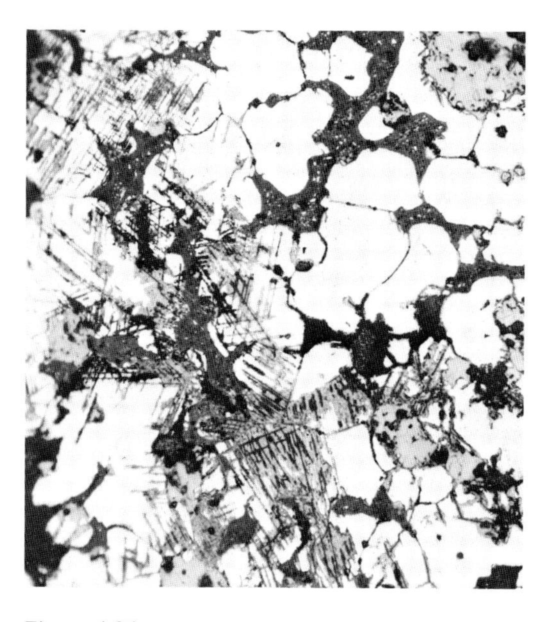

Figure 4.24
Corrosion along crystallographic planes in slightly cold-worked bronze. Microstructure of a bronze *pan* (Chou dynasty), area in vicinity of a chaplet. Etched with potassium dichromate. × 500.

The changes just discussed involve chemical reaction with the surface environment. Change can also occur internally in an alloy. The recrystallization and softening that occurs on annealing of heavily cold-worked metal or alloy is an example of purely structural atomic rearrangement. Another change, the basis of a twentieth-century discovery that is now an important way of hardening alloys, is the precipitation of crystalline particles that can be made to occur entirely within the matrix of another, previously homogeneous, crystal. Archaeologically, copper-silver alloys are the main ones that show such precipitation.

Structural Change in Silver Alloys Containing Copper

Silver was probably first used by man in electrum, its natural alloy with gold, but artifacts of pure metal occur very early since it is found native and is easily reduced from chloride or sulfide ores. Gold and silver alone among the common metals resist oxidation, and the refining of silver consists simply in heating it above its melting point in air to oxidize the impurities: at some point it was found that the process worked better if lead were present, for the lead oxide effectively dissolves more refractory oxides and accelerates the process. This process (known as cupellation from the little cups of absorbent refractory, commonly bone ash, which were used) was well known by Classical Greek times. Many lead ores contain small amounts of silver, and oxidizing large amounts of lead would produce small amounts of silver: the ores of the famous Laurion mines were actually of lead though, as is common, they were named for the more precious component. Cupellation was rarely carried to completion, and traces of lead usually remain in silver. Figure 4.25 is the microstructure of a little flat cake of Anatolian silver, which shows pools of lead between the silver crystals. Corrosion has resulted in the lead being completely replaced by a mineral, probably cerrusite, but the silver is untouched except at the extreme surface.

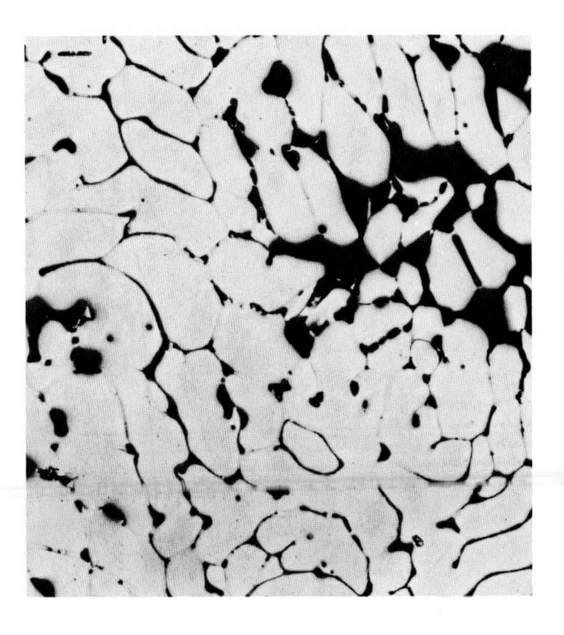

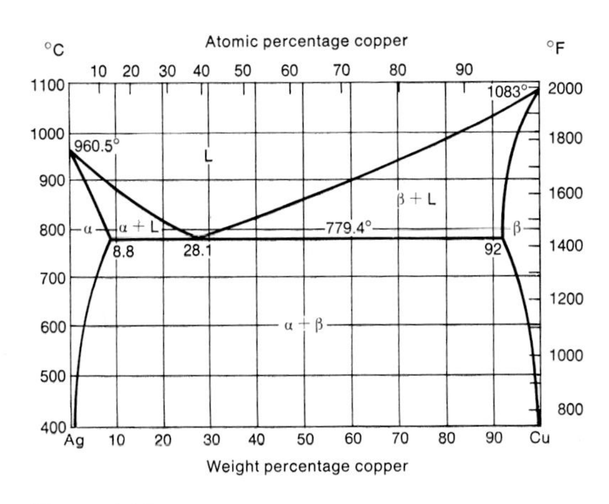

Figure 4.25
Microstructure of a flat cake of silver (Anatolia, fourteenth to thirteenth century B.C.). Etched with potassium dichromate. × 50.

Figure 4.26
Constitution diagram of silver-copper alloys.

In later times silver was alloyed with copper, partly as an inevitable impurity arising from repeated remelting, partly for reasons of economy, and partly because such alloys are stronger and harder than pure silver. The presence of copper introduces an unpleasant instability, and alloys of any antiquity containing over about 3 percent copper are invariably brittle. A glance at the constitution diagram, figure 4.26, shows that silver will hold in solution a relatively large amount of copper at the eutectic temperature, but progressively less at lower temperatures, where it will be precipitated—a fact that remained unsuspected until about 1925, despite centuries of work with the alloys. Figure 4.27, based on studies by Cohen[8] following the pioneering work of Norbury,[9] shows that this occurs at a relatively rapid rate, and by extrapolation from the laboratory experiments could occur at room temperatures in periods of a few centuries. The microstructure of some old silver-copper alloys appears in figures 4.28–4.30. Precipitation has occurred in continuous patches which start from the grain boundaries and grow out into the grains. (This is called "discontinuous" or "cellular" precipitation to distinguish it from general precipitation of particles throughout the grain such as occurs in Duralumin, for example.) This precipitation occurs relatively fast at intermediate temperatures, even—with relatively high copper-content alloys—during normal cooling following the silversmith's annealing operation, though it is easily suppressed by quenching. The microscope, unfortunately, cannot easily discriminate between the precipitate formed during cooling or on subsequent reheating. If the silver had not been cold-worked after annealing, the precipitate would be in the form of more or less parallel lamellae; if it had been cold-worked, pre-

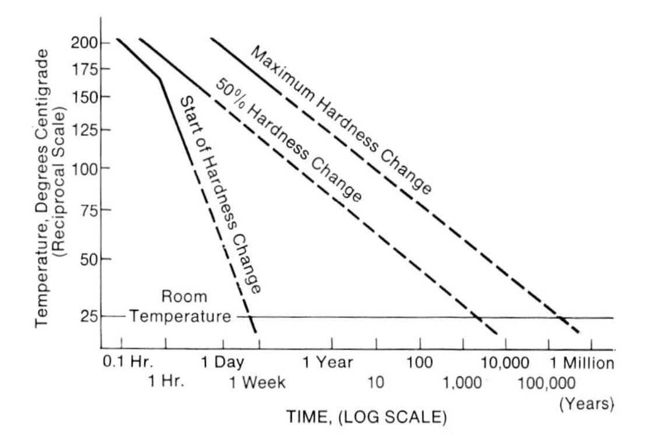

Figure 4.27
Curve showing the time required for change of hardness in a silver-copper alloy (8.75 percent silver) as a function of annealing temperature following quenching from 760°. (Based on data of M. Cohen, 1937.) The dotted portions of the lines are extrapolations and are very uncertain.

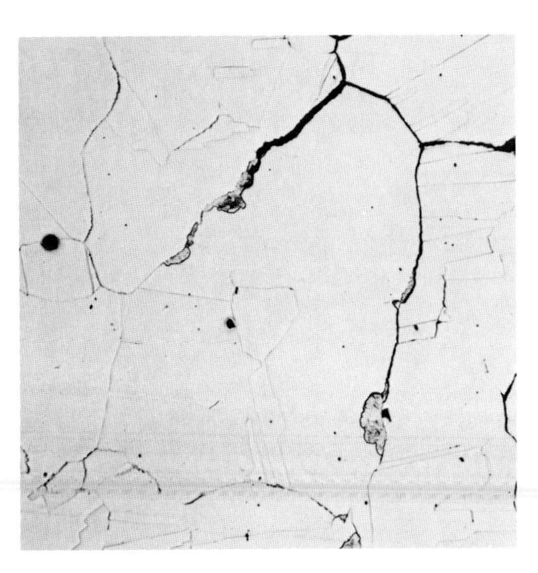

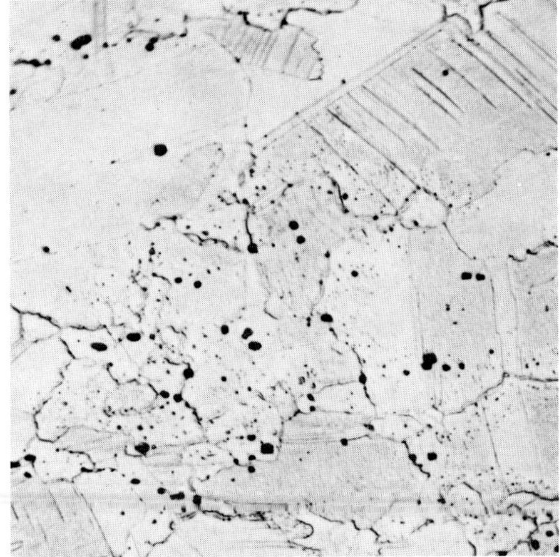

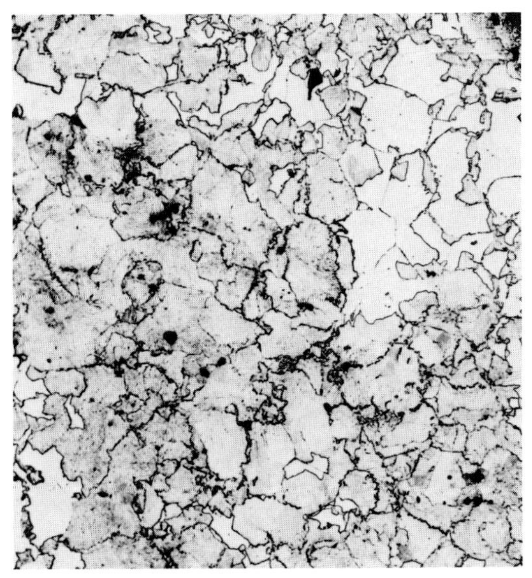

Figure 4.28
Silver repoussé head, Sasanian (Metropolitan Museum of Art). Etched with potassium dichromate. × 1000. Shows irregular grain boundaries associated with nearly invisible precipitation of copper. See Kate C. Lefferts, "Technical notes," *Bulletin* MMA, Nov. 1966, pp. 147–151.

Figure 4.29 (*right, top*)
Silver alloy sheet (Museum of Fine Arts, Boston). Etched with potassium dichromate. × 200. Shows beginning of precipitation accompanied by movement of grain boundaries.

Figure 4.30 (*right, bottom*)
Microstructure of ancient silver. Etched with potassium dichromate. × 200.

cipitation would be irregular, in spotty form accompanying local recrystallization.

In a silver-copper alloy, grain boundaries along which such precipitation has occurred seem to be highly susceptible to corrosion. Moreover there is a volume decrease accompanying the precipitation, and the combined effect is to form cracks even under mildly corrosive conditions. A frequent sight in the microstructure of old silver is the irregular grain boundaries shown in figure 4.28. Here the precipitation has barely begun, but the old boundaries have slightly moved, leaving behind the duplex precipitated structure. (A similar appearance of the boundaries is often seen in native copper, where the driving force is not precipitation but the relief of slight strain in the geologically deposited metal.) At low magnification such boundaries have a wiggly jagged appearance that is quite characteristic; at high magnification the patches of precipitation become visible.

No work has yet been done to show definitively the rate at which precipitation occurs at room temperatures in silver with different copper contents or to unravel the role of corrosion both accompanying or following precipitation of the copper.[10]

From the conservationist's standpoint, brittleness can be removed by heating the metal to a relatively high temperature (preferably in hydrogen), but even then it may not become really ductile unless it is hammered and reannealed. Such treatment will reconstitute a homogeneous alloy. Low temperatures of annealing (not in hydrogen) may coarsen a preexisting copper precipitate and if not in a reducing atmosphere, the corrosion products will gather into less harmful spheroidal particles. A silversmith aiming for permanence in his creations should not use copper-alloyed sterling. If not satisfied with the strength of pure silver, he might experiment with alloys of silver with zinc, cadmium, tin, or aluminum, all of which would probably be more stable than present sterling from the standpoint of both localized corrosion and metallurgical changes.

Copper-silver alloys provide good examples of another metallurgical phenomenon, that known as internal oxidation, which results in the formation of a layer of "subscale" beneath the usual oxide layer. During annealing (unless under reducing conditions) atmospheric oxygen diffuses into the alloy from the surface, oxidizing the copper in situ, to form innumerable tiny blobs of copper oxide embedded in a matrix of unoxidized metallic silver (figure 4.31). Silversmiths dislike this, of course, for it changes the surface appearance of the silver, and it is usually removed by scraping and grinding.

As with bronze, a copper-silver alloy in the cast condition is heterogeneous. The outer parts of the silver crystals are richer in copper, and if there is more than about 4 percent silver, some eutectic will usually form at the last stage even though it is not an equilibrium constituent below 8.8 percent copper.

A well-defined network of eutectic was used decoratively in Japan from the seventeenth century on. An alloy of copper with about 25 to 50 percent silver, known as *shibuichi*, when cast has a network of silver-rich eutectic on a scale just too small to be seen with the naked eye. When finished by polishing and chemically treated, it takes on a beautiful grey-brown-green patina with a delightful misty sheen that is very resistant to wear. Figure 4.32 shows the surface of a 200-year-old swordguard made by Masayuki. This is its original surface, not

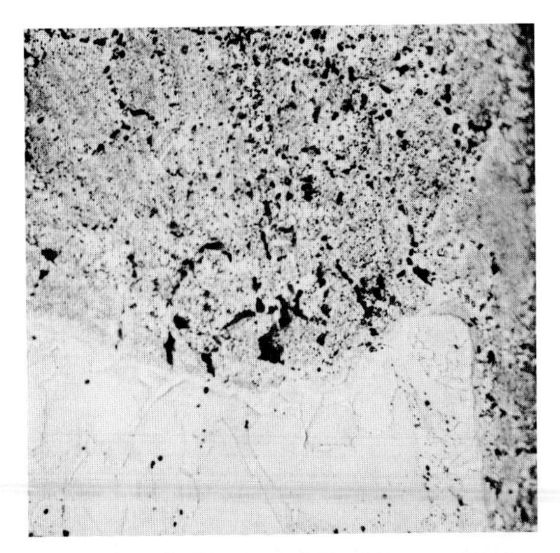

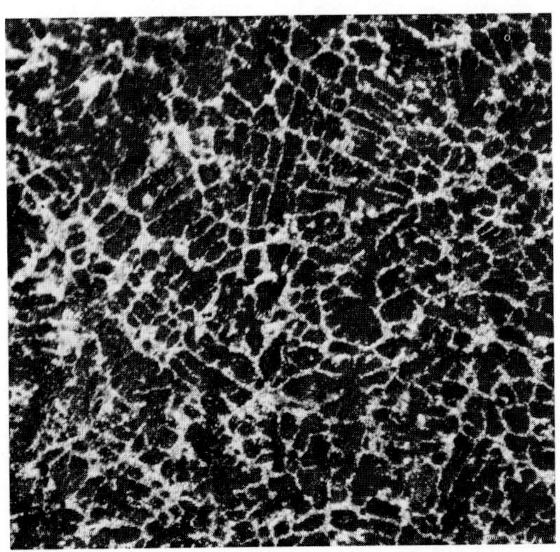

Figure 4.31
Zone of internal oxidation on the surfaces of a silver sheet. Etched with potassium dichromate. × 500. The dark particles are copper oxide formed by oxidation in situ.

Figure 4.32
Microstructure of a Japanese swordguard made of the silver-copper alloy *shibuichi*. Signed Masayuki (1696–1769). This is the original surface not repolished or treated in any way. × 85. Oblique illumination.

polished or retreated in any way, and would pass for a student's metallographic preparation. Both the sheen and the wear resistance of *shibuichi* come from the fact that the silver-rich eutectic stands slightly proud above the copper.

Wire, Filigree, and Granulation

Much of the beauty of even the earliest precious metal jewelry originates from the use of filigree work. The wire, however, was not drawn, and it is rather surprising that its extensive use should precede the invention of this process for so easily shaping metal. Die-drawing was perhaps used in making pure gold wire in Achaemenid Persia, but the conditions of the invention are not known and wire-drawing was certainly not common for the harder metals until very much later. Much mail armor, which so strongly suggests the drawing process, is actually not made of drawn wire. Figure 4.33 is a cross section of a link from a fourteenth-century European coat of mail: the transverse lamellar structure shows clearly that the wire had been rounded from a strip cut from a sheet. Turkish mail, even of the late eighteenth century, has a similar structure, but steel wire for European mail—and fishhooks—was being drawn by the fifteenth century.

An extraordinary and interesting example of metallurgical principles lies in the use of granules to decorate gold work, attached by a special kind of invisible soldering. The Etruscan goldsmith was superb at this, and the Greeks used it extensively. It is currently being used by John Paul Miller, the Cleveland goldsmith, whose work is the first to match that of the Etruscan heyday. Once the granules have been made by letting surface tension convert melted filings of gold into true spheres, the trick is to get a small amount of solder at the spots where it is needed, and not to flood other parts with the liquid metal.

Figure 4.34 shows a line of granules on an Etruscan gold fibula in the Boston Museum of Fine Arts; figure 4.35 a somewhat coarser display on the handle of the dagger from Tutankhamen's tomb.[11] The ancient technique is unknown, but it probably was similar to that described by Theophilus and Cellini, in which very finely ground copper oxide or carbonate is mixed with a little adhesive flux and painted thinly on the surface of the gold, then the assembly is heated to a temperature a little above 890°C where, although neither gold nor copper is melted, an alloy of about 20 percent copper is molten. In a reducing atmosphere, the copper oxide turns to copper, which forms a molten alloy by contact and diffusion with the underlying gold. The surface tension of the resulting thin film of liquid alloy draws the granules together and produces the joint. On further heating, the copper in the alloy at the surface will dissipate by diffusion leaving a perfectly solid joint, or it can be removed entirely by annealing under oxidizing conditions and pickling. This method is easier to work the purer the gold, and will not do for a low-karat gold for this already melts at a low temperature. The method would work equally well on silver, though for some reason silver has always been soldered,[12] as have all other metals, by the use of separate bits of a low-melting-point alloy that are applied somewhere to the surface and allowed to melt and run into places where they are needed. This is clumsy-looking compared with granulation, but just as effective mechanically.

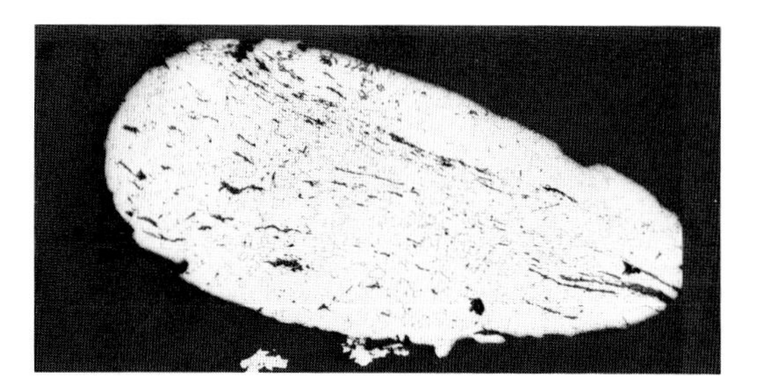

Figure 4.33
Cross section of link in mail armor. German, ca. 1500 A.D. Etched. × 3.7. The laminar arrangement of the slag shows that the wire had been made by cutting a rectangular strip from a sheet to approximately the final size, and it has only been slightly shaped since. Note the low carbon content.

Figure 4.34
Detail showing granulation work on an Etruscan gold fibula, sixth century B.C. × 25. The granules are 0.17 mm in diameter.

Figure 4.35
Granulation work on handle of Tutankhamen's dagger. × 10. Photo by W. J. Young.

The same principle of surface-tension shaping that was used in making these granules was also employed in making flans for coinage, which at first were simply large sessile drops of gold and silver. They were not cast in molds, which would have given variable weight, but were individually melted from a weight-adjusted collection of scraps to give a well-shaped flattened drop ready for striking.

The earliest solder to be used on copper seems to have been lead, which is a very poor material for the purpose. The earliest good soldering was that done with gold-copper or silver-copper alloys on jewelry. Common lead-tin solder and brazing solder (a copper-zinc alloy melting about 900°C or less if tin is added) appear much later. Early bronze workers, however, made innumerable joints and repairs by the running-in process, in which molten metal is made to flow through a space molded around the parts to be joined until the surfaces are fused. The related process of casting on, in which a precast part is incorporated in a mold so as to project into the space forming a main part of the casting, was used extensively in Chinese bronze ceremonial vessels, and in innumerable weapons and decorative objects in the Middle East and Europe.[13]

Iron and Steel

Figure 4.36 compares hardness of iron-carbon alloys (i.e., steels) and copper-tin alloys. Unquenched iron, or even steel, is no better as a tool or weapon than cold-worked bronze, and it is somewhat more difficult to fabricate. The importance of the first introduction of iron was primarily an economic one, for iron ores are widely available. Quenched steel, however, is spectacularly harder than bronze, and after quenching was

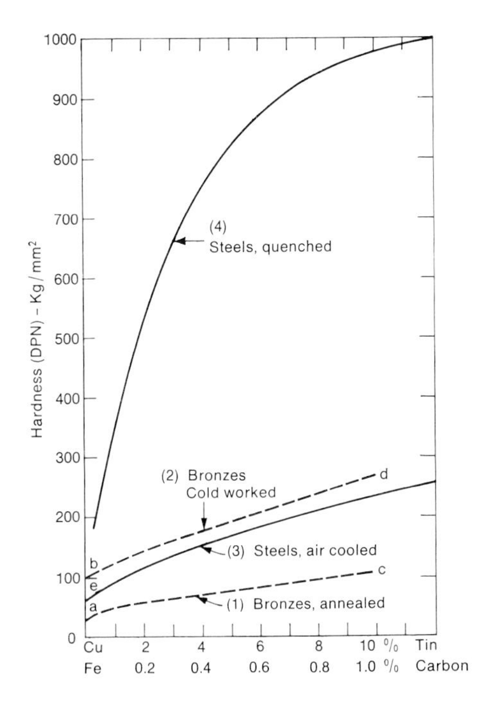

Figure 4.36
Curves illustrating the principal ways of hardening metals. Pure copper when annealed has a hardness (DPH, measured with the Vickers diamond pyramid indenter) of about 40 kg/mm^2 (point a). Cold working hardens copper progressively to about 100 kg/mm^2 (point b) after 70 percent reduction in thickness, and to a maximum of about 120 kg/mm^2, which is reached after 95 percent reduction in thickness. Alloying copper with tin in amounts up to about 10 percent (curve 1) progressively raises its hardness to about 110 kg/mm^2 (point c) if the alloys are annealed. The hardness can be further increased by cold work (curve 2) up to about 270 kg/mm^2 (point d). Pure iron, which has DPH of about 60 (point e), is hardened by the addition of carbon, the essential element in steel. If steels are heated and allowed to cool naturally, the range of their hardness (curve 3) is slightly below that of worked bronzes, but they become spectacularly superior if quenched (curve 4). The curves are approximate and the hardness varies considerably with impurity content, details of casting technique, prior annealing, and other factors. The brittleness of an alloy generally increases with its hardness.

introduced to harden the steel in a controlled way man could do previously impossible things with metal.

Apart from ceremonial arms and armor, the ferrous metals have been rather neglected by art collectors and curators.[14] The magnificent iron and steel objects that occur copiously in the graves that provide the world-famed Luristan bronzes are hardly known—mainly, I suppose, because time in destroying the surface has not, as on copper, replaced it with a corrosion product that is itself well formed and of pleasing color. A metallographic examination, however, is far more revealing of the history of an iron artifact than it is of one of copper. The macroscopic distribution of carbon and of slag inclusions shows how a piece had been welded and forged, and microscopic details of grain shape reflect intimately the final heat treatment. This comes mainly from three facts: (1) prior to about 1860 most iron and steel objects, except in China, were not melted but were consolidated without melting and therefore they retain a considerable heterogeneity originating in the mixture of sponge iron, steel, and slag that was found in the hearth of the bloomery or finery; (2) solid iron when heated in a charcoal fire readily absorbs carbon; and (3) iron undergoes an important change at 910°C to a high-temperature crystal form (called austenite by microscopists) in which carbon is significantly soluble and from which it crystallizes on cooling in a variety of forms highly characteristic of the cooling rate.

The making of iron has gone through two major changes. One was in the fifteenth century when the blast furnace was first used in the Western world to produce liquid cast iron from the ore for subsequent refining to wrought iron, at first in the finery hearth and later (1784) in the puddling furnace, both refining operations being carried out below the melting point of pure iron. The other change occurred around 1860 when, for the first time, it became possible to melt large quantities of low-carbon iron and cast it into relatively homogeneous ingots, so eliminating much of the slag that had characterized iron from the very beginning. Cast iron, which is iron containing around 3 percent of carbon and some silicon and is easily melted, was at first extensively made in Europe only for refining to wrought iron, though it was used increasingly for pots and pipes, firebacks, and occasional works of art before it became architecturally important in cast beams and machine frames at the time of the Industrial Revolution.

The constitution diagram of the iron-carbon system is shown in figure 4.37. We are mainly concerned with the area below 1 percent carbon. Quenching from the austenite area prevents the formation of equilibrium iron carbide and gives an intermediate phase known as martensite, the hardness of which increases with carbon content as shown for the quenched steels in figure 4.36. If the steel is slowly cooled, carbon-free ferrite (the low-temperature form of iron) will grow from austenite, concentrating the carbon in the latter (just as tin is concentrated in the liquid in a solidifying bronze, or alcohol in applejack) until it reaches 0.8 percent carbon, when the austenite is of the eutectoid composition and on further cooling breaks down to the fine mixture of lamellar carbide and ferrite. This structure is known as pearlite from its colored iridescent appearance after etching. The diffusion rate of carbon in austenite is high enough so that separation occurs fairly completely during air cooling, and the area of pearlite is therefore directly

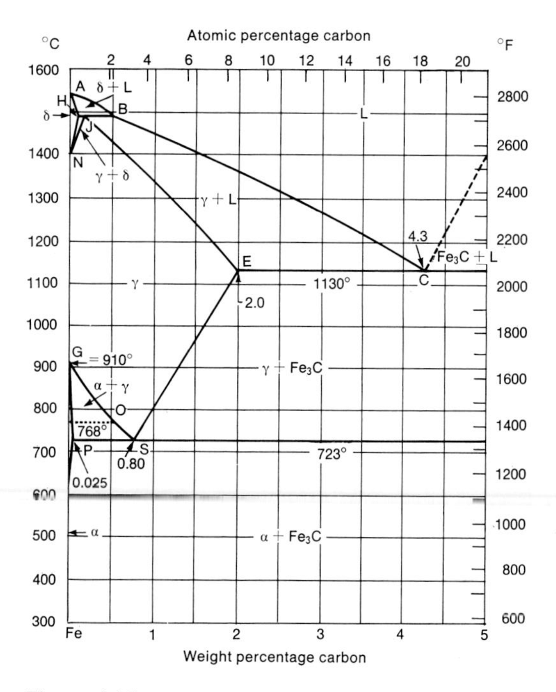

Figure 4.37
Constitution diagram of iron-carbon alloys. From *Metals Handbook*, A.S.M., 1949 edition. (Note: Under some conditions the carbide phase, cementite, will decompose into iron and graphite with a perceptible change in the solubility limits shown.)

proportional to the amount of eutectoid and hence of carbon.

Most archaeological samples show strong gradients in carbon content, with zones completely free from carbon not very far from those completely eutectoid and sometimes even containing free iron carbide.

Figure 4.38 shows an etched rectangular chunk cut from an iron currency bar from Sparta, fifth century B.C. Some sense of the flow of the metal during forging can be gained from the appearance of the different faces in relation to each other. Figure 4.39 shows the general appearance of a Luristan short sword of about 800–700 B.C.[15] It is a truly magnificent bit of smithing, made of several parts mechanically joined together after accurate preshaping. The center of the handle and the blade were forged integrally from a single piece, with shoulders or ridges formed so as to lock on the other pieces that were separately forged (perhaps using special swages or dies to give detail) and then attached by riveting or crimping. Holes were drilled or filed, not driven with a punch. Welding was used only in preparing the bloom for forging, not for final shaping.

I have examined only one sample of Luristan steel which had been hardened. This was a tiny fragment cut from an axe said by the Tehran Museum to be from the classic Luristan period. Its microstructure (figure 4.40) and its hardness (over 400 VHN in places) leave no doubt that it had been quenched. Hardening was certainly known at the time, for Homer refers to the quenching of iron (i.e., steel) in his story of the blinding of Polyphemus, but microstructures of 13 other Luristan weapons all indicate that no heat treatment had been given beyond such as would naturally accompany forging and air cooling. The carbon content is often high (though locally variable) and many of the

Figure 4.38
Rectangular prism cut from an iron currency bar, Sparta, fifth century B.C. Specimen from Lyle Borst. All surfaces have been polished and etched in nital to show the distribution of carbon in the metal. × 4.25.

Figure 4.39
Short sword from a Luristan grave, ca. 800 B.C. The blade and central part of the handle are forged in one piece, but the decorative features were separately made and mechanically joined together without welding. Photo courtesy of the Metropolitan Museum of Art.

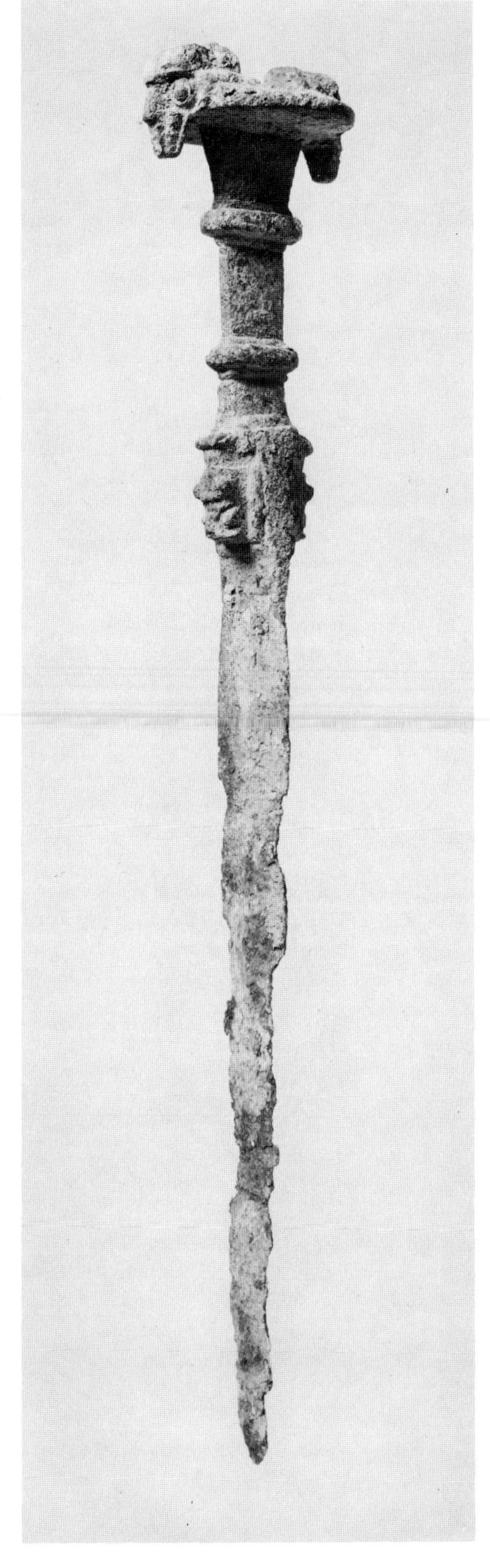

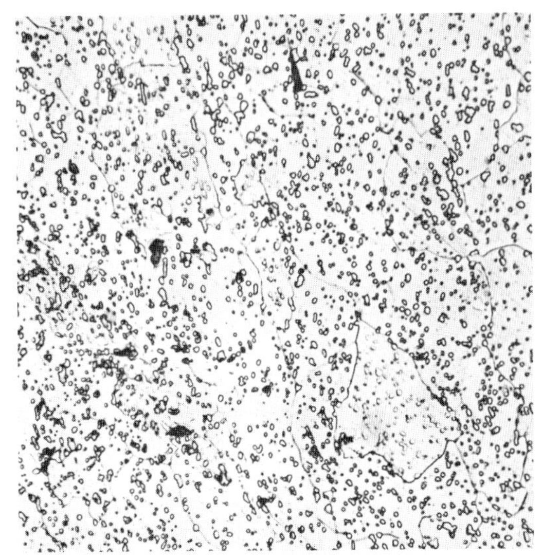

Figure 4.40
Microstructure of fragment of an axe said to be from Luristan. Etched in nital. × 500.

Figure 4.41
Microstructures showing spheroidized carbide in a Luristan sword. (This sword was a single-piece sword of simpler construction than figure 4.39, with a broad blade 50.3 cm long and 4 cm wide at the widest point.) Etched in nital. × 500.

blades have fully spheroidized carbide particles (figure 4.41), which indicates that the metal had been forged when it was hot but not hot enough to transform to the higher-temperature form—i.e., below 723°C.

In most early iron or steel objects, the distribution of the carbon content obviously originated in random variations in the bloom. In the absence of any knowledge that carbon was transferred from the fuel in the hearth to the metal, control was possible only by rule-of-thumb maintenance of tried operating procedure. Nevertheless, the fact that the metal became harder if it was kept hot for a long time well surrounded with charcoal must have been observed. By La Tène times (perhaps earlier) technique had advanced to the point where chunks of metal known to be hard were selected and welded onto a softer bloom in such a way as subsequently to provide hardness at the places where it was most needed, that is, at the cutting edges of a tool or weapon. Still later, local surface carburization of finally shaped tools came into use, using prolonged heat treatment of iron sealed in a box with the charcoal. This process gave a far more uniform and controllable carburizing effect than did the hearth, and its products are recognizable by the uniform diffusion of carbon in from all surfaces of the finished shape.

Iron has much too high a melting point to be melted by primitive means. The ore was reduced at temperatures of 1000° to 1400°C and consolidated by hammering when hot to weld the spongy mass of crystals together. Most of the rocky matter in the ore formed a molten slag or was fluxed by unreduced iron oxide (silica forms a eutectic with iron oxide that melts at 1180°C) to give a liquid that was squeezed out during the forging operation. Most of it, but not all; there remain inclusions of slag, the distribution of which

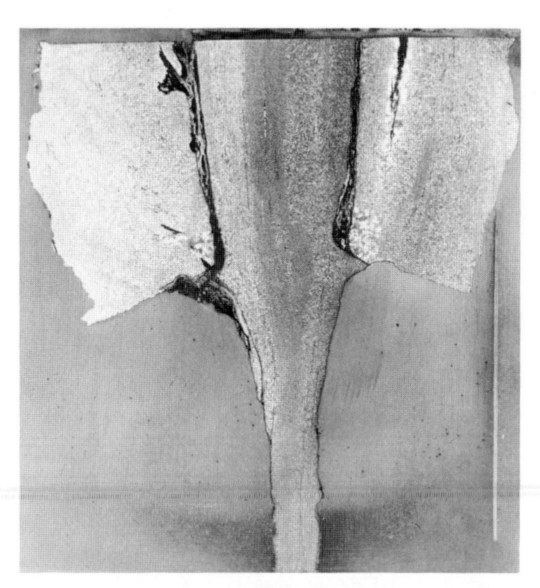

Figure 4.42
Longitudinal section through a Luristan sword similar to that shown in figure 4.39, at junction of blade and handle. Etched in nital. × 2.

Figure 4.43
Section through a transverse hole in the handle of a Luristan iron mace. Etched in nital. × 4. (The mace was 42.0 cm long and had a six-fluted head forged integrally with the handle. The hole was 9.2 cm from the end of the handle, which was 1.5 cm in diameter.)

provides clear evidence of the flow of metal during forging or other shaping. Who can doubt that the blade shown in section in figure 4.42 was forged integrally with the handle, or that the hole shown in section in figure 4.43 was cut by some means that did not distort the fiber of the metal as a hot punching operation would have? Moreover, if we could only read it, the composition and structure of the slag particles themselves provide evidence of the ore that had been used and of any fluxing material, such as limestone, that might have been mixed with the ore to facilitate its working. Except in the case of gross defects and poorly forged material, the slag particles are relatively small, and such methods as exist for studying them have not yet been extensively applied to archaeological material. It is a rich field for research. Note the obvious difference in appearance of the slag inclusions in figure 4.44, which shows two adjacent areas of an iron clamp which, with lead cast around it, was used to hold the stones of Persepolis together. The difference is partly related to the local carbon content of the steel, for the inclusions in the high-carbon area consist solely of the silicate $2FeOSiO_2$ (Fayalite), while in the more oxidized part there is still some unreduced FeO which, during cooling, has crystallized from the silicate melt in the form of rounded crystals of wustite. Not infrequently one encounters beautifully formed tiny spherical inclusions that are probably very high in silica, for they are colorless and noncrystalline as shown by their behavior under polarized light (figure 4.45). Other silicate inclusions, obviously once molten, show duplex crystalline structures that have not yet been deciphered (figures 4.46 and 4.47). Etching with hydrofluoric acid will often reveal detail in silicates (figure 4.48).

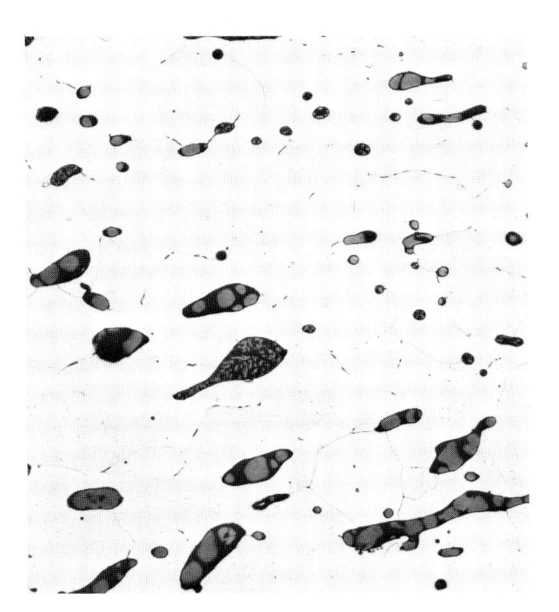

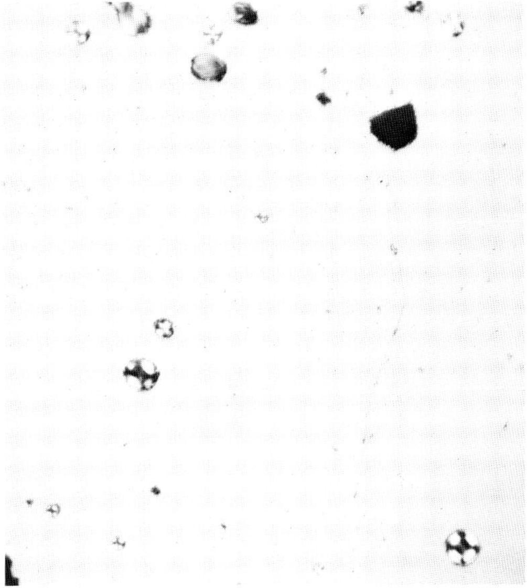

Figure 4.45
Glassy inclusions, probably high in silica, in iron. From Pandu Rajar Dhibi, India, ca. 1000 B.C. × 500. (Polarized light.)

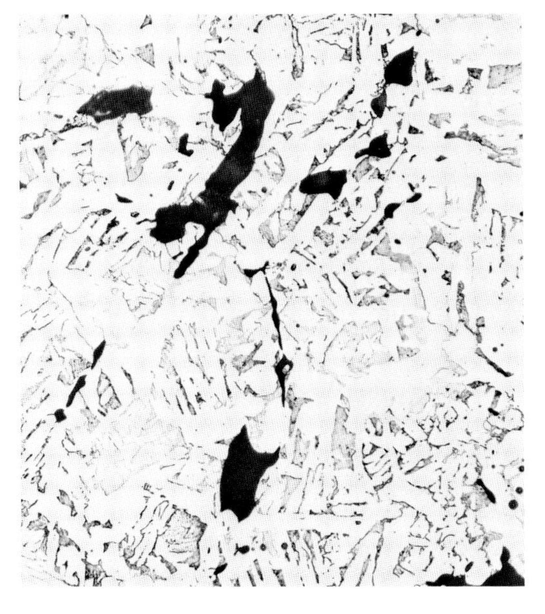

Figure 4.44
Slag inclusions in an iron clamp from Persepolis, ca. 530 B.C. *Top:* Slag in high-carbon area. *Bottom:* Slag in low-carbon area. × 500.

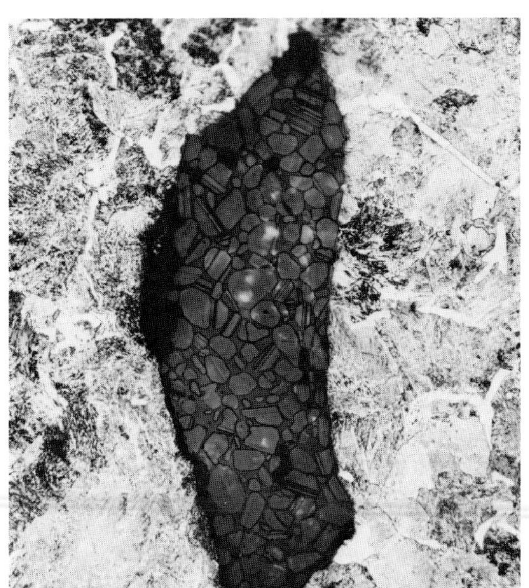

Figure 4.46
Slag inclusions in Luristan dagger, showing complex crystalline structure. Etched in nital. × 500.

Figure 4.47 (*right, top*)
Same sample as figure 4.46 showing different crystalline phases in slag. × 500.

Figure 4.48 (*right, bottom*)
Slag in Persepolis iron (cf. figure 4.44). Etched in hydrofluoric acid to reveal crystallization of the silicate phase. × 200.

The Japanese Sword

Slag is usually a nuisance for it weakens the metal and blemishes a polished surface. The Japanese, however, having developed methods of forging to distribute the slag so finely that it did not seriously weaken the steel, found that it conferred a beautiful texture to the surface of the blade after proper polishing (figure 4.2). The swords of different schools of smiths are identifiable by the resulting texture or pattern. The Japanese sword is remarkable in other ways also, and those made after the eighth century A.D. are beyond doubt of a quality toweringly above any other steel objects made prior to the introduction of modern science. It was made by extensive welding and forging to break up the slag and to enable the assembly in different parts of the blade to be of metal having the desired different qualities. Finally, it was quenched in water after being partly coated with clay (figure 4.49) so that the extremely hard metal at the cutting edge merged into a softer, heavier body, thus avoiding the brittleness that is inevitably associated with large pieces of extremely hard steel. Figure 4.50 demonstrates the variation of hardness down the center of a blade, showing the hard and soft zones and the transition between them.[16]

The center of the blade was usually of soft iron (often intentionally containing large slag inclusions to damp out vibrations) which was surrounded by hardenable steel of controlled (though not analyzed!) carbon content. Figure 4.51 is a cross section of a blade from edge to back.[17] Such patterns incorporate features depending on the grain size of the metal, inclusion content, the precise way in which the laminations of the metal are made, and the final surface cut, as well as the

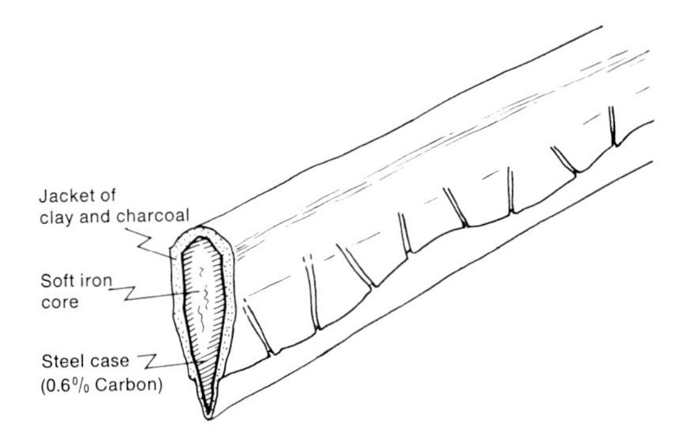

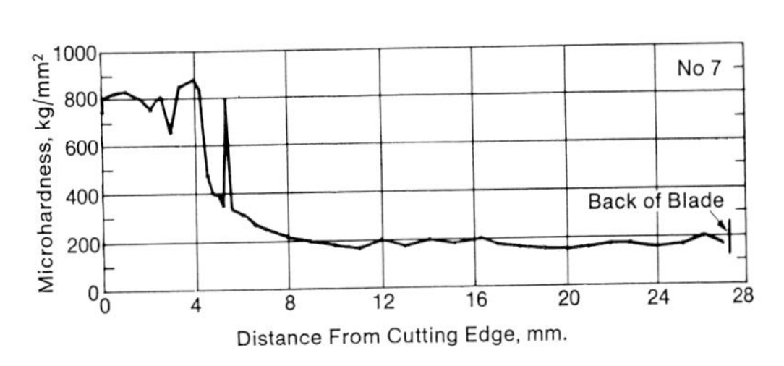

Figure 4.49
A Japanese sword blade prepared for heat treatment. Schematic.

Figure 4.50
Variation of microhardness along the center of a cross section of a seventeenth-century Japanese sword blade.

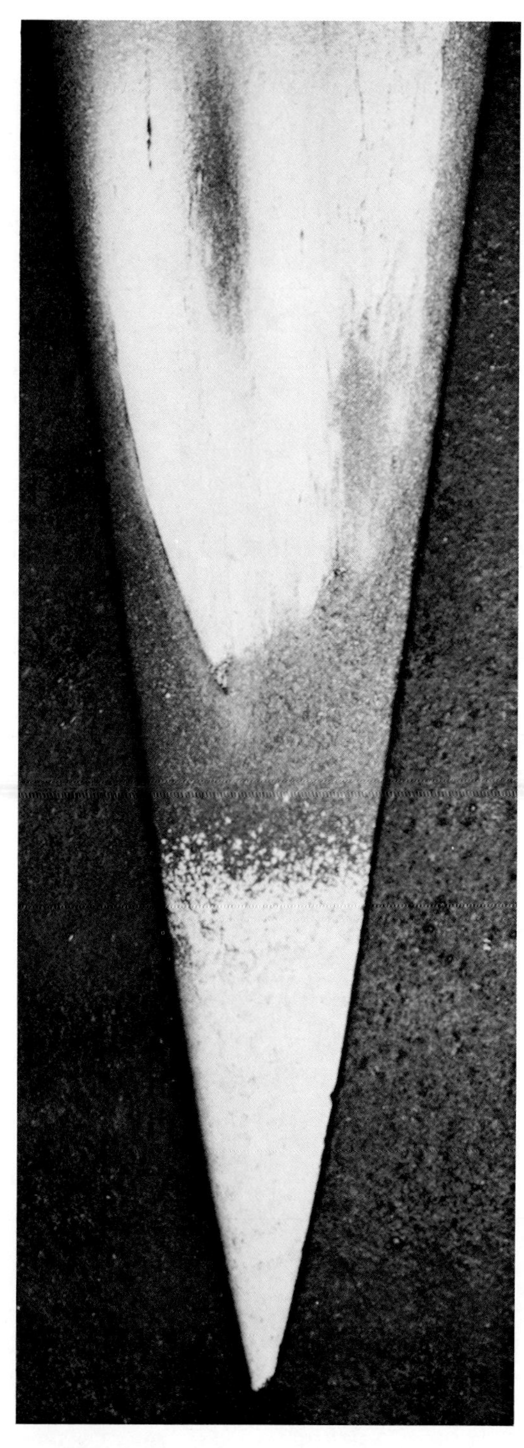

Figure 4.51
Cross section of a Japanese sword blade, eighteenth century. × 12. The white area near the cutting edge at the bottom is intensely hard martensite. This merges gradually into dark-etching pearlite which is of the same composition (0.69 percent carbon). The white-etching area above this is a soft iron core, surrounded by a jacket of the same steel as the edge.

details of the application of the clay jacket and the manner of quenching. They are often of great intricacy and beauty, and their innumerable variations provide the basis for the identification of the work of a swordsmith just as details of brushwork and style will identify a painter.

For aesthetic rather than metallurgical reasons, the Japanese also exploited the *mokumé* or wood-grain texture used decoratively in forged iron swordguards (figure 4.52). This texture resulted directly from the way in which the metal had been forged, and was not an artificially applied design. After the piece had been shaped, it was subjected to a rather deep etching which attacked metal of different composition in different ways and so revealed the flow lines. Extremely gaudy effects of this kind were employed in the *kris* of Southeast Asia (figures 4.53 and 4.54), which sometimes have welded-in layers of meteoric iron. A comparable effect in nonferrous metals was achieved by the Japanese by combining many thin sheets of different metals—including copper, brass, silver, *shibuichi*, and *shakudo* (the latter a dilute copper-gold alloy with a beautiful raven-black surface after pickling). A soldered multilayered sandwich was hammered down in such a way that the surface was slightly undulating, then a flat surface was cut, exposing rings of the different metals that could be distinguished by a final patinating pickle. An example in the Boston Museum (figure 4.55) has been cunningly worked so as to produce on its plane surface the illusion of cherry blossoms floating upon the wind-rippled surface of a pond.[18]

Other textured metals are the Merovingian pattern-welded blade—which, incidentally, was sometimes given a differential quenching to harden the edges as in the Japanese sword, though some centuries

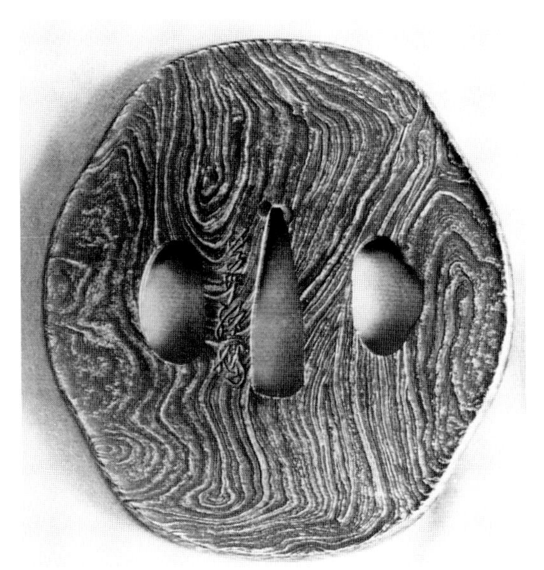

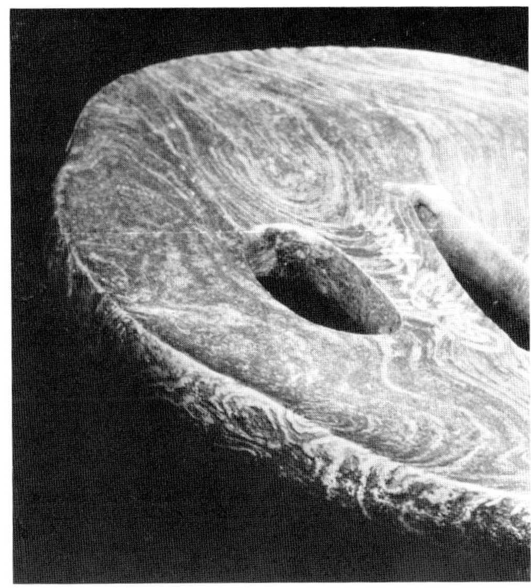

Figure 4.52
Swordguard of iron *mokumé*, Kisai Kyosai, eighteenth century, 8 cm diameter. Photo courtesy of the Seattle Art Museum.

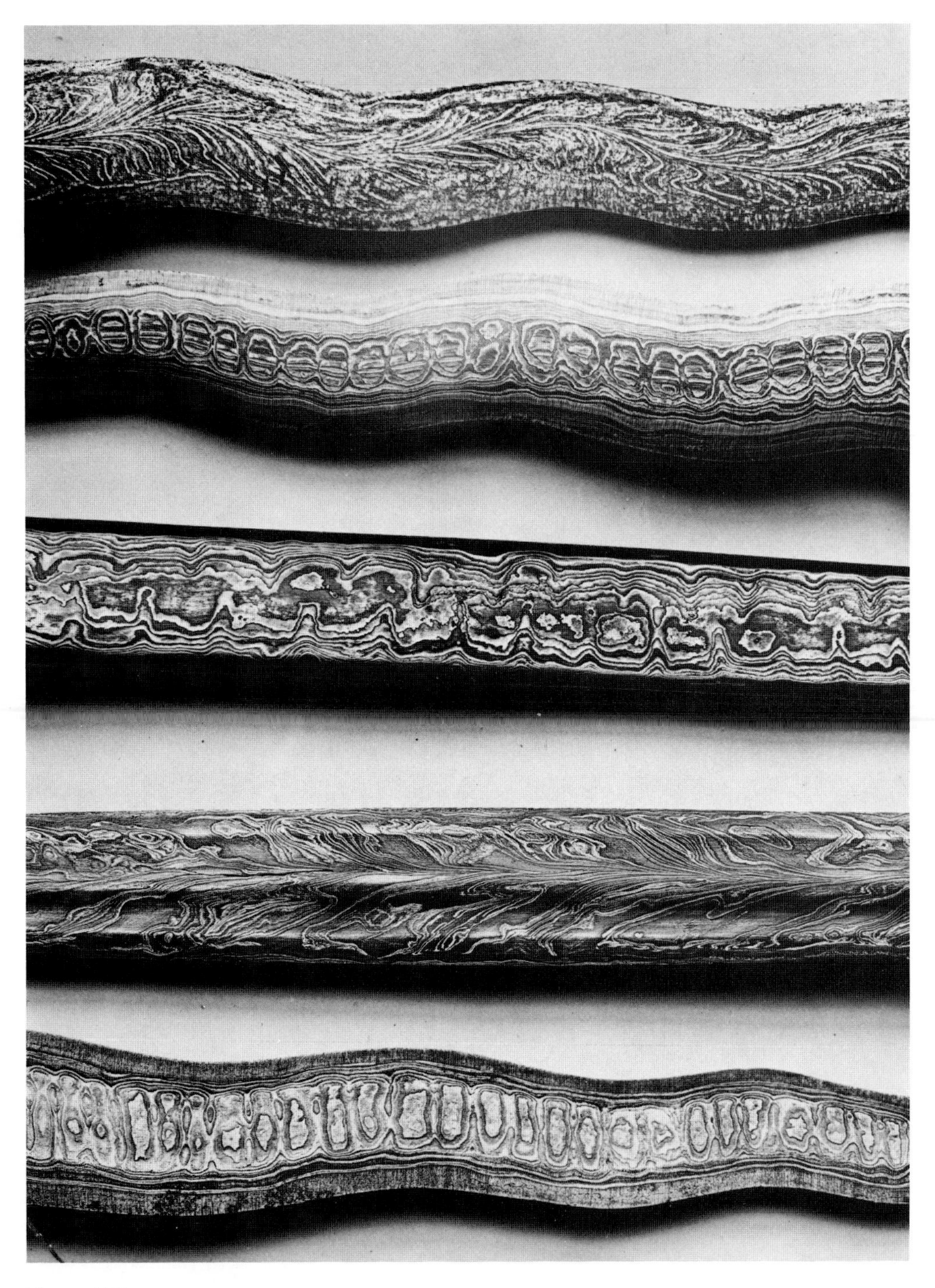

Figure 4.53
Blades of five Malayan *kris*. Natural size. British Museum, photo courtesy of Herbert Maryon.

Figure 4.54
Javanese *kris*, eighteenth century. This blade has been etched far more deeply than usual and reveals the laminae of acid-resistant metal (usually appearing in section as white lines on the surface) in three-dimensional splendor. Photo courtesy of the Seattle Art Museum.

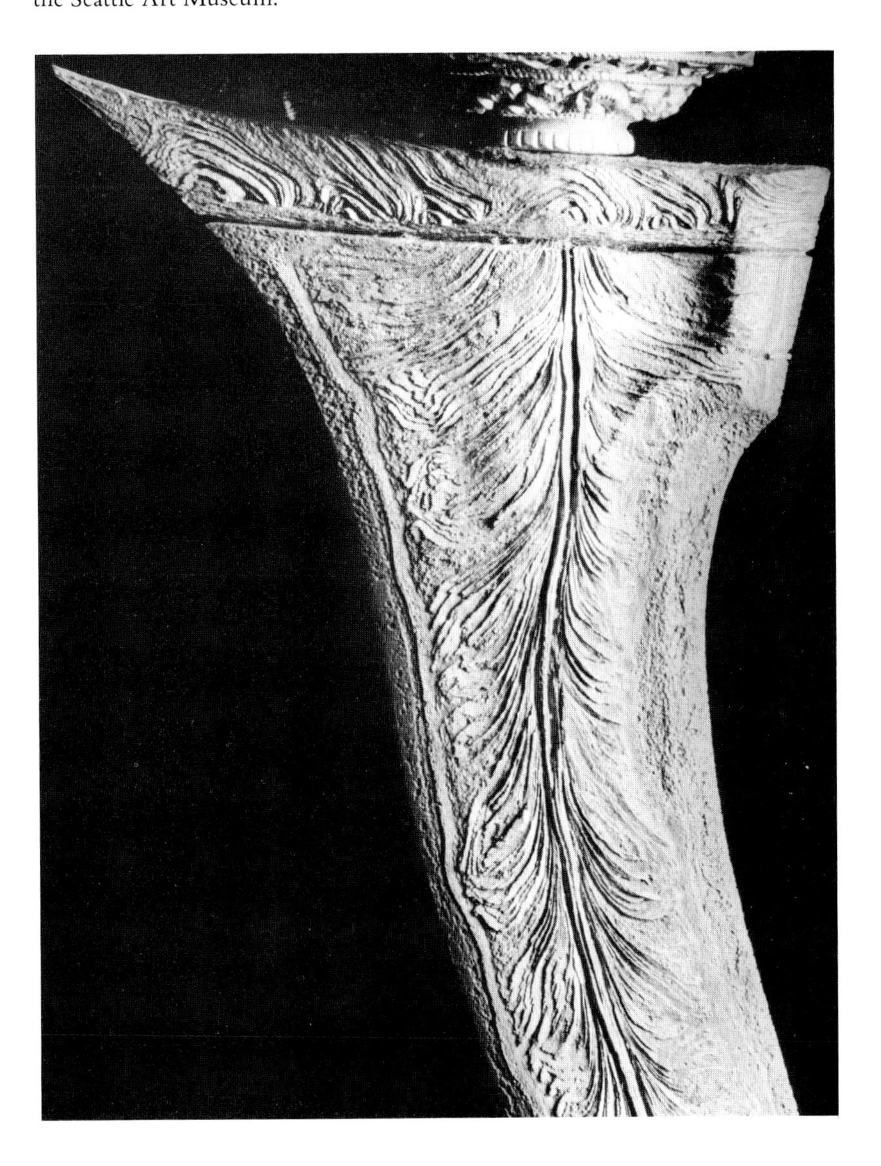

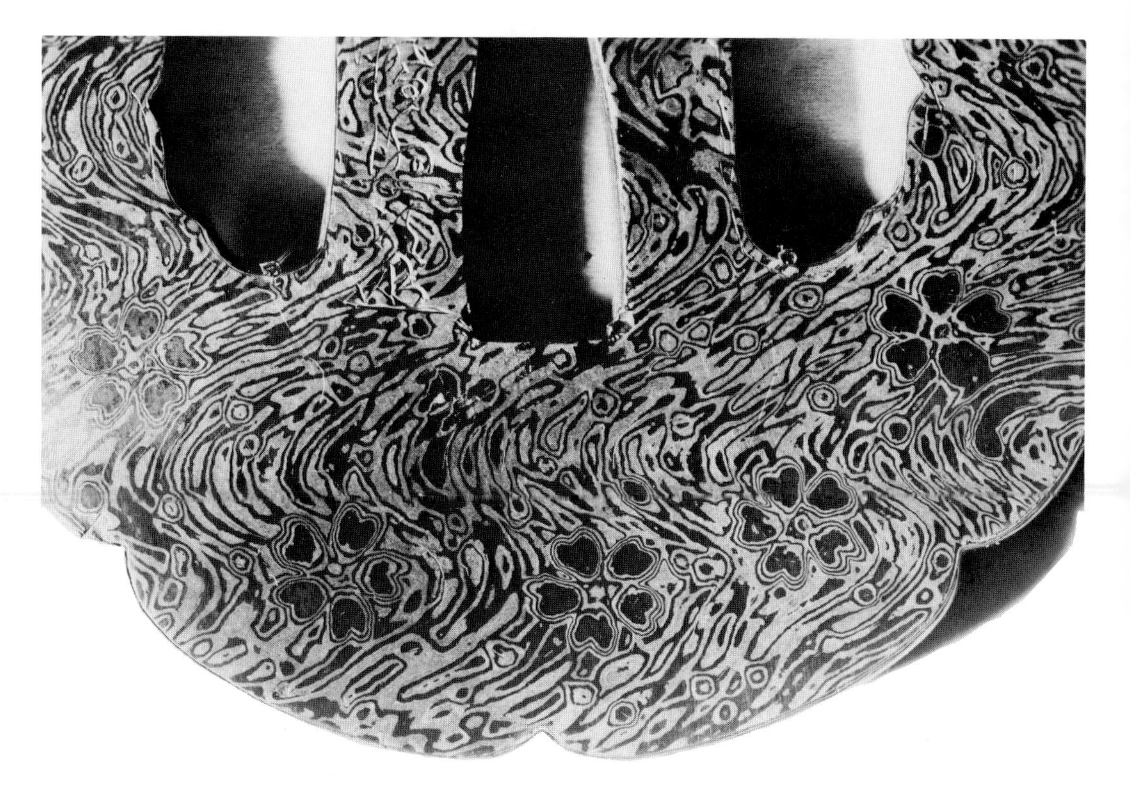

Figure 4.55
Japanese swordguard made by Takahashi Yogi (nineteenth century). Nonferrous metal *mokumé*. × 2.5. Note the cherry blossom design achieved by so punching the composite metal sheet that a layer of dark-etching metal would be revealed in appropriate shape when the final plane surface was cut. The guard is 7.3 cm in diameter. (Bigelow Collection, Museum of Fine Arts, Boston, No. 11.12445.)

earlier—and the Damascus sword (figure 4.1). The latter sometimes owed its pattern to the forging of different kinds of metal together, but in the best types the visible texture has its origin in the crystallization of a high-carbon steel during solidification from the molten state.[19]

On Quantitative Metallography

The discussion in the present paper has been almost all qualitative, yet I most strongly urge the museum metallographer to use quantitative methods wherever appropriate. It is desirable to measure microhardness at many points throughout a sample in order to find gradients in composition or treatment. Grain size should be measured by counting the number of grain boundaries intercepted by a traverse of known length. There are many occasions in which it is useful to know the volume fraction of a given constituent, which is most easily determined simply by counting the fraction of times that a given phase appears under the cross hairs of the microscopic eyepiece as the sample is randomly moved.[20]

Conclusion

The above discussion has shown that the internal structure of a work of art in metal can often throw as much, or more, light on its origin as can be derived from stylistic analysis. Moreover, the techniques employed can provide clues to the habits of mind of the people who originated them. The preference for casting shown by the Chinese artisans compared with the European fondness for forging must have human, or at least social, significance. The diffusion of specific forging and heat-treatment techniques can provide just as much evidence of cultural contact as the style of design on a sherd. Perhaps the most important reason for structural studies of museum objects is that the intimate knowledge so derived as to the way in which an object has been made adds so greatly to the aesthetic enjoyment of it. Very often some detail and sometimes the whole of an effective design arises directly in the exploitation of the merits and the overcoming of the difficulties of a specific technique, in the reaction between the artist's fingers and his material. It is regrettable that for the best results such metallographic studies require that a small piece of metal be cut from the object under study, or even that a complete cross section of it be made. Such vandalism, however, is compensated for by the knowledge gained. Moreover the accidents of time have frequently resulted in broken edges from which pieces can be removed for examination with no loss whatever, especially if the piece is destined to be restored anyway. Usually, however, well-authenticated objects in poor condition and of little aesthetic value are available, and these should be laid aside for such studies. Most archaeologists and many art museum curators are now willing to allow minor mutilations of the objects in their custody for the sake of major understanding.

Notes and References

1
See chapter 3 in this volume.

2
Two independent translations into English have been done: J. G. Hawthorne and C. S. Smith, *On Divers Arts: The Treatise of Theophilus* (Chicago, 1963); and C. R. Dodwell, *Theophilus: De diversis artibus* (London and Edinburgh, 1961).

3
For a discussion of the development of observations and ideas on the structure of metals, see C. S. Smith, *A History of Metallography* (Chicago, 1965).

4
The general history of metallurgy is well summarized in L. Aitcheson, *A History of Metals*, 2 vols. (London, 1960). For a history of castings with illustrations of works of art showing the relationship between design and technique, see chapter 6 in this volume.

5
See item 152 in the bibliography and D. L. Schroeder and K. C. Ruhl, "Metallurgical characteristics of N. American prehistoric copperwork," *American Antiquity* 33:162–169 (1968).

6
It seems that the first and most imaginative use of practically every material was, before quite modern times, in making something decorative. People are experimentally minded when looking for decorative effects, but they can't experiment with the established techniques on which their livelihood depends. The aesthetic urge has been the key to innumerable discoveries—witness the role of the precious metals and of colored dyes, pigments, and glazes in the history of chemistry. (See chapter 5 in this volume.)

7
A contemporary account of the process by a Jesuit priest, Sahagún, is given in the appendix to André Emmerich, *Sweat of the Sun and Tears of the Moon: Gold and Silver in Pre-Columbian Art* (Seattle, 1965).

8
Morris Cohen, "Aging phenomena in a silver-rich copper alloy," *Trans. AIME* 124:138–157 (1937).

9
A. L. Norbury, "The effect of quenching and tempering on the mechanical properties of standard silver," *J. Inst. Metals* 39:145–161 (1929).

The brittleness of antique silver has been examined especially by F. C. Thompson and A. K. Chatterjee, "The age embrittlement of silver coins," *Studies in Conservation* 1:115–126 (1952–54), and by A. E. Werner, "Two problems in the conservation of antiquities: Corroded lead and brittle silver," in W. J. Young, ed., *Application of Science in Examination of Works of Art* (Boston, 1967), pp. 96–104. Thompson and Chatterjee believe that the precipitation of lead is mainly responsible for brittleness, but lead does not form the wide duplex areas of precipitate that are commonly seen in the microstructure.

10
(*Note added 1980*) The mechanism underlying the changes in microstructure and brittleness is more complex than is indicated here. See F. Schweizer and Pieter Meyers, "A new approach to the authenticity of ancient silver objects: The discontinuous precipitation of copper from a copper-silver alloy," in the *Proceedings of the 18th International Symposium on Archaeometry and Archaeological Prospection* (Bonn, 1978), pp. 287–298. Could it be that corrosion is involved at a primary, not a secondary, stage? Once intergranular corrosion has been initiated, the boundary could be driven laterally by solution of the alloy on one side of a thin film of electrolyte, followed by redeposition of pure silver on the other side, the formation of copper oxide within volumetric limits, and the diffusion out of the remainder in countercurrent with oxygen atoms.

11
[For a recent study of this technique see D. L. Carroll, "A classification for granulation in ancient metalwork," *American Journal of Archaeology* 78(1):33–39 (1974).] Incidentally, the similarity between the geometric array of these granules and those of soap bubbles in the crystal model of the frontispiece will be recognized. One of the mysteries of the history of science is the delay in accepting the simple concept that the basis of crystal form lies in the development of geometric shapes having directional properties from the stacking of isotropic (spherical) parts. It was proposed by Kepler, Hooke, and others in the seventeenth century, and thereafter was forgotten while crystallographers packed little polyhedral unit crystals together: only in the present century did it become a useful textbook picture.

12
Since writing this, I have seen some fine earrings and other jewelry of the ninth century from Pohanskol, Czechoslovakia, which have invisibly soldered silver granulation that appears to have been made by a true granulation technique.

13
H. Drescher, *Der Überfangguss . . .* (Mainz, 1958).

14
See, however, the exhibition *Made of Iron* organized by Mrs. J. de Menil for the Art Department of the University of St. Thomas in Houston, Texas, September–December 1966. An illustrated and well-annotated catalogue was published by the Department in 1967.

15
For detailed metallographic studies of Luristan iron objects see item 161 in the bibliography and A. France-Lanord, "Le fer en Iran au première millénaire avant Jésus-Christ," *Revue d'Histoire des Mines et de la Metallurgie* 1:75–127 (1969).

16
The most complete scientific study of Japanese sword blades is by Kunichi Tawara, *Nippon-to no kagateki kenkyu* (Tokyo, 1953); in English there is only item 97 in the bibliography and H. Tanimura, "Development of the Japanese sword," *Journal of Metals* 32(2):63–73 (1980).

17
For more detail on the Japanese sword see chapter 9 in this volume and the references cited therein.

18
For a detailed study of the techniques used in Japanese swordguard manufacture see item 193 in the bibliography.

19
Textured metals are the subject of five chapters in C. S. Smith, *A History of Metallography* (Chicago, 1960). On Damascus swords, see Carlo Panseri, "Damascus steel in legend and in reality," *Gladius* 4:4–66 (1965), and Jerzy Piaskowsky, *O Stali Damascenskiej* [On Damascus Steel] (Warsaw, 1974). On similar textures in materials in general, including lacquer, ceramics, and wood, see item 144 in the bibliography.

20
These methods are summarized in E. E. Underwood's book *Quantitative Stereology* (Reading, MA, 1970). This also gives methods for estimating the shape and size distribution of three-dimensional particles on the basis of measurements made on a two-dimensional section.

5
Matter versus Materials: A Historical View

Not many years ago, I was a practical industrial metallurgist, and it is with some surprise that I find myself delivering a lecture in honor of a great historian. George Sarton pioneered in the application of the techniques of the historian to the then-neglected area of science. His immense energy, his proper regard to rigorously checked detail, his respect for the boundaries of his chosen period, and his insistence on comprehensiveness within these boundaries set standards for two generations of scholars in the United States and for the entire discipline on a world scale. I have done detailed research in both science itself and its history, but I want to use this opportunity to make some general remarks on man's attitude toward materials (in contrast to matter) throughout the whole of history. These derive from the fact that I happen to have lived at the time of some rather exciting developments in materials science—in fact even its formation as a recognizable area of knowledge—and have had a moderately intimate (if one-sided) look both at the recent history of science and at archaeologists' findings of the earliest uses of materials of many kinds. I see science reversing the trend toward atomistic explanation that has been so triumphant in the last 400 years, and I predict a more human future based on the symbiosis of exact knowledge (which is by its very nature limited) and experience. This I do hesitantly, certain only that this is an important area for discussion at this particular stage of history. Materials provide a good illustration of the difficulties of applying exact knowledge to a complicated world.

Much of the history of materials has been rather dull, for man has usually been satisfied to make do with what he had, but there are three periods at which sharp changes occurred. These correspond to the first discoveries of the principal alloys and ceramic materials, the beginning of scientific explanation, and the very recent realization that, by the control of their structure, materials that possess almost any property in high degree can be designed and produced for special applications.

The Discovery of Materials

What Peter Drucker has called the first technological revolution began more than 7000 years ago in the Middle East, where there arose an appreciation of the possibilities of technology combined with a pattern of social organization that both allowed the necessary specialization and provided the necessary superstructure for its exploitation and its control.[1] Anyone who studies the reports of the major archaeological excavations, or, better, exposes his senses to the magnificent objects in clay or metal displayed in the museums of the world, cannot help but be impressed with the extent of real if untheoretical knowledge of materials displayed by early craftsmen. The well-shaped and decorated pottery from Chatal Huyuk (7th millennium B.C.), the gold jewelry from the royal graves at Ur (2600 B.C.), and the efflorescence of ceramic, stone, and metal art in Egypt and in Persia provide plenty of evidence that man knew, if he did not understand, a vast amount about the behavior of materials under chemical, thermal, and mechanical treatment.

Practically everything about metals and alloys that could have been discovered with the use of recognizable minerals and charcoal fires was discovered and put to some use at least a millennium before the philosophers of classical Greece began to point the way toward an explanation of them. It was not intellectual knowledge, for it was sensually acquired, but it produced a range of materials that continued to serve almost all of man's needs in warfare, art, and engineering continually until the end of the nineteenth century A.D. It is of basic significance for human history that, from the cave paintings on, almost all inorganic materials and treatments of them to modify their structure and properties appear first in decorative objects rather than in tools or weapons necessary for survival. Aesthetically motivated curiosity, or perhaps just play, seems to have been the most important stimulus to discovery. The men who first produced basic changes in the very nature of common earths and stones by heating and mixing them must have felt an almost godlike power. Rare materials of many sorts are intimately associated with magicians' practices. They have long been used in the decoration of temples and churches to promote religious feeling and in palaces to inspire awe.

Through most of European history, since the Renaissance, both connoisseurs and historians of art have reserved their highest praise for painting and sculpture, and the most marvelous artistic uses of materials have been designated "decorative" or "minor" arts. However, the recent trend toward nonobjective art has been accompanied by a new appreciation of the aesthetic richness in textures, colors, and other physical properties of materials. Such properties and the technologies derived from them have been appreciated far more in the Orient, both in the Middle East where it all began, and especially in the Far East. Chinese, Korean, and Japanese craftsmen in stone, wood, clay, and metals have sensitively used the subtlest properties of plastic and viscous flow, of crystallization, surface-tension differences, and color changes resulting from ions in various states of oxidation and polarization. They have enjoyed the beauty conferred on a surface by chemical degradation and the irregularity that comes from fracture, deformation, and sectioning of polycrystalline materials. This sensual awareness of the properties of materials long preceded the Taoist and Ch'an (or Zen) philosophies into which it was formally incorporated.

The main characteristic of today's science of materials is a concern with properties and the dependence of properties upon structure. This is exactly where the story began. The history of materials has been a long journey in search of knowledge in strange and difficult terrain, finally to return to the familiar scene with vastly better understanding. Yet most of the histories of science are quite unconcerned with the structure of atomic aggregates, but deal rather with the basic philosophic question of the existence of matter, and later, as chemistry evolved, with questions of composition. Historians of science commonly regard as their central theme the throwing over of the Aristotelian principles and their replacement with analyzable elements and with weighable atoms. What a triumph it was to discard earth, air, fire, and water and to find atoms of silicon, carbon, hydrogen, and oxygen! Matter cannot be understood without a knowledge of atoms; yet it is now becoming evident that the properties of materials that we enjoy in a work of art or exploit in an interplanetary rocket are really not those of atoms but those of aggregates; indeed they arise in the behavior of electrons and photons within a framework of nuclei arranged in a complex hierarchy of many stages of aggregation. It is not stretching the analogy much to suggest that the chemical explanation of matter is analogous to using an identification of individual brick types as an explanation of Hagia Sophia. The scientists' laudable striving to eliminate the evidence of the senses has sometimes produced a senseless result. But if exact science is used to illuminate empirical experience, and if experience is used to temper the extrapolation of the simple ideal systems of the scientist, then indeed we have real knowledge. Some materials scientists and materials engineers (both very recent professions) are, I think, beginning to see this. Their concerns are more than interdisciplinary, for they add a measure of art to discipline.

The Philosophy of Matter

The simple direct approach to materials is evident in the writings of the earlier Greek philosophers.[2] Democritus (400 to 357 B.C.), following Leucippus, held that the distinguishing characteristics of materials depended upon three distinguishing characteristics of the aggregation of their parts—shape, order, and orientation. Such a relation, if not a theoretical dependence, must have been obvious to every stonemason, smith, or foundryman who used texture (revealed on a fractured surface) as a criterion of quality of both his raw materials and his products. These simple truisms, however, disappeared on further elaboration. Concern with real states of aggregation got lost in the search for the ultimate nature of matter. The Pythagoreans and Platonists seem to have regarded the numerical aspects of form as more important than form itself (as perhaps they should be to a philosopher).

It is the same with qualities. Aristotle in discussing his predecessor Empedocles (455 to 395 B.C.) says—quoting the most recent translation of Gershenson and Greenberg: "Empedocles was also the first to say that the elementary constituents of the matter in the universe are four in number, although he does not treat them as if they were really four separate substances. Instead he treats them as if they were only two—heat substance on the one hand, and on the other hand dry dust, colorless gas, and clear liquid[4] (whose properties contrast with those of heat substance), all dealt with like a single substance."[3] There could hardly have been a clearer statement that energy and the three main states of aggregation of matter—solid, liquid, and gas—are the important things to consider. Later philosophers gave special meanings to these constituents and disguised their simple physical meaning in almost occult principles.

In Aristotle's own treatment of matter, the four material elements of Empedocles are derived from various combinations of four primary qualities—hot, cold, dry, and moist. Thus earth is cold-dry; air, hot-wet; fire, hot-dry; and water, wet-cold. Structurally, Aristotle does no more than distinguish between visibly homogeneous and heterogeneous bodies and invoke the presence or absence of pores to account for some properties. However, Aristotle's reference to the distortion of ceramic ware in the kiln and his puzzling discussion of the melting and solidification of wrought iron in the steelmaking hearth leaves no doubt that he had observed workshop processes in some detail, and the 18 qualities of homoeomerous bodies that he chose to explain in detail in his *Meteorologica*, are just those fine points of behavior that would be noticed in a workshop. They are: solidifiable, meltable, softenable by heat, softenable by water, flexible, breakable, fragmentable, capable of taking an impression, plastic, squeezable, ductile, malleable, fissile, cutable, viscous (the converse of which is friable), compressible, combustible, and finally, capable of giving off fumes. He gives examples of materials possessing each of these qualities and the converse ones, and explains them in terms of the relative content of his four elements.

This redundant list of properties is not the neat classification of a philosopher. It reads more as if it were based on a conversation with a workman whose eyes had seen and whose fingers had felt the intricacies of the behavior of materials during thermal processing or as they were shaped by chipping, cutting, or plastic deformation. And the attributions of the proportions of the elements were an attempt to assign a measure of solidity or fluidity: it was more physics than chemistry, and it related more to real materials than to the fundamental nature of matter. The very word used by Aristotle for matter in general, *hyle*, was simply the word for wood or lumber, the common material of construction with real tangible properties.

The Chinese philosophy of matter, as it began to take definite form in the time of Tsou Yen (350 to 290 B.C.?) was based even more tangibly on the properties of materials. At the beginning the five elements—earth, wood, metal, fire, and water—were associated with phases of temporal cycles, but it will be noted that they also constitute a tribute to, and a classification of, the materials familiar to the workshop of the potter, the carpenter, the smith, and the dyer. As in the West, later Chinese philosophers got away from the artisan's sensual approach to matter and they developed an elaborate series of sequential relations between the elements and a complex system of symbolic correlations with seasons, tastes, smells, and much else—even including politics.[5]

Histories of philosophy are full of discussions of the development of the concept of matter, yet hardly at any point do they touch on the nature and properties of materials. Atoms and the qualities that accompany their aggregation became pure exercises in thought, with the significance of monism and pluralism more important than the visible, tangible aggregations that, in the craftsman's hand if not in the philosopher's mind, were directly relatable to useful properties. Through most of history, matter has been a concern of metaphysics more than physics, and materials of neither. Classical physics at its best turned matter only into mass, while chemistry discovered the atom and lost interest in properties. Only in the last few decades was it possible for solid-state physics to mature and to merge with a growing technology as attention turned again to materials and to qualities, which had now become measurable properties.

For 19 centuries after Aristotle, virtually all thinking about matter was expressed in terms of his elemental qualities: then came a period in which all advance arose from demolishing the misunderstanding that had accumulated in his name. During both these periods sensitivity to the wonderful diversity of real materials was lost, at first because philosophical thought despised the senses, later because the more rigorous experimentally verifiable thought patterns of the new science could only deal with one thing at a time. It was atomistic, or at least simplistic, in its very essence.

The practical world, of course, continued to exploit materials regardless of the state of science. The Greeks with all their sensitively shaped ceramics, sculpture, and buildings, and the Romans with their large-scale military and engineering enterprises, made good use of the materials that had been discovered one or two millennia previously, but they added few new ones. Development was mainly in the economy and scale of production. In the Middle East, however, the ancient techniques were elaborated to some extent. In the working of steel, artisans of this region were particularly effective, as the crusaders who encountered the Sword of Islam learned painfully. Still farther east, in China and especially in Japan by the thirteenth century A.D., techniques of steelmaking rose to unsurpassed heights, but this had no influence on European technology or science.

Alchemy and Iatrochemistry

Greek philosophy—Stoic, Gnostic, as well as Aristotelian—had a lusty if deformed child in alchemy, which reached its height in the sixteenth and seventeenth centuries A.D. Much of alchemy seems to belong to the history of religion and psychology more than to the history of the physical sciences. The alchemists' attempts to relate macrocosm to microcosm, their extensive symbolism of sacrifice, corruption, death, and resurrection had an integral beauty that one must admire, but they helped mysticism more than metallurgy. Nevertheless, though chemists can legitimately scoff at the alchemists' attempts at transmutation, physicists should not, even those who are not concerned with nuclear reactions. Transmutation was a thoroughly valid aim, a natural outgrowth of Aristotle's combinable qualities, and its truth was demonstrated by every child growing from the food he ate, by every smelter who turned green earth into red copper or black galena into base lead and virgin-hued silver, by every founder who turned copper into gleaming yellow brass, by every potter who glazed his ware, by every goldsmith who produced niello, by every maker of stained glass windows, and by every smith who controlled the metamorphoses of iron during its smelting, conversion to steel, and hardening. Such changes of properties, seen physically, *are* transmutations, but they are not chemical in the purified modern sense, and the chemistry had to be clarified before the physics could be studied. The impossibility of making real gold lay in the necessity of duplicating all of its properties simultaneously, but taken separately the malleability, reflectivity, color, thermal conductivity, in fact practically everything but the density of gold, could be singly matched by suitable operations upon common materials. It must have been tantalizing and frustrating. There were many examples of the validity of the aim, and theory taught of the combination of qualities but gave no reliable way of achieving it. Many wonderful things must have been seen by the alchemists in their mixings and heatings, more perhaps even than by the old craftsmen who sought only enjoyable aesthetic effects, but they added little to transmittable knowledge. Their symbolic language had the effect that any security system has in hampering initiates as well as outsiders, and their theories, too firmly believed, closed their eyes to many phenomena and made visible what was not there.

Through the entire alchemical period, the workshop transmission of practical knowledge continued, of course. Many superb pieces of jewelry and other metalwork were produced in the so-called Dark Ages. Early in the twelfth century there appeared, for the first time in all history, a practical metalworker who wrote extensively of his craft. This was the pseudonymous Theophilus, a Benedictine monk, who gave a superb factual description of all the techniques of churchly decorative art and felt no need for a word of theory or for speculation about ultimate causes.[6] The sixteenth century saw, among the flood of new works encouraged by the printing press, a sudden growth in practical literature at various levels. The most notable are the extensive books by Biringuccio and Agricola, who between them deal with all aspects of the winning of metals from their ores and their application to man's use.[7] Both reflect the organized industrial framework that had replaced the craftsman's shop of Theophilus's day, but no new materials had appeared, and neither author felt the need of theory either to guide practice or to organize the description of it. Theory had, however, begun to change by this time, and significantly the change came from a man who had practical aims. The turbulent annoying medico Paracelsus (about 1491 to 1541) wanted to turn the main body of chemistry to the service of medicine, but, unlike contemporary metallurgists, he felt the need of theory and applied it both to the treatment of patients and to the preparation of his medicines, which he thought should be simple pure substances. Dissatisfied with the Aristotelian elements, he superimposed on them a new set of active principles—salt, sulphur, and mercury. These *tria prima* were not, of course, the materials known by those names, any more than the Aristotelian principles were really earth, air, and water, but they were to be combined in almost an Aristotelian way.

Paracelsus's principles were clearly suggested by three classes of real materials that have basically quite different characteristics: mercury, primarily metallic but also liquid; salt, the ionic compound that gave its name to the whole class of salts; and sulphur, soft, easily melted, its molecules held together by van der Waals forces. These three materials exemplify three of the four types of interatomic bonding in today's quantum-mechanical theory of solids. Paracelsus in providing a more sophisticated version of the Aristotelian qualities showed a great insight into the nature of solids. From the physicist's point of view, if not from the chemist's, it was an important advance, but it was not one destined to develop, and only one of his principles survived to the eighteenth century. This was sulphur, which became a general principle of inflammability, a reducing principle, eventually phlogiston.

Corpuscular Philosophers of the Seventeenth Century

A few decades after Paracelsus had redefined the chemical principles, the monopoly of qualities was challenged by a revival of interest in the structure of matter. The rebirth of atomism and the growth of Descartes's corpuscular philosophy have generally been treated in terms of the philosophical question regarding the ultimate divisibility of matter: I see them more as premature but well-based attempts to unravel the significance of larger structural units, the microcrystals, subgrains, and precipitated particles that the materials man today observes and controls. By the end of the seventeenth century, virtually every scientist took it for granted that matter was particulate in nature, but Newton had cast physics so strongly in the mold of quantitative mathematics that complex aggregates were beyond his reach, interest even. During the seventeenth century there was much delightful if unproductive speculation, but by its end the best scientists again abandoned real materials and settled for the study of only those properties of matter that are insensitive to structure. Theories of mass, acceleration, hydrostatics, elasticity, and kinetics found their way into every textbook of physics, but the promising studies of the strength of materials done by Musschenbroek in 1729 had few followers except among engineers, and then only after a delay of three quarters of a century.

During the seventeenth century, however, very many of the structural ideas that lie at the basis of today's approach to materials were suggested in a qualitative, conjectural way. It began with the rediscovery of Greek atomism, which was effectively used in the attacks on Aristotelian orthodoxy by Giordano Bruno and others, and slowly passed from philosophy to natural philosophy. Johannes Kepler in 1611 (perhaps following a suggestion of Thomas Harriot) described the various ways of stacking spherical particles into crystals.[8] Hooke in 1665 showed how such a model would account for the various angular facets on polyhedral crystals. Huygens in 1678 related growth, cleavage, and optical properties to the stacking of spheroidal parts in calcite.

Descartes had an immense following, especially on the Continent, for his picture of the world based upon the aggregation of elementary (but not indivisible) corpuscles. These particles resulted, he proposed in 1644, from the fragmentation of primary matter in his universal vortices and were shaped by attrition into polyhedra and rounded particles of various sizes, to leave a still finer form, his subtle element, the circulation of which served to compress the others together. The continental Cartesians were, shall we say, more imaginative than the proponents of the "mechanical philosophy" who precisely because of their restraint and disregard of metaphysical principles became so influential in England. Curiously, the Cartesian speculations contain very little about geometric crystallinity and a very great deal about the fitting together of irregular parts. There were spherical and polyhedral molecules; springy wire balls; denticulated parts which could slide over each other only when heat had separated them, thus accounting for the softness of iron at high temperatures; little needles of acid which could insinuate themselves between the parts of a metal; clumps of particles tied to each other tightly so that those on the surface were indifferent to external attack; felted and carded aggregates of fibers; loose structures that can bend, and tight structures that must crack; parts that can slide over each other without losing adherence; aggregates of more than one shape of particle that are stronger than aggregates of a single shape.[9]

Every one of these structural concepts can be found in one form or other today, but they mostly relate to complex groups, not the atoms themselves, and are part of a coherent doctrine,[10] not a collection of ad hoc assumptions to account for individual phenomena. Cartesianism was rightly discarded. Yet, just as the early craftsmen had intuitively felt the nature of their materials, the intuition of the corpuscular philosophers rightly suggested that the variety of behavior of matter was in some way related to the way in which its parts were put together. One of the last to use Cartesian structural concepts was the great Réaumur, who in 1722 made excellent practical use of the theory to account for the properties of iron and steel in terms of changes in contacts between different kinds of parts. Though he used the theory as a guide to develop two eminently practical materials—malleable cast iron and devitrified glass "porcelain"—he had no followers. Knowledge advanced in another way. Not structure, but composition was to be the center of understanding materials for the next two centuries.

Eighteenth Century and Chemical Revolution

It has often been said that the revolution in physics preceded that in chemistry by a century. This is true only of part of physics, for the physics of solids lagged more than a century behind an equivalent level of understanding of their chemistry.

The eighteenth century, rather uneventful in physics, was one of the most exciting periods in chemical history, for it saw the change from principles to clear-cut chemical elements determinable by quantitative analysis. The sulphur principle of Paracelsus had become associated with the qualities of oiliness and unctuousness, with combustibility, and with organic matter in general. Because such things usually contain carbon, it became also the reducing principle involved in metallurgical smelting operations. One of the various classes of earths postulated by J. J. Becher was turned by G. E. Stahl in 1703 into phlogiston, which was putatively responsible for the profound physical effect of producing metallicity.

From Meissen in Saxony came not only the most successful European answer to the challenge of Oriental porcelain but a book, *Lithogeognosis* by J. H. Pott (1746), which experimentally classified earths into calcareous, gypsumlike, argillaceous, vitriable, or siliceous—still associating chemistry with qualities. But the main development in eighteenth-century chemistry was in the field of analysis. Even in the sixteenth century, the quantitative analysis of precious metals in ores and objects had been in an advanced state, for rather simple pyrotechnical operations (involving molten slags, sulphides, and the oxidation of metallic lead) would produce beads of pure gold and silver from almost anything that contained them. But the extension to other materials needed wet methods of analysis. These arose mainly in Sweden and eventually led to the new definition of an element. The growth of pneumatic chemistry led to the realization that the metallurgist's ancient charcoal fire had been a source of carbon and oxygen for chemical reaction as well as of heat. The subsequent story of the filling out of the list of chemical elements, of the quantification of the chemical atom and simple molecules, is a magnificent one, but it is too well known to need development here; it is of greater interest to consider what was not done.

Though the discovery of oxygen and the true chemistry of reduction was a triumph, in achieving it a physical feel for metals was put aside. It was not just stupidity that made a few chemists reluctant to abandon phlogiston. They were trying to preserve some of the quality beyond the composition. Though the oxygen atom is a rather big thing to overlook, it is nevertheless true that its importance lies in its physical effect of removing an electron from the state in which it confers metallic properties upon matter. In a way, the outer valence electrons of atoms correspond almost tangibly to the phlogiston postulated in the eighteenth century. Metals are metallic not because they do not contain oxygen but because they do have their valence electrons in the conduction band. Such internally free electrons confer ductility, conductivity, and other metallic properties. Their ready availability in carbonaceous and hydrogenous materials is responsible for the chemically reducing properties of supposed phlogiston sources. The electropositive elements soak them up. Bergman in his classic paper on the analysis of iron used as a quantitative measure of the amount of phlogiston in various forms of iron the volume of hydrogen that was evolved on their solution in acid.[11] Was he not physically right, for the yield would depend upon the electrochemical equivalent? Today we can handle phlogiston by itself; indeed we pump phlogiston through a resistor to generate heat and through an electrolytic cell to make metals. All of these possibilities were temporarily thrown aside when, after Lavoisier and Dalton, the determination of atomic ratios became the main aim of chemistry, and the role of the electron in solids had to be laboriously discovered by men of a different stamp of mind, quite unaware of its background in outmoded phlogiston.

Empirical chemical experimentation on materials continued, of course, long after chemical theory gave the main direction of research. Many problems that are of interest to solid-state physicists today were noted by practical chemists early in the nineteenth century, for example, the thermodynamic basis of elasticity in rubber, the catalytic effect of alumina surfaces, and various phosphorescent, thermoluminescent, thermoelectric, and photoelectric phenomena. Most of the electrical properties of materials were uncovered by simply observing, by new means, substances that had long been available.

Molecules and Crystals

Although the physical ideas toward which the corpuscular philosophers and experimental chemists like Boyle had been straining were slow to develop, the chemistry became clear and quantitative with Dalton's atomic theory. Here, too, an immense gain was accompanied by a not insignificant loss. The confirmation of Dalton's hypothesis was possible only by ignoring the large fraction of substances that do not conform to the law of definite proportions. Because of this, the whole relation between chemistry and metallurgy began to change. In the nineteenth century, metallurgists were foremost users of analytical chemistry, for they used composition, the presence of both major and minor alloying elements, to explain old mysteries such as brittleness in iron and copper, as well as to control the efficiency of production processes and to find new types of ore, but metallurgy lost the close association with the forefront of chemical research that it had had in earlier centuries, indeed from the very beginning. The separation, reduction, and refining of metals had provided the chemist with most of the reactions that taught him about the behavior of matter. For a whole century after Dalton, chemists' eyes were closed to all but molecules of stoichiometric compounds, and most alloys, to say nothing of sulphides and slags, are not of such definite composition. The bias toward reactions in aqueous solution was reinforced, although the fact that composition was not everything was shown by the newly discovered chemical identity of the different physical forms of carbon and sulphur, and the existence of isomorphism between crystals of chemically different substances. The simple concept of the molecule was reinforced by every new discovery of the physical and the chemical behavior of gases and triumphantly vindicated by the kinetic theory of gases. In organic chemistry, the concept of the molecule was of utmost importance. Though it began as a notational scheme, molecular structure became a physically real model with Kekulé's benzene ring and with the development of stereochemistry. The very success of the molecule in gases and in carbon compounds, however, effectively limited any serious thought about higher levels of organization of either atoms or molecules in solids. Arguments about the crystallization of metals by vibration were carried out by engineers, not physicists.

Through most of the nineteenth century, the crystallographers, the physicists, and the chemists talked little to each other. Though Haüy's polyhedral boxes had become mathematical unit cells to contain molecules, the properties of crystals were attributed to the shape of individual molecules and to their orientation in space rather than to the manner in which they were stacked in the crystal. The concept of a crystal of a simple substance as a stack of balls is so familiar today that it is hard to account for the earlier disinterest in this model, especially since it was the first one actually to be proposed. Though the ball model was common among physicists, if not philosophers, in the seventeenth century, it had been replaced by little polyhedra (suggested by cleavage fragments) early in the eighteenth century. The mathematics of crystallography was, at first, the simple analysis of steps in such polyhedral packing and, later, of point-group symmetry, both done with careful avoidance of any suggestion that the mathematical units were chemical atoms or molecules. Wollaston said so clearly in 1813 (unaware at first that he was reviving the seventeenth-century view), but he had few followers until 1883 when Barlow (also an Englishman with an empirical bent) brought chemistry and crystallography together.[12]

In the twentieth century, chemical crystallography in both England and Germany was beginning to make great advances based largely on Barlow's concept, when the discovery of x-ray diffraction in 1912 suddenly gave a superb new tool for studying crystalline units and brought physics and chemistry together to take a look at real solids. By coincidence this discovery occurred within a few months of the publication of Niels Bohr's concept which gave physical meaning to the energy levels within the atom that had been revealed by optical spectroscopy, and it at once provided the means of extending spectroscopy to the shorter wavelengths and higher energies needed for its full confirmation. From our viewpoint, x-ray diffraction also marks the beginning of a new concern on the part of physicists with the structure of atomic aggregates. Indeed, it marks the beginning of a reversal in the movement toward the ever smaller that had characterized physical science since its beginning. Structure had become measurable, and with it arose an interest in those properties that were so sensitive to structure that they had previously been beyond the possibility of good "scientific" treatment.

To the physicist, even more important than the structure of the atomic framework is the structure of the electronic energy levels interacting within it, so elegantly treated by quantum mechanics. The electron, once discovered, quickly joined with old electrochemical theory to become the material basis of chemical valence. The Drude electron gas theory had some success in explaining metallic conductivities but otherwise had the same difficulty that the Rutherford atom had. The success of the quantum theory within the atom was soon followed by Schrödinger's generalized equation for aggregates, and for the first time in four centuries the fundamental approach to the nature of matter began to move upward in scale and complexity. A science of materials as distinct from matter became possible. Like molecular structure earlier, quantum mechanics began almost as a notational device, and even today physicists tend to ignore the rather obvious spatial structure underlying their energy-level notation, but the theory has the important basic quality of showing the dependence of energy on the entire structure: the structure within the atom—its outer shells at least—being dependent on its environment, and vice versa. Its very essence is hierarchical. It soon led to the explanations in terms satisfactory to the physicist of the various properties of solids that had been sensed and used so long before. It is still an ideal picture explaining matters in principle rather than in full detail, but the difficulties lie in the complexity of the calculations, not in the simplification of the model itself. The reasons for the ductility and optical reflectance of metals, the hardness of diamond, the softness of sulphur, and the qualities which the early chemists had seen in the vitriols and salts were now apparent. All derive from the different patterns of the interaction of electrons and photons within the fields of the positively charged atomic nuclei, stabilized in a particular morphology by the interaction of the levels themselves. Matter is a holograph of itself in its own internal radiation.

Not the least important part of the new approach, forced by the obvious inadequacies of the ideal structure of crystalline matter that resulted from the first crystallographic studies by x-rays, was the focusing of attention on the role of disorder and imperfections, both mechanical and chemical. It was only after 1940, however, that physicists discovered the importance of the metallurgists' older naive concepts and empirical data on the behavior of grain boundaries and on the work-hardening of crystals (both currently explained by the interaction of dislocations within the crystals); on diffusion, which metallurgists had used for millennia in making brass, steel, and gilded surfaces (now seen to require vacant sites in the crystal lattice); and eventually electrical imperfections both in the form of ionic substitutions, vacancies, and local charge anomalies called excitons. Models based on each of these unit imperfections could be mathematically treated in the approved way. A field that particularly attracted physicists in the decades after World War II was that of semiconductors, and here it must be noted that the knowledge of the old craftsman contributed little: it was physics of a new kind, theoretical and practical men working together for complex objectives in which the physicist's passion for understanding was influenced by an acknowledged desire to be useful, perhaps even rich and influential, and this led him to a keener awareness that understanding things "in principle" was not always enough.

Metallography as a Harbinger of Solid-State Physics

The rapid advance of solid-state physics in the 1950s would not have been possible had there not existed a rather well-developed body of knowledge relating properties to the level of structure visible under the optical microscope. This realistic concern with structure was the metallurgists' particular contribution to science, for it brought back into view, via empirical but intelligent observation, some of the more complicated aggregates about which the corpuscular philosophers had speculated in the seventeenth century. This has been discussed at some length in my *History of Metallography* and is only outlined here.[13] For some reason the early microscopists had failed to find significant structure in metals, and even Réaumur's masterly study of fracture, mentioned above, had no followers, for most scientific metallurgists in the eighteenth and nineteenth centuries were involved in exploiting the application of chemical analysis.

Some interest in structure remained, however, on a practical level. Fracture tests continued to be useful especially in controlling the quality of iron and steel. Some disastrous railroad failures in the 1840s precipitated fierce arguments over the crystallization of metals by vibration. After a hint from geologists who had developed coarse crystalline structure in iron-nickel meteorites, the true microstructure of steel was at last disclosed. This was done in 1863–64 by Henry Clifton Sorby who was the first European to prepare metal surfaces by methods delicate enough to avoid the obliteration of the significant structure. Studies of metallography, as this branch of physical metallurgy became misleadingly called, took on renewed meaning after 1900 when Gibbs's thermodynamic principles were shown to be simply applicable to the analysis of the existence of phases. Scientists—mainly German chemists—undertook to determine constitution diagrams of innumerable binary alloy systems by mixed thermal and microscopic means. No rules of alloy formation were uncovered (ordinary valence seemed not to apply), and so much unrelatable data had been discovered that the chemistry of alloys was beginning to lose interest in 1912 when the new technique of x-ray diffraction opened up structure on a different level for exploration.

The relation of the visible microstructure of metals to useful properties, however, continued to be a popular activity among scientific metallurgists. Theories of deformation, of the nature of intercrystalline boundaries, of transformation mechanisms, and many other subjects popular today were advanced and discussed by metallurgists decades before physicists discovered that there was any interest in this scale of matter. But x-ray diffraction inevitably led the physicist into contact with the whole range of solids, and made imperfections unavoidably visible. By 1930 there had been postulated several different types of imperfection—those resulting from gross polycrystalline heterogeneity and various types of mechanical and chemical imperfections within an ostensibly homogeneous single crystal. These models provided satisfactory explanations of many age-old phenomena. An extremely fertile period of interaction between metallurgists and physicists resulted, now, fortunately, extending to those who work with ceramics and organic materials as well.

The new viewpoint is so potent that it has, perhaps, caused too many metallurgists to forsake their partially intuitive knowledge of the nature of materials to worship at the shrine of mathematics, a trend reinforced by the curious human tendency to laud the more abstract.

Nature of Materials Science

Even in the field where he was once supreme because he alone could make or build, the engineer is currently losing status to the scientist. Personally, I think this is temporary, partly because many people who are called scientists are simply users of computers with no more understanding than the old unimaginative users of engineering handbooks, but more because I see in materials engineering the germ of a new and broader kind of science, an attitude of mind, a method and a framework of knowledge applicable to many areas. The materials engineer, no longer the specialized smelter of ferrous or nonferrous metals, is now beginning to look at all materials competitively or rather comparatively. He is as likely to be interested in ceramics and synthetic organic polymers as he is in metals. His job is to find, to invent, and (or) to produce materials having the particular combination of properties (mechanical, magnetic, optical, electrical, and others, including economic) that is needed for a given service. The materials engineer's complex knowledge of what it is possible to achieve, involves him in the very center of discussion of most new projects, whether scientific, engineering, or social in nature.

The materials scientist has in large degree recaptured in more definite form many of the discarded intuitions of the past. He has returned to a direct and intimate concern with the qualities that fascinated man from the beginning, and the explanation of these properties is now seen to depend directly upon structure, that is, form. But significant structure is a mixture of perfection and imperfection. The imperfect aggregates are, on one scale or another, not much different from some of the aggregates that were postulated with undisciplined enthusiasm by the Cartesians. Phlogiston, which the chemists had to discard as an inadequate compositional reason for the difference between a calx and a metal, has become the quantum theorists' conduction-band electron, quite literally responsible for metallic qualities and in other associations for pretty nearly everything.

The whole story of man's relation to materials involves the interaction between the simple and the complex, with all of the triumphs of science up to the present being in the direction of the atomistic (or at least simplistic) and all of the realities of matter being complex. The laws of science apply under definable circumscribed conditions. The transistor shows that simple things matching the mathematics can be made in practice, but most things that human beings deal with are complex systems that are the result of a long succession of single events, recorded in the emergent structure but in combination are essentially uncomputable. In practice it is necessary, therefore, for what exists to be measured grossly, using mainly sensual experience to reveal the cooperative effect of innumerable factors that are computable only in isolation.

The enormous success of the rigorous atomistic approach in the last three centuries has led us to expect continued illumination from the same approach. In the limited high-energy world below the atom it seems as if there are always particles below particles, but nothing so fundamental lies in the realm of concern to us aggregate humans, where the need is, now, for the study of real complexity, not idealized simplicity. In every field except high-energy physics on one hand, and cosmology on the other, one hears the same. The immense understanding that has come from digging deeper to atomic explanations has been followed by a realization that this leaves out something essential. In its rapid advance, science has had to ignore the fact that a whole is more than the sum of its parts.

Polanyi has strongly argued that biology is not reducible to physics and chemistry since the existing morphology of an organism, which provides the boundary conditions within which the physical or chemical laws operate, are physically and energetically indistinguishable from other no less probable morphologies that have not happened to come into existence.[14] This argument is valid and applies equally to the much simpler aggregates of the materials engineer.

Science now relates to the two extremes of elementary atomistic physical chemistry on one hand and averaging thermodynamics on the other. But why cannot science develop a new approach encompassing the whole range? I am not as pessimistic as Polanyi, for I see in the complex structure of any material—biological or geological, natural or artificial—a record of its history, a history of many individual events each of which did predictably follow physical principles. Nothing containing more than a few parts appears full panoplied, but it grows. And as it grows, the advancing interface leaves behind a pattern of structural perfection or imperfection which is both a record of historical events and a framework within which future ones must occur. Deoxyribonucleic acid is simply a mechanism to save time in reaching higher levels of organization, though, of course, with severe limitation of possible structures. It is neither possible nor necessary to study all structures that might have existed, but there is need for studying more than a statistically averaged structure. Is there not possible an intermediate science using the structure that exists—important for no other reason than that it does exist—both as a key to history and as a framework for continuing process?

Hesitantly, in my ignorance, I predict the development of some new principles of hierarchy that will enable the effective resonance between molecule and organism to be explored: possibly the way to this may be pointed by the emerging science of materials, so incredibly simple beside biology but complex enough to demonstrate a kind of symbiosis between scales, the interwoven importance of both atoms and aggregates. Such things can be appreciated and understood only by a parallel aggregation of viewpoints, one intellectual, atomistic, simple, and certain, the other based on an enjoyment of grosser forms and qualities, but somehow the two must join as they do in matter. A few men will be in touch with both levels, but human capacities are such that most must specialize, and the liaison man will be far more important than he has been in the past when the greatest intellectual opportunities lay at the frontier. The scientist need not despair. He must become a little more of a whole man. He must restore his senses to a position of respect, though not in domination over his intellect, for each must supply something lacking in the other.

This approach would bring together fields that because of their special complexities have been unrelated; it would minimize the difference between the scientist and those who try to understand the human experience. It would incorporate the historian's interest in the past as the basis of the present and the artist's feeling for the complicated interrelatedness of things. Encouraging diversity but controlling disruption, it would suggest more viable political structures. Using man's mind, hand, and eye in coordination, it would be a thoroughly human activity. This conclusion, if not perhaps some of the details of my historical approach to it, George Sarton would have approved.

Notes and References

1
P. Drucker, "The first technological revolution and its consequences," *Technology and Culture* 7:143 (1966).

2
A good summary of the Greek philosophy of matter is in S. Sambursky's two books, *The Physical World of the Greeks* (London, 1956) and *The Physical World of Late Antiquity* (New York, 1962).

3
D. E. Gershenson and D. A. Greenberg, "The first chapters of Aristotle's *Foundations of Scientific Thought (Metaphysics Liber A)*," *Natural Philosophy* 2:5 (1963).

4
In all earlier translations these elements appear as "earth," "air," and "water," respectively. Gershenson and Greenberg's version rightly emphasizes the state rather than the material, but the old terms are so deeply embedded in the literature that they are probably ineradicable.

5
J. Needham, "The origin and development of the five element theory," in *Science and Civilisation in China* (London, 1956), vol. 2, pp. 232–253.

6
Theophilus, *De Diversis Artibus* (ca. A.D. 1125). There are two recent English translations, one by C. R. Dodwell (London, 1961) and one by J. Hawthorne and C. S. Smith (Chicago, 1963). The latter includes technical notes.

7
V. Biringuccio, *De la Pirotechnia* (Venice, 1540; English translation by C. S. Smith and M. T. Gnudi [New York, 1942]); G. Agricola, *De Re Metallica* (Basle, 1556; English translation by H. C. Hoover and L. H. Hoover [*Mining Magazine*, London, 1912]).

8
J. Kepler, *Strena Seu de Nive Sexangula* (Frankfurt, 1611; English translation by C. Hardy [Oxford, 1966]).

9
The best sources are R. Descartes, *Principia Philosophia* (Amsterdam, 1644); J. Rouhault, *Traité de Physique* (Paris, 1671); C. Perrault and P. Perrault, *Essai de Physique* (Paris, 1690); and N. Hartsoeker, *Principes de Physique* (Paris, 1696). For further discussion see items 104 and 107 (chapter 8) in the bibliography.

10
My colleague Nathan Sivin points out that whatever else was wrong with Cartesianism, its aim was to be a coherent doctrine. However, despite this coherence at the highest levels, and despite the belief in the importance of the shape and arrangement of particles at the lowest levels, when explaining the properties of materials both Descartes and especially his followers invented innumerable particles purely on an ad hoc basis. The modern "coherent doctrine" at least uses the same units in all of its discussions of aggregates a level or two above the atom.

11
T. Bergman, *Dissertatio Chemica de Analysi Ferri* (Upsala, 1781; English translation in item 150 in the bibliography).

12
W. Barlow, "Probable nature of the internal symmetry of crystals," *Nature* 29:186 (1883). The best history of early crystallography from the present viewpoint is J. G. Burke's *The Origins of the Science of Crystals* (Berkeley, CA, 1966).

13
C. S. Smith, *A History of Metallography* (Chicago, 1960). For a listing of the various types of microstructures that are known and their origins see item 129 in the bibliography.

14
M. Polanyi, "Life's irreducible structure," *Science* 160:1308 (1968).

15
(*Note added 1980*) The present state of materials science and engineering was the subject of a four-volume report to the National Academy of Science: M. Cohen and W. O. Baker, eds., *Materials and Man's Needs* (Washington, D.C., 1975); a partial summary has been issued as a special number of the journal *Materials Science and Engineering* 37(1)(1979), also available in hardcover.

16
The author is grateful to the Sloan Fund for Basic Research at MIT for support of his historical activities in general, and to St. Catherine's College, Oxford, for a visiting fellowship in the spring of 1968 during which time this manuscript took its final form.

6
The Early History of Casting, Molds, and the Science of Solidification

"In its natural state metal is enclosed in earth: likewise when it is being made into useful implements it is also enclosed in earthen molds."

—Sung Ying-Hsing

The Beginning of Casting

I propose to discuss three stages of history—the very beginning of some techniques of mold-making; the theories of solidification that arose when scientists first became intrigued with the properties of matter that had long been known to artisans; and the period when empirical theories of solidification and segregation laid the groundwork for the more mathematical approach of today.

Metallurgy did not begin with casting and, if the powder metallurgists have their way, it may not even end with it. The main bulk of metal (iron) was not melted at all until the sixteenth century A.D., and even now casting is done mainly to facilitate the production of something that can be wrought.

It is perhaps fitting that man's first use of any metal depended on the most notable metallic property, the ability to be plastically deformed without loss of cohesion. Little bits of native copper were being hammered into decorative forms at least as early as the ninth millennium B.C. in the Middle East. In North America, where native copper lumps were abundant, large useful tools and weapons were shaped by hammering (with intermediate annealing) by about 3000 B.C., but in the Middle East extensive use of metals for tools and weapons had to await the discovery of smelting from abundant ores. The fact that the earliest cast copper is of higher purity than most subsequent metal prior to electrorefining suggests that smelting was preceded by a stage in which native metal was melted without chemical change. Whether or not this is so, the first castings were probably no more than lumps formed at the bottom of a container in which metal had been (s)melted.

There are good discussions of prehistoric casting methods by Coghlan,[1] Tylecote,[2] and Aitchison.[3] Wertime[4] has analyzed "Man's first encounters with metallurgy."

All available evidence points to copper, which needs a temperature of about 1100°C, as being the first metal to be melted, not common low-melting lead. It is neither certain nor improbable that the first melting occurred in a potter's kiln—for potters were the leading users of pyrotechnic means to modify materials—though an annealing fire could also have provided the condition necessary for the discovery. In both cases unusual circumstances would have been needed to give a temperature high enough to melt copper. The first cakes of melted and solidified metals were, supposedly, simply a supplement to the supply of native metal being hammered to useful shape. They would, nevertheless, quickly demonstrate that molten metal would take the shape of the surface against which it lay and would preserve this after solidification. Simple cast copper objects appeared by 4000 B.C. From here on, the development of casting is partly the history of mold materials and techniques, and partly the discovery of new compositions that could be more easily cast or that would give sounder, stronger, cheaper, or more decorative castings.

There is a most intimate connection between metallurgy and ceramics; from the earliest days the techniques of the potter in exploiting the unique combination of plasticity and refractoriness of clay have been the basis of the mold-maker's art. Sand, so

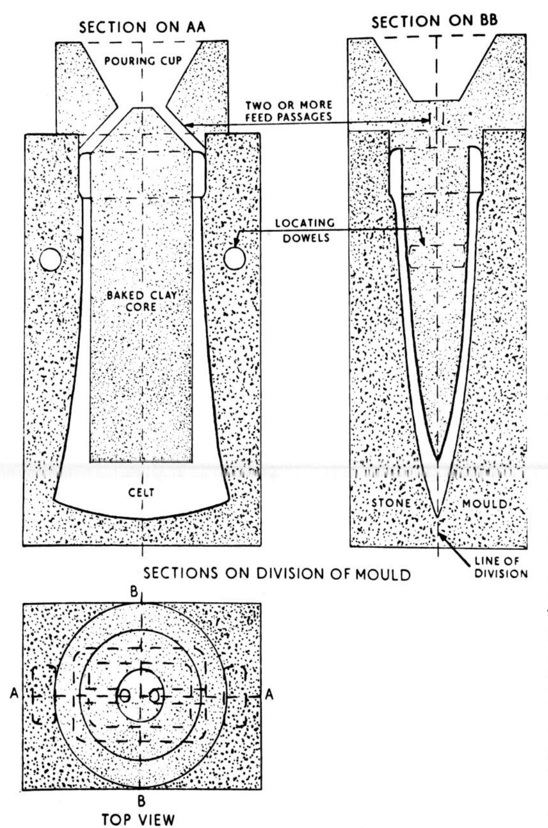

Figure 6.1
Probable construction of stone mold with baked clay-sand core for casting a socketed celt. Drawing courtesy of H. H. Coughlan.

popular today, is a very recent material for molds.

The best records of early casting techniques, as of most technological processes, are to be found in art museums. The skill of the artist lies largely in relating his aesthetic aims to the requirements dictated by the properties of his materials and the processes of shaping them.

The simplest mold is the open groove pressed into clay or, for semipermanence, cut into stone, leaving one face of the casting horizontal, exposed to air, and unfinished. Next follows the bivalve mold and the bivalve with a core either transversely as in the Luristan axes or axially as in the Celtic socketted axe (figure 6.1). An aesthetically pleasing bilateral symmetry results from the joints in the mold, which are necessary to allow the shaping of interior surfaces, regardless of whether this is done, as in stone molds, by direct cutting, or, when using clay, by pressing against a preshaped positive.

Chinese Ceremonial Bronze Castings

The jointed mold becomes very complex if anything but simple bilaterally convex shapes are to be produced, yet it was used to produce some of the most magnificent castings the world has known—the Chinese ceremonial bronzes of the Shang and Chou dynasties (figure 6.2). Until quite recently it has been casually assumed that these vessels with their intricate detail "must have been" made by the lost-wax method, but the researches of Li Chi, O. Karlbeck,[5] Barnard,[6] and Gettens[7] leave no doubt that they are made by the piece-molding technique, and, indeed, that much of the strength of the design comes from the exploitation of the in-

Figure 6.2
Chinese ceremonial bronze vessel, type *tsun*, Shang dynasty (twelfth or eleventh century B.C.). Mold joints can be seen on the flanges. Photo courtesy of the Smithsonian Institution, Freer Gallery of Art, Washington, D.C.

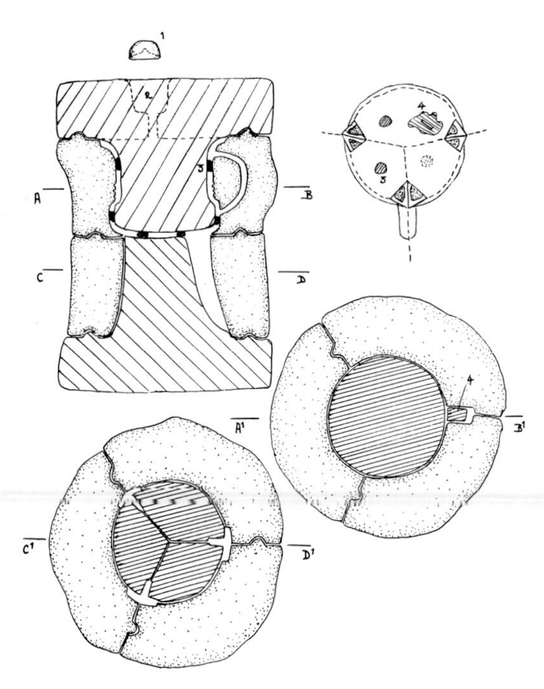

Figure 6.3
Sketch showing probable mold assembly for casting a ceremonial bronze *chia*. Spacers (3) keep mold and core apart. The gate (2) is left as part of the finished casting, the metal in it being shaped with a plug (1). A repair patch with a rectangular sprue can be seen in the bottom (5). Drawings courtesy of Noel Barnard.

evitable defects associated with the many joints in the mold. Figure 6.3 illustrates Barnard's ideas on the mold assembly for casting one of the three-legged vessels called *chia*.

The molds for the earliest castings seem to have been made by molding around a smooth pattern or model, the mold being in sections to allow easy removal from convex surfaces. The impressions of the joints can often be clearly seen as defects running over the surface of a finished casting (figure 6.4). The pattern may itself have been made of mold clay, and separately scraped down to serve as a core. Separate pieces were incorporated to shape handles and knobs. The smaller clay cores were often left embedded in the finished casting. It seems likely that the decorative detail was at first carved or incised directly on the mold surface. More elaborate designs soon followed, a characteristic feature of which is that some symmetrical figures are molded in moderately high relief, while the area between these is covered with elaborate decoration, apparently incised on the mold surface. Particularly effective is the contrast between a common feature, the monster mask called *t'ao t'ieh* with its relatively plain convex surface relieved with a few lines, and the background which is covered with the fine intricate squared spirals of the *lei-wen* (figure 6.5). The modern foundryman cannot but be amazed at the perfection of the *lei-wen*, with almost no tool slips or mold fractures to give clues to the molding technique. It is frequently impossible to decide which decorative details were carved positively on the pattern for impression into the mold, and which were carved negatively directly on the mold surface. Figure 6.6 shows a fragment of a mold from the early period of casting in Anyang, one of many that have been found

Figure 6.4
Bronze *chia*, early Chou dynasty. Detail showing fins and pattern disregistry resulting from the movement of mold parts. Photo courtesy of the Boston Museum of Fine Arts.

Figure 6.5
Cast bronze *kuei*, early Chou dynasty. The fine detail was probably carved directly into the mold surface; the bolder features in relief are impressions from a pattern. Photo courtesy of the Boston Museum of Fine Arts.

Figure 6.6
Fragment of a clay mold from Anyang. Scale in mm. Photo courtesy of the Sumimoto Collection.

in recent archaeological excavations. The material seems to be a high silica loess, with relatively little clay.[7] It cannot have been a highly plastic material.

As experience grew, the molds became more elaborate. In Eastern Chou times, eighth century B.C., pieces such as legs and handles were cast separately and incorporated in the mold of the main vessel. Examination shows that these rarely fused into the metal in the body of the vessel: this would have required metal that was too hot to give a good casting and, of course, was unnecessary in a decorative object not subject to high stress (see figure 6.7). Later, precast parts were attached to a precast body in a separate joining operation that involved casting an intermediate little ring locking into both—the beginning of autogenous welding. Evidence on the use of true brazing solders in China is still uncertain. Drescher has examined at length the use of the casting-on process in the West, where it became of even greater importance than in China.[8]

Most of the Chinese ceremonial bronze vessels are only about 2 to 4 millimeters thick, and the positioning of the core in the mold offered a considerable problem. Studies at the Freer Gallery have shown that this was commonly achieved by the introduction of chaplets consisting of small, approximately rectangular though broken pieces of previous vessels, placed rather liberally around the mold.[7] They are sometimes distinguishable on the surface of the finished casting by a local difference of patination (figure 6.8) and can be found by radiography. Often, instead of a chaplet, a little pyramidal or roof-shaped projection was left on the clay core to rest against the mold body, leaving a fine unfilled line on the casting. This was a refinement of earlier cruciform clay spacers, which often

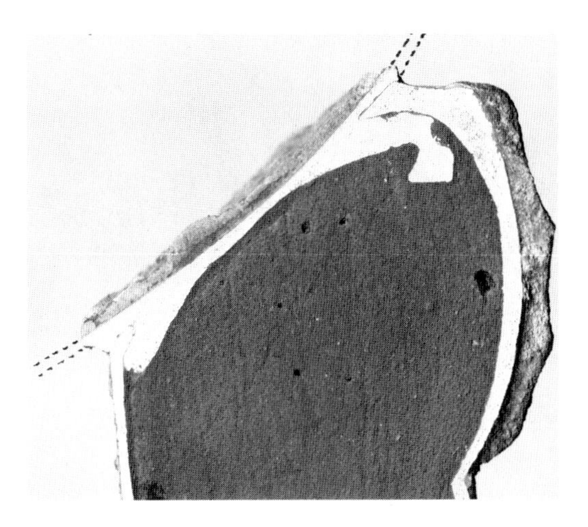

Figure 6.7
Section of a cast bronze *ting* showing a joint with cast-on leg. Photo by R. J. Gettens, Smithsonian Institution, Freer Gallery of Art, Washington, D.C.

Figure 6.8
Cast bronze *hu*, Western Han dynasty (202 B.C.–9 A.D.). Detail (× 1.5) showing square chaplets revealed by their lighter green corrosion product. The displacement of some of the chaplets during casting resulted in the clustering of this group. Photo courtesy of the Boston Museum of Fine Arts.

left holes as much as 8 millimeters across.

By Eastern Chou times (770–221 B.C.), much of the detail in the vessels was provided by premolding clay against a flat master pattern of the decor, then cutting out little slablets that were joined into a larger assembly which in turn formed a section of the main mold. The master pattern was itself made by repeating some units, evidently by physical molding of the details. It is possible—for instance, in the Freer bell (figure 6.9)—to see repeated detail of both pattern units and clusters. Many of the surfaces of castings from this period contain detail that is obviously molded and yet could not possibly draw from a pattern. The mold parts were probably bent over a smooth curved surface when soft immediately after molding. Throughout, there is a fascinating interplay between the mechanics of mold preparation and the very nature of the design. As art history, this is a field that is only just beginning to open, but every metallurgist and potter will immediately appreciate the significance of the impressions of the joints and the other technical details that form the design elements in these ceremonial objects.[9]

The making of these castings in China begins rather suddenly shortly before 1600 B.C., and there is very little archaeological evidence of an experimental period. The techniques are so different from the earlier bronze casting in the Middle East as to suggest independent invention, a view that is strengthened by the manner of their first occurrence in one place with progressively widening circles of diffusion thereafter and no detectable line of transmission from the West.[10] However, the lack of any evidence of a period of experimentation makes it more likely that there had been a transfer, at least of a nucleating idea if not a rudimentary technique, from the West. One suspects that the rapid and elaborate development of mold-making in China was in large measure attributable to the local potter's long and successful acquaintance with the technology of clays.

Lost-Wax Casting

In the West, long before there were any Chinese bronze founders, the more elaborate castings were being made by the lost-wax (*cire perdue*) method. In this, a wax model is completely invested in clay or other refractory material and is then melted and burned away, leaving its space to be filled with molten metal. Castings bearing no evidence whatever of mold joints and possessing such a freedom of shape that they could hardly have been made other than in invested wax first appear around 3000 B.C. in Mesopotamia. There is, of course, no evidence whatever of what "wax" was used. Any easily shapable material that was fusible or volatile, or that left a light removable ash on combustion, would serve—for example, a hard animal fat, beeswax, rosin, some of the harder forms of bitumen, and, on occasion, the bodies of insects or a light low-ash wood. With this technique the artist is free from any restraint arising in the negative nature of the mold, for the pattern does not have to be withdrawn and the mold surfaces, however contorted, reflect all the details of the wax and do not have to be accessible to further tooling. The sculptor could, indeed, model things as freely as he had done in clay for millennia. At first these castings were solid, but later the art of molding the wax over a sandy clay core was introduced. Pierced areas in the design were necessary to

Figure 6.9
Cast decoration on top of a large bronze bell, late Chou dynasty. × 0.5. Photo courtesy of the Smithsonian Institution, Freer Gallery of Art, Washington, D.C.

Figure 6.10
Section of mold for casting a large equestrian statue by the lost-wax process. Note the heavy iron armature and the numerous channels for metal ingress. From Volume VIII of plates to Diderot's *Encyclopédie* (1771).

keep the mold and core properly spaced, or metal spacers (chaplets) were used, to remain embedded in the finished casting. The superb life-size bronze statues of Classical Greece were probably made in this way, although they were not cast integrally. A large and elaborate lost-wax mold for an equestrian statue is shown in figure 6.10, taken from the French *Encyclopédie* (1771).

It is interesting to compare the products of three schools of founders who have used lost-wax casting—those of Syria (figure 6.11), of Nigeria (figure 6.12), and of Colombia (figure 6.13), each of which in a different way exploits the versatility of the medium.

Japanese founders in the nineteenth century used the lost-wax technique to obtain the most exquisite detailed surface finish that had ever been achieved by casting. Gowland[11] and Wilson[12] together provide an excellent account of Japanese lost-wax casting methods. The importance of a hot mold is particularly stressed if fine detail is to be reproduced.

Lost-wax casting, for centuries kept alive for works of art and later in the precision-shaping of dental metalwork, has had a modern renaissance in the making of accurately shaped castings, often of highly refractory metals, for engineering use.

A most intriguing modern variant of the principle of the disposable pattern is the use by the sculptor Duca of a styrofoam model which is embedded in a sand mold into which molten metal is directly poured, evaporating the low-density foam and taking its place.[13] An example of such casting is shown in figure 6.14.

Figure 6.11
Figurine cast in bronze, with hammered bronze and silver helmet. From Tell Judeideh, Syria, ca. 2900 B.C. This is a very early example of the lost-wax technique. Photo courtesy of the Boston Museum of Fine Arts.

Figure 6.12
Bronze (that is, brass) figure of a "flute player," cast in Benin, Nigeria, during the classic period, 1500–1700 A.D. Photo courtesy of the Brooklyn Museum.

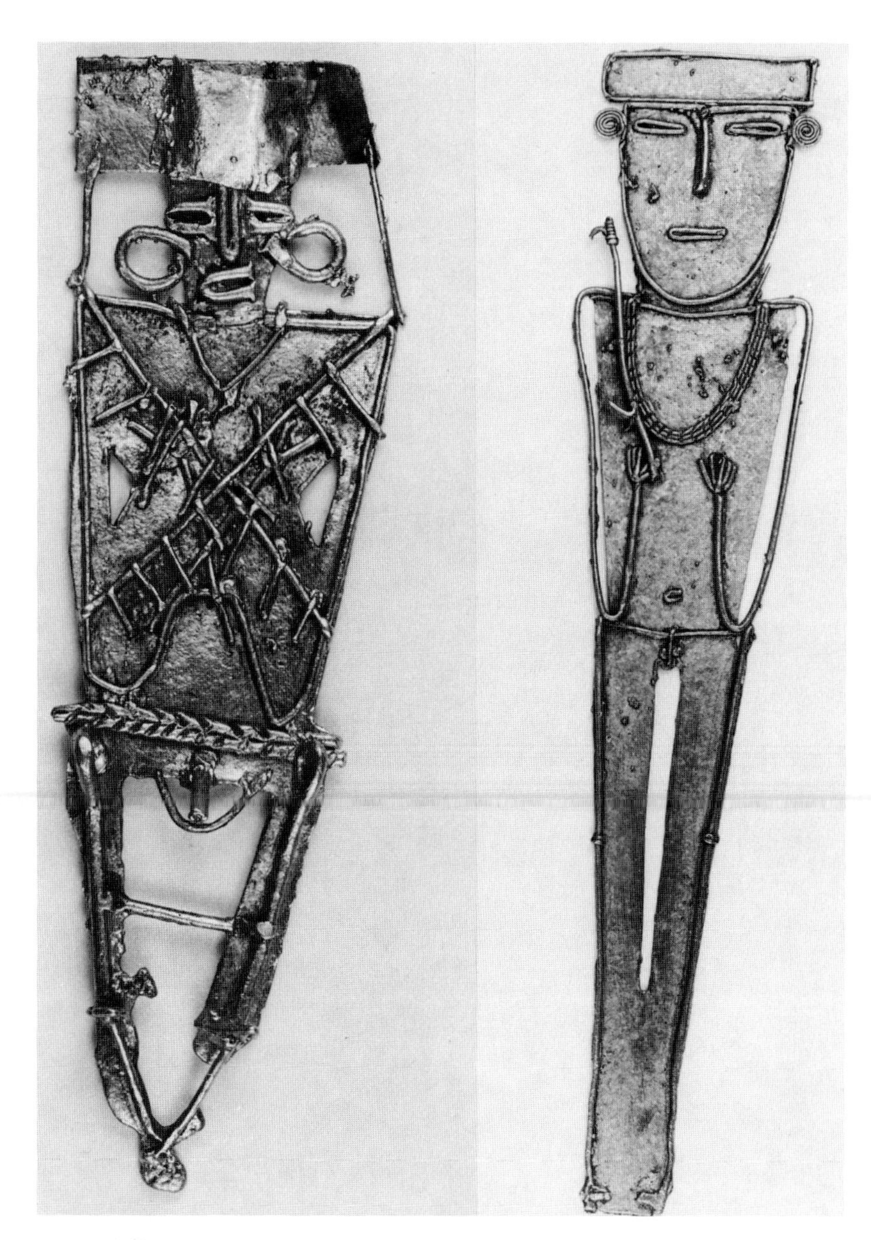

Figure 6.13
Precolumbian votive figurines made of gold by lost-wax casting. Muisca culture, Colombia. Photo courtesy of Clemencia Plazas, Museo del Oro, Bogota, Colombia.

Figure 6.14
Casting made by the foam-vaporization technique, ca. 1960. Alfred Duca, sculptor.

Bells and Guns

Before the sixteenth century (and to a large extent even today) the best records of technical matters are provided by objects, "not written in words but wrought by labor." Nevertheless, no one interested in the subject of casting should fail to read the vivid discussions of bell founding by the pseudonymous German monk, Theophilus, in ca. 1125 A.D.,[14] and the meticulous discussion of metalworking operations of all kinds by the Italian metallurgist, Vannoccio Biringuccio, in his *Pirotechnia*, published in 1540.[15]

Shapes of rotation (produced by the potter on his wheel by about 3000 B.C.) became important in the foundry in the making of bells, for the lathe permits the accurate generation of the shape of both core and cope (the inner and outer parts of the mold) without the use of a pattern or with the use of only an outline cut into a strickle-board. Theophilus, who is the first in all history to record practical details of metalworking on the basis of his own experience, has a magnificent account of the making of bells. He first turned a clay core, then applied tallow or wax and turned it to give the shape desired in the metal, and finally completed the mold in clay with iron bindings. This was heated in a special furnace in a pit to remove the wax and to prepare the mold to receive the bronze, which was melted in a kind of cupola furnace with bellows. Theophilus also used a lathe for preparing pewter pots and for the core of an elaborate cast censer, though he carved this in such a way that the final object was far from a simple rotational shape.

Some of the best records of casting technique are those on ordnance production. Although the first large guns to be used in warfare were made of longitudinal bars of wrought iron held together with rings, these were soon replaced by nonferrous castings. Enormous bronze guns were made in the fifteenth century, using techniques that were appropriated—as often was the very metal—from the bell founder. Several are still extant, one of the most interesting being the one with a screwed-on breech, now at the Tower of London, that dates from the time of the siege of Constantinople in 1453. A description of the casting process written only a few years later has been preserved,[16] but it adds few metallurgical details beyond those given by Theophilus some centuries earlier.

Biringuccio has more detail on the casting of guns than on any other single topic. Figure 6.15 shows the building up of a pattern for a gun and a section of a complete mold taken from the manuscript by Prado y Tovar written in 1603.[17] Essentially the same drawings appear in the better-known later descriptions by St. Rémy[18] and Diderot.[19] The iron core support near the breech should be noted. This was left in the gun. The remains of such iron rods can be found embedded in most surviving guns of the sixteenth century. Stronger core-free castings were made by the end of the eighteenth century, being bored from the solid when better machines became available.

The Casting of Coins

Though coins have at various periods in history been cast, the earliest and the best ones were finished by striking between engraved dies. The earliest real coins are those made in Lydia by a predecessor of the famed King Croesus about 630 B.C. (figure 6.16). Much

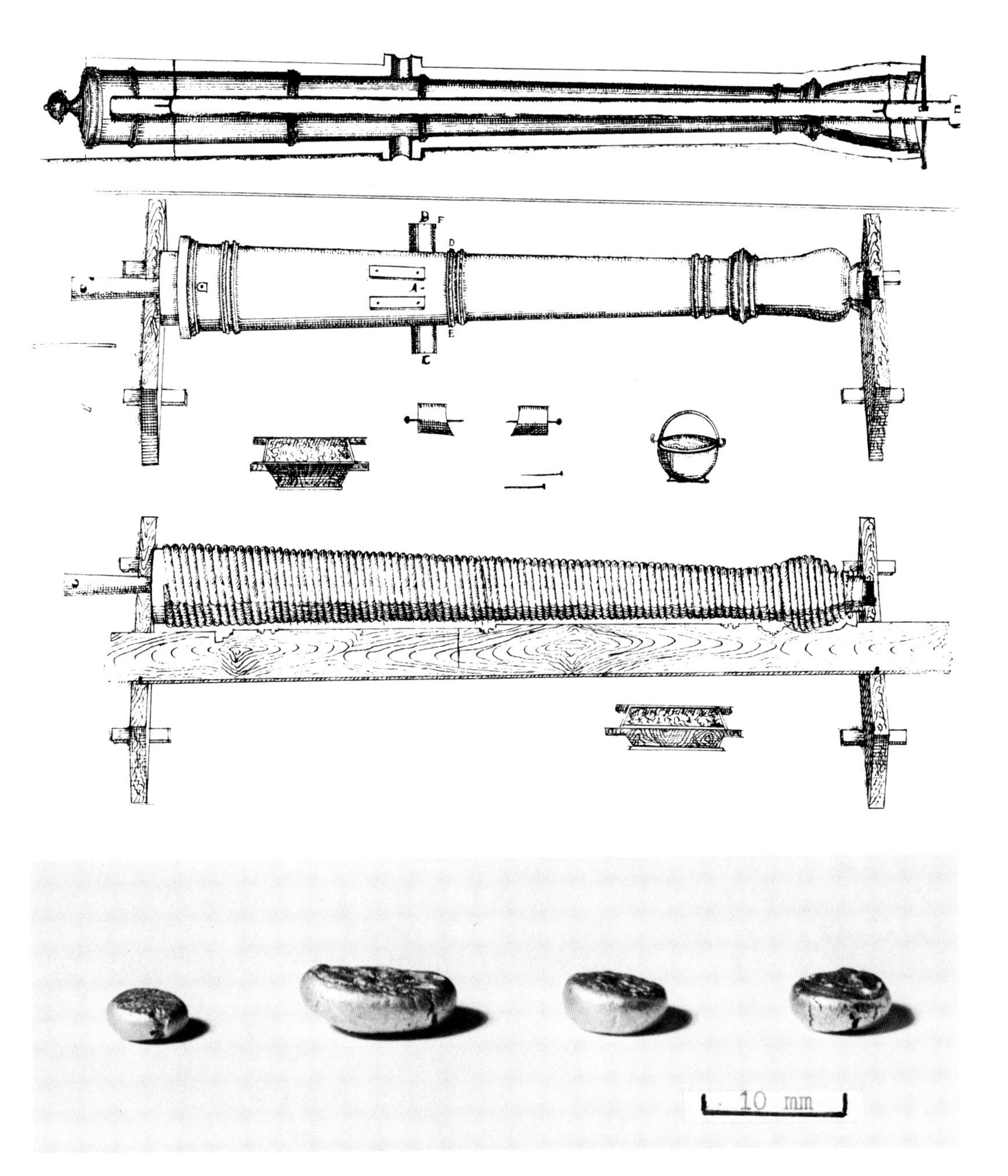

Figure 6.15
Building up a clay-and-wax pattern for a gun, and a section of a completed mold. From Diego de Prado y Tovar's *Encyclopaedia* . . . (1603), courtesy of Cambridge University Library.

Figure 6.16
Examples of the earliest metallic coinages. All are Electrum. Asia Minor, mid to late seventh century B.C. Photo courtesy of the American Numismatic Society.

has been written about these coins, but a fact that is obvious to a metallurgist has not been commented upon: this is that they are simply surface-tension shapes, slightly flattened. The flans were castings made without a mold, using surface tension as the shaping force. There were antecedents to both the physical and economic aspects of these coins: the use of surface tension to shape the minute droplets that were so effectively used in granulation work in gold jewelry (figure 6.17), and the custom of assembling hoards of small metal lumps and fragments to facilitate exchange and trading.[20] It was an inspiration to standardize the metal pieces in weight and in fineness, and to give them a characteristic appearance and mark to indicate authenticity.

It would have been virtually impossible to achieve accurate adjustment of weight by pouring metal from a crucible or ladle into a mold, but easy to do so by simply assembling a small pile of fragments cut and selected to give the exact weight, and then heating until the metal melted and was drawn together by surface tension. Figure 6.18 illustrates what happens when fragments of coin silver are so melted. Such shapes, formed by a balance between the spheroidizing force of surface tension and the flattening force of gravity, are known to anyone who has ever spilled molten metal and are seen in perfect form on the cupel at the termination of every fire assay of gold or silver. When such droplets are hammered flat they duplicate almost exactly the appearance of the early coins, both the general curvature of the edges and the local irregularities in deformation arising from the large grain size. This is not only the easiest way of getting an exact weight adjustment, it is aesthetically superior to the squarer edges of a cut or mold-cast blank. The obverse of the early coins was finished by striking with a die bearing the effigy of the king, but the incuse was unmistakably made with a punch of a brittle crystalline material (bronze, steel, or stone) which was simply fractured so that the impression conveyed an unduplicable guarantee of authenticity of the sort provided by a fingerprint, the halves of a broken clay tablet, or a split wooden tally.

The details of the melting operation are unknown. One can imagine a horizontal plate heaped over with clean charcoal, blown with bellows, or possibly even covered with a muffle to prevent contact with the fuel. Shallow depressions may have been made to render exact horizontality unnecessary and to prevent the individual drops from running together. Several clay "molds" are known from Central Europe which would permit the easy melting of several flans at once (figure 6.19). It has hitherto been assumed that metal was poured into these cavities from an external ladle or crucible, but it seems far more probable that these indented plates were indeed multiple crucibles in which preweighed quantities of metal were melted and allowed to solidify under surface forces alone.[21] The process was eventually displaced by mass-production methods of cutting the flans from flat strip (hammered or rolled), but most Persian, Greek, and Roman coins that were not cast to the final shape have the rounded edges that would result from the suggested process.

Metal granulated by surface tension has continued to this day to be important both for decorative purposes and for ease of handling in the assay laboratory and workshop. The earliest record of a blast furnace in Europe is of one producing cast iron granules, perhaps for use in the production of steel.[22] In the seventeenth century, Prince

Figure 6.17
Detail showing fine granulation on a gold pendant. Indian, Shunga period (185–72 B.C.). The area shown is 2.3 cm wide, and the smallest granules are about 0.2 mm in diameter. Photo courtesy of the Cleveland Museum of Art, Purchase, John L. Severance Fund.

Figure 6.18
Reconstruction of early coin-making technique. Left: Pile of metal pieces adjusted to weight, 2.7 g. Center: The same, melted into a single sessile drop, 10.3 mm in diameter. Right: Drop flattened with a hammer.

Figure 6.19
Multiple crucible for coin-drop melting, found at Manching, Bavaria. Celtic, ca. 100 B.C. Photo courtesy of the Prehistorische Staatssamlung, Munich.

Rupert developed his famous method[23] for casting spherical lead shot by allowing the metal to solidify in free fall in air or water after the addition of arsenic to give a strong skin of oxide to reinforce the natural surface tension. (Parenthetically, what interesting studies could be done on the solidification of spheres of molten metal under the gravity-free conditions of a satellite laboratory!)

Typecasting

The greatest influence of the casting process upon the development of civilization has probably been through printer's type. Although the earliest typography in China, ca. 1045 A.D., used carved and fired ceramic letters, cast tin was mentioned in 1313. Several fonts of different sizes of cast bronze type were made in Korea, at least three—those of 1403, 1420, and 1434—before the introduction of printing in Europe.[24] The type was cast in a conventional way using beechwood models and sand (? clay) molds. It was of rather rough workmanship and far too irregular in shape to permit its locking together (figure 6.20).

Figure 6.20
Cast bronze type from Korea, early fifteenth century or later copies. Photo courtesy of the American Museum of Natural History.

The full potential of printing from movable type was realized in Europe shortly thereafter. Although there is no description whatever of European typecasting for a century after its appearance, it seems probable that Gutenberg employed all of the basically significant principles that appear later, that is, the use of a low-melting-point metal hardened by alloying, a mold with sliding metal parts to give rectangular type bodies that would pack rigidly together, and replaceable matrices in the mold containing the forms of the letters impressed with an accurately shaped punch. The first description of all this is in 1540, by Biringuccio. After a few incidental accounts, there follows in 1683 the highly detailed and well-illustrated book on the whole art of printing by Joseph Moxon.[25] His mold is illustrated in figure 6.21. There was little change thereafter until the introduction of machine casting in the nineteenth century.

Casting in permanent molds, of course, long preceded Gutenberg's precise use of it. Not only were many of the earliest weapons and tools cast in stone, bronze, or (in China) iron molds, but trinkets and jewelry were also so cast in quantity. The designs are usually characterized by recognizable joints in the object. Theophilus has a description of the casting of lead cames in a split iron mold which has all of the flavor of later accounts of typecasting, except for the "shake" by which the typecaster threw the molten metal against the matrix, employing the principle that later was used more reproducibly in centrifugal casting. In the Middle Ages and later, religious shrines sold innumerable cast lead pilgrim's tokens made in molds of soapstone or similar material, one side being flat. Three-dimensional figures in multipiece molds were made of lead and pewter. Cast pewter plates and pots were common; a pewterer would have a stock of molds cut in stone, and some of these undoubtedly had replaceable parts containing undercut parts or decorative detail. The use of a punch was common in bookbinding and in every goldsmith's shop, and there would be no mystery in the inverted typographic impression. Nevertheless, though all the roots of the invention existed, it took Gutenberg's genius to combine them and, above all, to realize the need for a new invention.

Nothing is known of Gutenberg's alloys. One may guess that he first used simply pewter, and then hardened it with larger amounts of antimony and bismuth than were

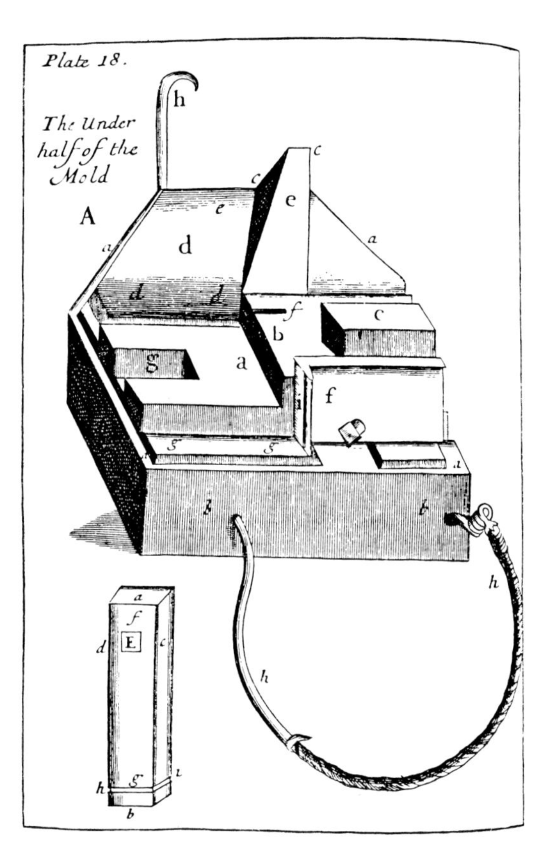

Figure 6.21
Mold for casting type. The artist has inverted the slope of the back part of the mold (d). In use, the matrix (E) is placed against the guide (i). The other half of the mold is nearly identical and forms a rectangular space for type at (a). From Moxon's *Mechanick Exercises* (1683).

normally present in the alloy. Biringuccio describes a tin-rich alloy, "a composition of three parts of fine tin, an eighth part of black lead, and another eighth part of fused marcasite of antimony." His marcasite was supposedly the metallic regulus of antimony, though it may be a confused reference to bismuth, then a relatively new metal which was sometimes called marcasite. Cheaper, harder alloys of lead with antimony came into use in the seventeenth century (if not before), and are described by Moxon.[26]

Slush Casting

Slush casting is the process of making hollow castings without using cores, by filling a mold and pouring out the still-molten metal after a layer of the desired thickness has solidified. It is perhaps best known from its use in casting tin soldiers, though it has been widely used for cheap decorative objects in pewter, lead, and zinc. I have not yet traced its beginning. The slip casting of ceramics in porous plaster molds is a somewhat analogous process of higher status.

The physical principles involved in slush casting were used long ago in the casting of thin sheets of tin and lead for roofing and for organ pipes.[15] An inclined board was covered with a half-inch layer of sand or ashes, spread out, and made exactly flat. A bottomless wooden box was then fitted to slide over the board. When the molten metal had been poured into it, the box was moved slowly down the foot of the board, leaving behind a layer of solid metal of uniform thickness and width. Its thickness, of course, would be determined by the thermal diffusivity of the sand bed and the time that the molten metal was in contact with it. When thicker ingots of lead were needed for rolling into sheet (a seventeenth-century development, using most ingenious reversible mills), the molds were made horizontal and the sliding part became a mere skimmer. Félibien[27] is not the first to illustrate this process, but his illustration of both types of mold bed is most instructive (figure 6.22).

Organ pipes are still made of sheet cast in open horizontal beds. Of great metallurgical interest are those made of "spotted metal." These are reputedly of superior tone and are easily identified by their characteristic surface pattern (figures 6.23 and 6.24). The "spots" occur only with lead-tin alloys containing between 35 and 65 percent lead: the metallurgist at once suspects that the two-dimensional cellular pattern is produced by equiaxed growth of crystals from random nuclei. Microscopic examination of the organ pipe material shows, however, that the lead dendrites are very much smaller than the pattern spots, and, in fact, that lead dendrites irregularly standing proud in shrinkage troughs are responsible for easy visibility of the structure. The cells are eutectic colonies, nucleated at random but infrequently, which have grown through the liquid interstices of a spongy mass of lead dendrites, concentrating all the solidification shrinkage to form depressions along the lines where they meet.

Casting in Sand

It is one of the mysteries of technical history why so simple a process as green sand molding was so late in being developed. Virtually all molds until about 1500 A.D. were made of metal or stone or, if complicated, of foundryman's sandy clay that was dried and thoroughly baked before use. This did not

Figure 6.22
The two methods of casting thin sheet metal for roofing and pipe organs. From Félibien's *Principes de l'Architecture* (1676).

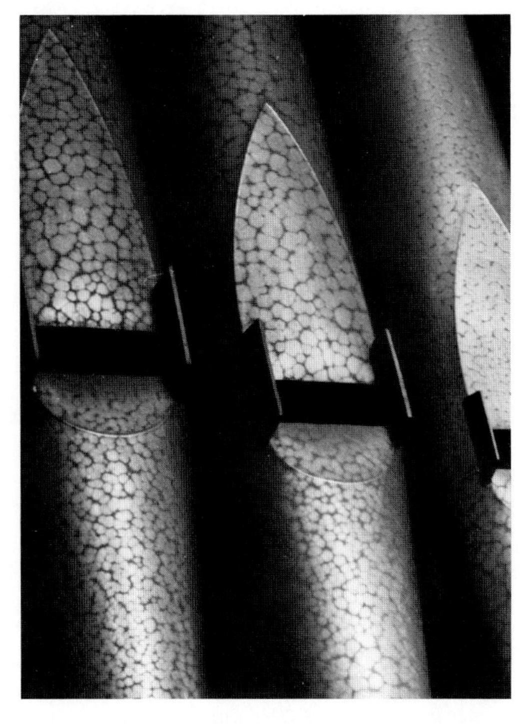

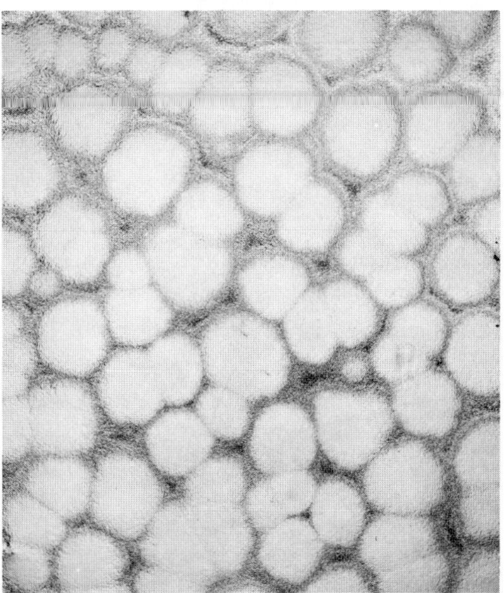

Figure 6.23
Spotted metal organ pipes in a church at Witney, Oxfordshire. Photo by T. M. Newton & Co., Ltd.

Figure 6.24
Modern spotted metal organ pipe. Surface of cast sheet of 52/48 lead-tin alloy about 1.5 mm thick. Natural size. Sample courtesy of William Bunch, Aeolian-Skinner Organ Company.

prevent mass production. Roman coin-counterfeiters' molds of clay in multiple layers have been found. Biringuccio describes a foundry in Milan in which 1200 small castings of everyday objects were made in a single mold composed of 20 superimposed layers of clay, each junction surface in succession being molded against a metal pattern complete with gates and runners. A century and a half later, a foundry at Islington was making 6000 thimbles per day by casting brass into molds, made of "a sort of sand" mixed with red ochre, each of which contained six gross thimbles that were finished by turning and stamping on special machinery.[28]

Medallions and small objects generally were molded in a fine powder made from ground pebbles, sand, brick, ashes, or similar refractory material, bonded with sodium silicate solution (made by heating salt in a clay crucible). The powders were pressed against the pattern in split metal or wooden flasks and thoroughly dried before use; often they were hard enough to be used for more than one casting. Biringuccio concludes his account by remarking that "it is truly a quick and easy method." Art museums are full of Renaissance medallions made by such methods which show remarkably fine detail. (Parenthetically, the use of a vacuum followed by air pressure for improving the surface detail of a casting was demonstrated to the Royal Society by Papin more than two centuries before vacuum casting was commercially employed [figure 6.25].)[29]

The technique of casting in green sand is perhaps old, but the well-informed Biringuccio, writing shortly before 1540, certainly regarded it as unusual, for he begins his description of it by the statement that "It has been discovered, contrary to the natural order of art, how to cast in moist earth in order to save labor and expense. This is truly a thing that many desire and few practice, because it is not as smooth a way or as easy to effect as it apparently seems." He describes frames or boxes, much like the flasks of today's foundryman, though his mixture is a ground yellow tuff or fine washed sand mixed with ashes and flour, moistened with urine or wine, and the mold surface was smoked. Leonardo da Vinci had previously mentioned the use of a box of river sand moistened with vinegar to make castings rapidly and simply. Cellini, in 1568, writes:

> By the bye, as I write I am minded of a very rare kind of this tufa which is found in the bed of the Seine in Paris. While there I used to take what I wanted from hard by the Sainte Chapelle, which stands on an island in Paris in the middle of the Seine. It is very soft, and has the property, quite different from other clays used for moulding purposes, of not needing to be dried, but when you have made from it the shape you want, you can pour into it while it is still moist, your gold, silver, brass, or any other metal. This is a very rare thing, and I have never heard of it occurring anywhere else in the world.[30]

The great Réaumur accompanied the report of his discovery of the method of annealing white cast-iron castings to make them malleable (1722) with a good account of the making of dry sand molds, which he insisted should be as hot as possible. He considerably amplified this part in the second edition, which was published posthumously in 1762 as part of the *Descriptions des Arts et Métiers* of the Academy of Sciences.[31] This same series included a comprehensive section by Courtivron and Bouchu covering most aspects of the iron industry, which is particularly interesting for its discussion of foundry work.[32] Plates illustrate the making

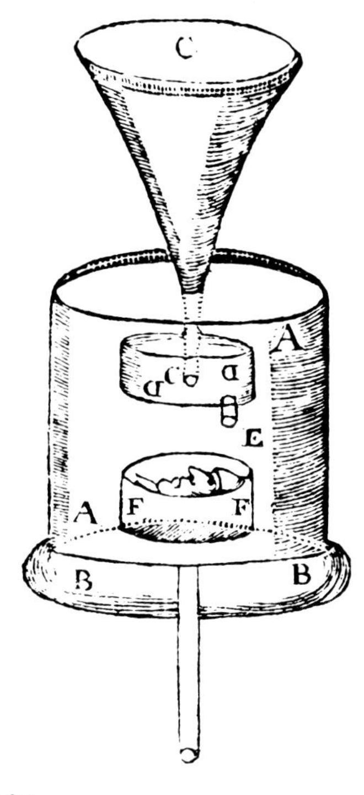

Figure 6.25
Device for casting medals in vacuum. Papin (1684) explains: "The shank of the funnel CC, ought to be stopt with wax or cement, and the receiver being well exhausted of air, the melted metal is to be poured quickly into the funnel CC, so it will melt the wax, and run into the vessel DD, and falling through the hole E it will fill the mold FF. Then the outward air following the metal immediately, will press it into the mold with such force, that it must take the impression of any little stroke that is printed therein much better than if the mold had not been exhausted of air. I have therefore cast two medals, one in the open air, and the other *in vacuo*, and I find much difference between them."

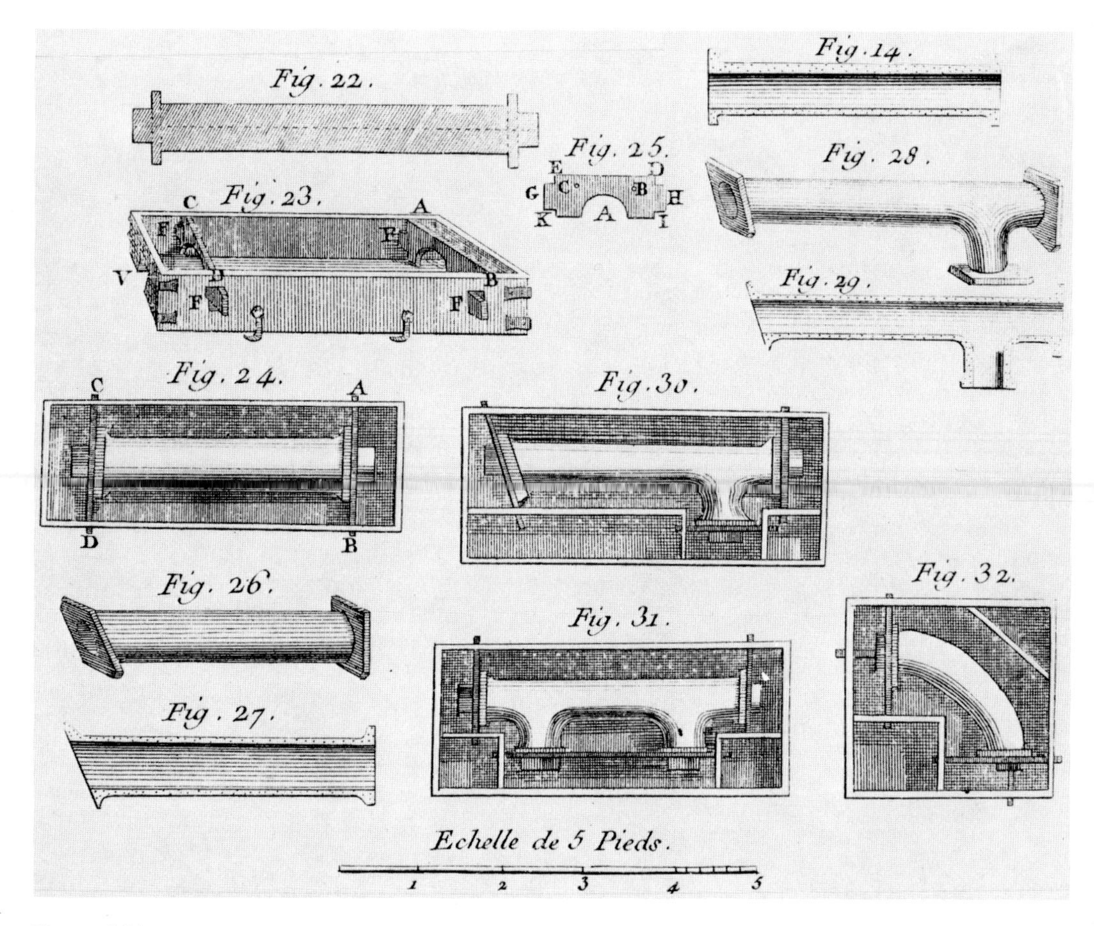

Figure 6.26
Cored sand molds for cast-iron water pipes. From Courtivron and Bouchu's paper in the *Descriptions des Arts et Métiers* (1762).

of cast iron waterpipes—a very important technical basis of the development of cities—in cored sand molds (figure 6.26) and cooking pots in both loam and sand. The same processes appear in the somewhat better plates of the *Encyclopédie*.[33] From this time on the making of sand molds improved in skill and became more and more mechanized, but the basic principles were little changed. A revolutionary change in the production of cast bronze cannon was the introduction of jointed sand molds by Gaspard Monge, who played an important role in modernizing armament production for the French Republic.[34]

Ingots of metal for further working were from the earliest times commonly cast in permanent molds of stone or metal. The earliest description of an ingot mold is by Theophilus, who also tells how to make a long iron mold for casting H-section lead cames for glaziers' use, and a clamped-together mold of plates and rings to give a disc of silver for hammering to his chalice. In the sixteenth to eighteenth centuries, brass slabs to be hammered into sheet and cauldrons were cast between thick slabs of stone (figure 6.27). At the same time, bars of precious metal for working into coinage were being cast in sand molds in conventional frames. The best contemporary description of such casting methods is that of Boizard in 1696.[35] A remarkable process formerly used in Japan for casting copper ingots is illustrated in figure 6.28.[36] The metal is shaped in the depressions in cloth draped over a wooden rack immersed in a tub of hot water. It acquires a beautiful pink oxidized surface that is extremely resistant to tarnish.

Many and wondrous have been the compounds used for dressing metal molds to prevent sticking and to give a good surface. Theophilus used wax and clay on different molds, but most subsequent descriptions of casting of brass ingots use some organic matter such as tallow or oil with some filler such as lampblack or graphite. The smell of burning lard oil was the characteristic smell of a brass ingot casting shop until very recently.[37]

Alloys for the Foundryman

Space does not permit a discussion of the important question of the selection of alloys for the foundryman. One of the great events in the history of metallurgy is the transition from copper, first to copper-arsenic and copper-antimony alloys which were easier both to melt and to cast and harder than pure copper, and then the discovery of bronze which dominated the casting scene for four thousand years.

The introduction of cast iron in China occurred about 400 B.C. With the long tradition of casting in China, it was natural that the casting properties of high-carbon iron would be appreciated and exploited. In Europe its brittleness rendered it unappreciated until the late fourteenth century A.D. It was made principally as a stage in the production of "useful" wrought iron simply to get more efficient removal of the metal from the ore. By mid-fifteenth century, however, it was used for simple decorative objects and by the early 1500s for cannon, though bronze remained the ordnance master's preferred material until mid-nineteenth century.

The introduction of cast iron for building began in England toward the end of the eighteenth century—the Coalbrookdale bridge over the Severn river in Shropshire (erected in 1779) being a famous mark in both landscape and history. The desire to

Figure 6.27
Casting brass slabs in stone molds for battery work. From Galon's *L'Art de Convertir le Cuivre Rouge en Laiton* (Paris, 1764).

Figure 6.28
Casting of copper ingots in a cloth mold under water. From Masuda's *Kodô Zuroku* (1801).

achieve the most economical section of cast-iron beams lent impetus to studies of the "strength of materials"—the most notable early work with cast iron being that of the English engineer, Tredgold[38]—and to the development of the mathematical theory of design. The relative brittleness of cast iron enabled the foundryman to break an occasional casting and obtain visible evidence of defects that he could relate to design and gating practice.

The conclusion to be drawn from the foregoing discussion is clearly that there was ingenuity enough in the castings industry long before the advent of modern science: advancing technology has increased production and cheapened the product, but there has not been any basically different way of mold-shaping or producing the transfer from liquid to solid state. The properties of castings are still usually inferior to wrought metal. The balance of this paper deals with the growth of scientific understanding, which may, in the future, open up quite new possibilities.

Early Scientific Speculations on Solidification

Scientific attempts to understand the structure of castings and the mechanisms of solidification have in the past involved three essentially different aspects, all of which are combined in today's theory: (1) the understanding of the basic mechanism of transition from the liquid to the solid state, which today we take for granted as being synonymous with crystallization; (2) an understanding of the local changes of chemical composition that accompany such crystallization; and (3) the analysis of heat flux in mold and metal. In addition there is the physics specific to the casting itself as distinct from the material, that is, the manner of the growth and impingement of the crystals to give the actual structures and compositions that are found in different parts of castings having different thermal and mechanical histories.

I have discussed elsewhere the curious delay in appreciation of the idea that regular internal order, not external shape, was the essential characteristic of a crystal.[39] The earliest fracture tests showed that metals were composed of grains, but the fact that the grains were always crystalline does not clearly appear until the nineteenth century.

In 1644 Descartes discussed crystallization at some length in terms of his corpuscular theory, which greatly influenced all subsequent scientific thinking.[40] He observed the "coming to nature" of iron in a refining hearth, and rightly surmised that the subparticles that composed one "drop" (grain) of iron were more closely bound to each other than to those in other grains. He attributed this to differing directions of coordinated movement rather than to orientation, as we would today. To Descartes the shape of the crystal was quite incidental to the shape of the parts. For example, he believed that the cubic shape of salt crystals originated from rodlike parts being brought together under the influence of the local curvature of the surface of the evaporating solution, just as two greasy needles floating on water would be. The parts collect into square patches, but mutually incompatible arrangements appear, particularly at the edges, and he gives, unwittingly, the first illustrations of crystalline imperfections (figure 6.29).

Robert Boyle took the particulate structure of matter for granted, though he did not adopt Descartes's ideas on its ultimate constitution. In 1672 he experimented with the

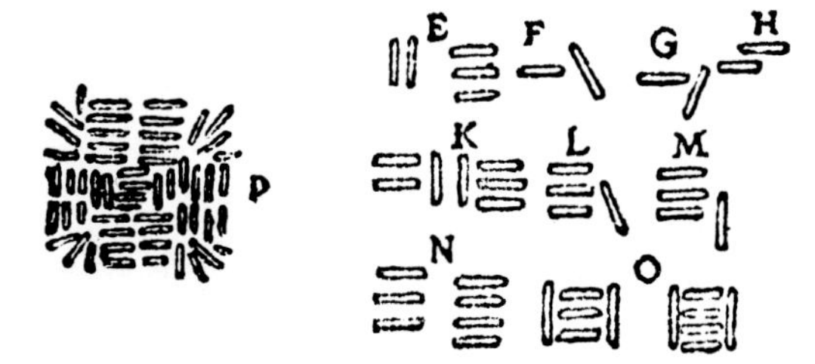

Figure 6.29
The aggregation of saline particles to form an imperfect cubic crystal. From Descartes's *Principia Philosophia* (1644).

solidification of bismuth in a spherical bullet mold and saw that the parts, however disordered they may be in the liquid state, rearrange themselves into smooth planes on cooling, the arrangement being "more uniform and regular than almost anyone would expect in a concretion so hastily made; notwithstanding which their internal contexture will be much diversified by circumstances as particularly the figure of the vessel or mold wherein fluid matter concretes." He concluded that the bullet was made up of

> a multitude of little shining planes so shaped and placed that they seemed orderly to decrease more and more as they were further and further removed from the superfices of the globe; and they were so ranked, that they seemed to consist of a multitude of these rows of planes reaching every way, almost like so many radii of a sphere from the center or middle part. . . . The orderly composition of planes in our bullet (which some curious persons . . . looked on as a not unpleasant sight) may be derived from this that the matter was cooled first on the outside by contact with the cold iron mold . . . and that the coagulation being thus begun, the parts of the remaining fluid as they happened to pass by this already cooled matter with a motion which . . . was now slackened, they were easily fastened on the already stable parts . . . and the refrigeration still reaching further inwards until it came last of all to the middle of the globe, that being remotest from the refrigerating agents; and the apposition was successively and orderly made until the whole matter was concreted . . . .[41]

Réaumur, whose work on steel is known to every metallurgist with an interest in history, wrote a paper in 1724 which is central to our present topic.[42] He saw that the fracture of a casting did not reveal the shape of the truly elementary parts because the arrangement varied so much with the manner of cooling: the needles characterizing the

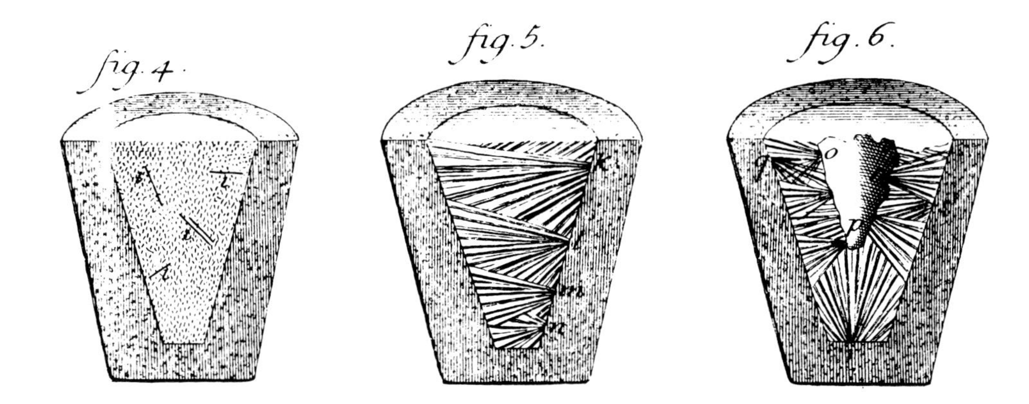

Figure 6.30
Fractured ingots of antimony sulphide made under various conditions to study the mechanism of solidification. From Réaumur's paper of 1724.

fracture resulted simply from the order in which the molecules were able to arrange themselves as the fiery particles which agitated them in the liquid were removed. He experimentally varied the direction and rate of cooling of melts of antimony sulphide, with the results shown in figure 6.30. No. 5 shows the contents of a crucible rapidly cooled in the air, which started to solidify in needles radiating from one side only; in No. 6 the direction of growth was locally reversed by the insertion of a cold poker. The upper left (No. 4) shows the appearance of the fracture of an ingot that was allowed to cool slowly in the fire. It was evidently a single crystal, but Réaumur did not know it, and he explained the featureless surface result as indicating that the molecules were preserved in liquidlike disorder because of the lack of the ordering tendency associated with a temperature gradient. He assumed that a strong temperature gradient was necessary to form a nucleus as well as for it to grow: "The molecules nearest to the molecules already congealed are the next ones to congeal, and so on. Thus the more that contact with a fixed molecule tends to fix another and to take away its movement, the better does each congealed molecule attach itself to its neighbor and the more inevitably does it take the same direction."

The eighteenth century saw many attempts to grow good crystals of all kinds from liquids, and the first observations on metal crystals growing in shrinkage cavities in large castings. Much of the scientific interest and most of the foundryman's practical problems originate in the fact that metals solidify in the form of dendrites. Though alchemists had earlier observed their mystic Star of Antimony, the scientific history of the dendrite begins, as does crystallography itself, with Johannes Kepler's *New Year's Letter on the Hexagonal Snow Flake* (1611). There is much observation and speculation on snowflakes in meteorological literature thereafter. It is not entirely fortuitous that the first illustration of a metallic dendrite (figure 6.31) should be contained in a book on "iron snow," a crystalline salt that Zannichelli was investigating for medicinal purposes and which led him to a broader study of iron-bearing crystals in general.[43]

In 1775 the scientific ironmaster, Pierre C. Grignon, published drawings of dendrites from a shrinkage cavity in a huge mass of cast iron in a blast furnace that was allowed to cool over a period of 15 days (figure 6.32).[44] Grignon describes the geometry of dendrites in great detail and discusses the role of heat and of slag as solvents for metallic crystallization. He also studied the directions of solidification of balls of white cast iron which he explains essentially in the same way that Boyle and Réaumur had done with their materials: "To the degree that the heat retires to the center, the molecules accumulate one on the other, following the progress of cooling right up to the center, which is the point toward which they all tend." Asymmetrically cooling the balls predictably displaced the center (figure 6.33).

Guyton de Morveau seems to have been the first, in 1777, to state specifically that the shrinkage lines, the little trees or whole dendrites on the free surface of a frozen lump of metal, were related to the structural details displayed on fracture.[45] He regarded the crystallization of metals from the liquid as being exactly analogous to the crystallization of salts, both being due to the evaporation of a solvent, for the igneous matter in melts acted in the same way as water in other solutions. Mongez (1781) first showed how to produce well-shaped crystals by draining a partly solidified melt.[46] David Mushet collected crystals in great abundance from the feeding heads of guns, and gives a good verbal description of growth directions and successive branchings of dendrite arms.[47]

The belief that solidification of a metal was somehow analogous to the crystallization of a salt continued to grow in the nineteenth century, although the idea that metals were always crystalline was not appreciated until the very end. The radiating texture of a casting required only a unidirectional aggregation of parts. There was, however, a remarkable paper by Savart in 1829 which touches on almost all of the questions that are of concern to present-day structural metallurgists.[48] With no more tools than a brilliant mind and Chladni figures to reveal elastic anisotropy, and though he rarely saw a crystal, Savart concluded that metals do indeed consist of groups of crystals in which each one considered individually presents a regular structure while the mass as a whole is confused. Cast discs of lead were always but variably anisotropic, which suggested the presence of a few crystals formed randomly. The discs became isotropic if, when solidifying, they were vibrated to produce many

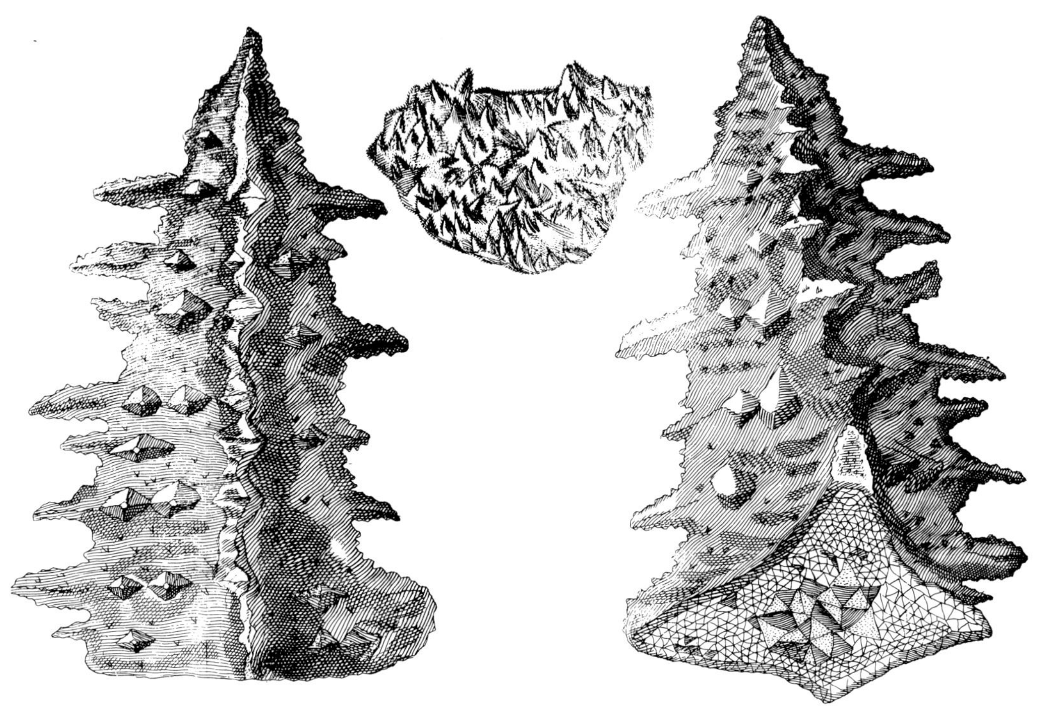

Figure 6.31
The earliest known representation of a metal dendrite (iron). From Zannichelli's *De Ferro . . .* (1713).

Figure 6.32
Dendrites formed in a shrinkage cavity in a large block of iron that was slowly solidified. From Grignon's "Mémoire . . ." (1775).

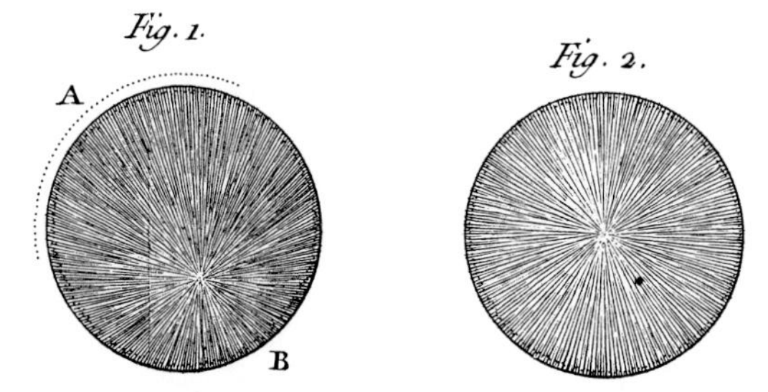

Figure 6.33
Cast-iron balls, fractured to show the displacement of the last parts to solidify by asymmetrical cooling. From Grignon's "Mémoire . . ." (1775).

random crystals. He also showed that a cold rolled strip was anisotropic and that annealing restored isotropy.

Robert Mallet's work (1856) provides a good picture of advanced, if slightly eccentric, thinking on problems of metal solidification at that time.[49] He misunderstood the nature of the orienting force in crystallization, but fully appreciated its geometry. He remarked that "It is a law (though one which I do not find noticed by writers on physics) of the molecular aggregation of crystalline solids that when their particles consolidate under the influence of heat and motion, their crystals arrange and group themselves with their principal axes, in lines perpendicular to the cooling or heating surfaces of the solid."

He clearly related the planes of weakness in a casting to the meeting of opposing zones of crystallization moving in from different surfaces, and he showed how it could be obviated by attention to mold shape. Figure 6.34 shows his sketch of a broken hydraulic press cylinder, as well as a new cylinder redesigned by Robert Stevenson, who appreciated the true nature of the failure. Mallet also clearly described the effects of casting section on shrinkage and has a good account of the principles of feeding. Some gunmakers (the most successful being Whitworth) attempted to produce sound castings by the application of pressure to the solidifying metal.

D. K. Tschernoff, in a paper on the making of steel ingots published in 1878, dealt with virtually every aspect of solidification —shrinkage, blowholes, cracking, columnar and equiaxed granular crystallization, and segregation.[50] He reported finding some dendrites so small that they could only be seen under a 150-magnification microscope, and saw that dendritic growth caused innumerable locked-in spaces in which contraction

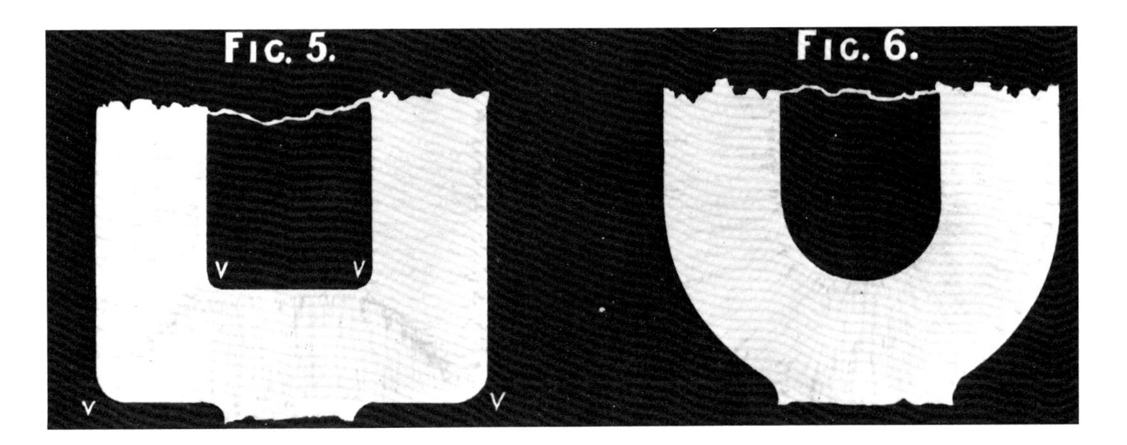

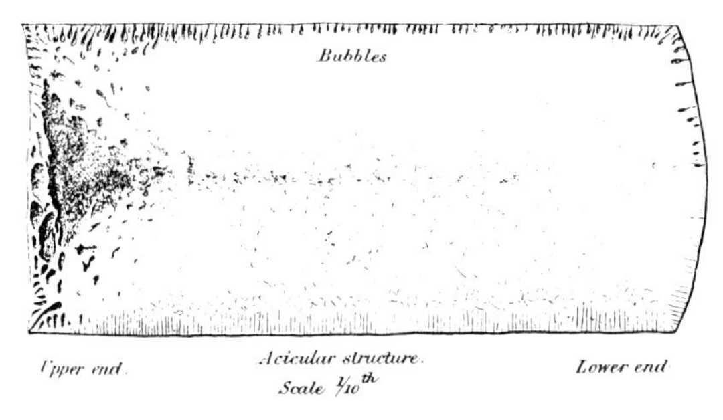

Figure 6.34
Fracture of iron castings for a large hydraulic jack. The one on the left shows a plane of intercrystalline weakness at the square center, eliminated in the design at the right. From Mallet's *On the Physical Conditions Involved in the Construction of Artillery* (1856).

Figure 6.35
Longitudinal section of a cylindrical steel ingot about 107 cm high. The drawing shows blowholes, a shrinkage pipe, and other defects. From Tschernoff's paper of 1878.

cavities would form. Tschernoff observed the difference between the columnar and the equiaxed grain portions of an ingot, but thought that the latter resulted from recrystallization of the solid metal under the influence of shrinkage stresses. He thought that an amorphous structure would come from rapidly oscillating the mold during freezing. He advocated forging to improve the grain size of a steel casting, as well as to close up the innumerable gas bubbles and shrinkage cavities (figure 6.35) that were present in an ingot unless silicon had been added. A few years later Tschernoff published a picture (figure 4.4) of a 15-inch iron dendrite, the largest that has appeared in the technical literature.[51]

One of the more interesting nineteenth-century books on the structure of matter is O. Lehmann's *Molekularphysik*, published in 1888,[52] which is an important forerunner of D'Arcy Thompson's more famous and philosophical *Growth and Form*. Lehmann describes an enormous range of phenomena and emphasizes the diversity of form in which inorganic matter can appear. He is the first to discuss the growth of dendrites in terms of the enhanced availability of material by diffusion or by convection at apexes.

Miers's work on crystal growth, in which he revealed concentration gradients in the vicinity of crystal faces by optical studies, also influenced metallurgists.[53]

A most interesting example of the way in which wrong theory can lead to the advance of science is provided by the foam-cell hypothesis of Quincke.[54] He had done extensive studies on the nature of emulsions and really pioneering work on liquid crystals, both of which proved that there could be some degree of structure in a liquid system. In 1903 he postulated that the formation of a foam in the liquid state was an essential preliminary to solidification. He pointed out that the crystals in some solidified liquids such as benzene can be seen to form a foam with a network of boundaries meeting always at an angle of 120 degrees exactly as in a soap froth. (The significance of the 120-degree angle had been beautifully shown by J. Plateau,[55] whose work is truly a classic mixture of physics, geometry, higher mathematics, and simple eye-catching experiments.)

Quincke was, of course, fundamentally right in insisting on the nearly universal cell-like structure of matter. Polycrystalline matter *is* a foam. If, in retrospect, Quincke seems to have been somewhat unobservant and unimaginative in attributing the origin of the foam to the liquid, consider for a moment the difficulties in conceiving an identifiable and indestructible interface between two particles of matter that are identical in all respects except in orientation. Grains are referred to in the first speculations on structure, but in the more natural form of granular matter like so many peas or grains of wheat with an individual and identifiable skin. It involves a whole inversion of a pattern of thought to appreciate a grain boundary of purely geometric origin and nature.

At one point, Quincke says that the plasticity of a material undergoing solidification "proves the presence of jelly, that is, of oily visible or invisible foam walls, over this range of temperature." All liquids, he thought, form a turbid solution, and "the walls and contents of the foam cells consist of heterogeneous substance," for absolute purity is impossible to attain.

Quincke's theory had a strong impact on English metallurgists, being championed especially by Beilby—he of the amorphous-surface-layer theory—who donated the sum of £100 for research leading toward its proof. This yielded two excellent papers by Cecil H. Desch, in which he failed to support the theory but provided an excellent analysis of the shape of crystals.[56] Surface tension, he said, appeared to be the main shaping force. Following the presentation of the papers, there was a long discussion which reflected strongly the views of physical chemists a few decades earlier on the differences between crystalloids and colloids. The great metallurgist Rosenhain, who a few years earlier had championed the amorphous metal theory of grain boundaries,[57] must have thought them to be without energy, for he criticized Desch, saying that he found it hard to believe there could be significant surface forces between two crystals of almost identical composition. Even Desch, after replying that the significant interface tension was that between solid and liquid, went on to say that this would disappear when the two growing crystals came into contact. The understanding of grain boundaries is one of the most fascinating aspects of metallurgical history!

Concern with the grosser problems of solidification of ingots has continued to the present day. Important studies on the general question of ingot design and ingot solidification were those of Brearley and Brearley, who made very effective use of model castings in stearin to duplicate crystallization, shrinkage, and gas evolution in steel ingots (figure 6.36).[58] Their work was of great value in educating the practical steelmaker but did not directly inspire more detailed scientific study. Much was learned by dumping or bleeding the liquid from partly solidified ingots. When thermocouples became common, temperature measurements were made in different parts of solidifying ingots and the gradient and lack of it related to columnar and equiaxed crystallization. The use of mathematical models, which have greatly aided the study of heat flow in castings, has been summarized by Ruddle[59] and will not be further discussed here. Let us return for a moment to a chemical aspect of casting, that of deoxidation.

Deoxidation

The earliest use of a minor element to prevent oxidation of a major one and the evolution of gas by reaction during solidification is undoubtedly the unconscious use of arsenic and iron in smelted coppers at the beginning of metallurgy. Zinc appeared in bronze as a minor, probably unconscious, addition before it became a major component in brass. The efficacy of phosphorus in promoting fluidity and soundness in bronzes was not discovered until 1870.[60] The production of "tough pitch" copper involves a delicate balance in the refining process to leave in just enough oxygen and hydrogen to cause the evolution, at the last moment of solidification, of a volume of gas exactly equal to the shrinkage, and so produce ingots with a flat top. Somewhat the same was later done with steel ingots, though the porosity was not so evenly distributed.

The deoxidation of cast iron was no problem, for smelting invariably reduced enough silicon to serve as a deoxidizer. When crucible steel was introduced, however, sound ingots were obtained only by "killing," which involved holding the molten metal for an additional hour or so in the furnace. Later analysis showed that this resulted in the reduction of 0.1 to 0.3 percent silicon from the clay crucible. The intentional addition of silicon to molten steel in commercial castings was done at the Bochum Works in Prussia around 1860. The additions of manganese to crucible steel, in which it served both as a deoxidizer and as a neutralizer of hot-shortness due to sulphur, was the subject of a disputed patent awarded to J. M. Heath in 1839.[61] Deoxidation with both silicon and manganese was quite crucial to the commercial success of the new methods of making low-carbon steel by the Bessemer process in 1856 and the Siemens-Martin process shortly thereafter.

The belief that Oriental wootz owed its superiority to the presence of aluminum goes back to erroneous analyses by Faraday in 1819, and many workers experimented with aluminum in steel once the metal became available. In 1885 a factory was successfully established in Sweden for the production of the so-called Mitis castings, which were ductile castings with thin sections and fine detail made simply by melting carbon-free wrought iron in a crucible and adding aluminum before casting in dry sand molds.[62] Aluminum was later used in the production of pure "ingot iron" from metal melted in open hearth furnaces.

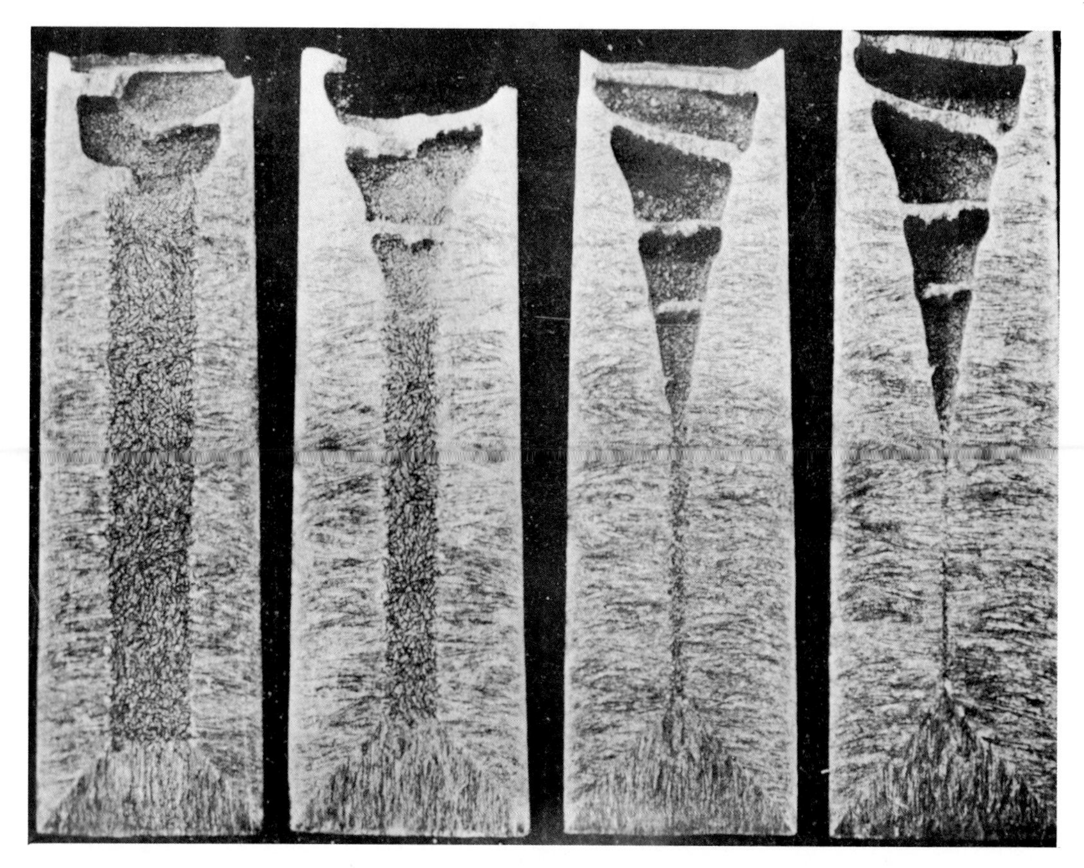

Figure 6.36
Sections of ingots of stearin used to study mold design. From Brearley and Brearley's *Ingots and Ingot Moulds* (1918).

Liquation and Segregation in Castings

The first bronze founders would have noticed that their alloys passed through a long mushy range in the course of solidification, and later, when men's minds were becoming chemically oriented, it would not have required much imagination to suspect that this was accompanied by local variations in composition.

The purification of salts by crystallization from aqueous solution is an extremely old example of chemical separation during phase change, and so is the separation of liquid smelting products into slag, matte, and metal, as well as the purification by drossing of tin and lead.

Segregation by partial solidification is the basis of the liquation process for the desilverization of copper, which was a source of immense wealth in Europe from the fifteenth century on. This involved the preparation of large cakes of an alloy containing about three times as much lead as copper, and then reheating these at a dull red heat to allow the lead to drain off, carrying most of the silver with it. Among the many descriptions of the process, none is better than that by Lazarus Ercker in 1574.[63] By this time it was fully developed, and it underwent substantially no further change until it was displaced by a sulphuric acid parting process in the mid-nineteenth century, and eventually by electrolysis. The Pattinson process for desilverizing lead, introduced about 1833, utilized partial solidification and separation of the crystals from the more silver-rich liquid in an ingenious countercurrent operation of high efficiency.[64] (In the 1940s the same principles were used in the zone melting process of W. G. Pfann, at first to produce silicon of extremely high purity that made the transistor possible, and later applied to many other materials, organic and inorganic, metals and semiconductors.)

Biringuccio (1540) was the first to refer to the phenomenon of segregation in a casting—he mentions the "leanness" of the bronze at the top of a gun casting, and tells how this can be corrected by adding more tin to the metal for the feeding head, the last that runs into the mold. Cellini's autobiography has a dramatic description of the household pewter being thrown into the molten stream, though he implies that it was done only to increase the fluidity of the bronze.

It is natural that the first quantitative measurements on segregation should have been done on alloys of the precious metals. Segregation complicated the sampling of raw materials being purchased and made it hard for the minter to produce a coinage of legal fineness without endangering his profit. Ercker, in 1574, says that the sample for assaying copper matte should be cut neither from the center of the cake nor from near the rim, "for silver follows the cold . . . so that the rim of a cake of matte is richer than the center," and in casting black copper, "the mold has to stand level lest the copper ingot become thick at one end and thin at the other; for you may be sure that wherever the copper flows with a rush and the ingot becomes thicker, there will be more silver in the ingot, especially when the copper is rich."

Most subsequent books on assaying carefully instruct on the selection of samples to compensate for segregation, and some of them attempt to explain it. J. A. Cramer, for instance, in 1739, knew that silver, gold, and copper were not evenly distributed throughout every part of a mass of lead, and that the effect depended upon the casting temperature.[65] There could be composition differences due to insufficient stirring of the liquid, especially if there was partial immiscibility, as with lead in alloys of gold, silver, and copper. A copper ingot containing lead should not be quenched in water, for the copper in growing cold contracts and "repels the lead still in fusion to the outside, and toward that part of the ingot which is immersed the last into the water. Likewise, the ingot is always richer in silver and gold in the [thicker] place where the [mold] has been inclined toward the horizon than it is in the opposite extremity, which is especially true of those mixtures in which lead and copper enter together." This idea that the contraction of the outer part of an ingot squeezed liquid metal out of the interior reappeared in the twentieth century.

That peripatetic metallurgist Gabriel Jars, in a 1769 memoir on the reworking of base silver-copper coinage, said that the common belief that the heavier metals sank toward the lower part of an alloy ingot could not be true whenever the degree of heat was strong enough to hold the two metals in fusion: the separation must have occurred during cooling.[66] Jars attributed segregation to differential repulsion of the cold mold:

Every chemist and metallurgist knows that metals in fusion have a strong tendency to jump up in the air with a lot of commotion whenever they come into contact with cold or humid bodies. Lead is less sensitive than silver, and the latter still less so than the copper. From this I conclude that, on pouring an alloy of silver and copper into an ingot, the mass, meeting a body less hot than itself, seeks to withdraw from the walls as soon as it touches them. Copper receives a stronger impulse by virtue of its greater sensitivity, and tends to approach the center from all sides.

The introduction of the Daltonian atomic theory had a profound effect on chemistry, but confusion followed when chemists tried to apply the new theories of simple combining proportions to metallic solutions and mixtures. A. Levol, an assayer in the Paris mint, thought that alloys that were true chemical compounds would not segregate, and published a paper on copper-silver alloys in 1852–53 that was to answer the question: "Are alloys simple mixtures in any proportion whatsoever, a kind of reciprocal solution of metals, or do the metals that constitute them form combinations in definite proportions?"[67] He assayed samples taken from many positions in cubes and spheres cast in iron molds. Table I summarizes his results. Levol noted that the alloy Ag_3Cu_4 (actually Ag_3Cu_2) was the only one to show no liquation and concluded therefrom that it was a compound, although he thought this to be an extraordinarily complex formula. A comparable compound, red lead, Pb_3O_4, was explained by chemists as

Table I Segregation in copper-silver alloys (Levol, 1852)

	$AgCu_2$	Ag_3Cu_5	Ag_2Cu_3	Ag_3Cu_4	AgCu	Ag_2Cu	Coinage alloys	
Intended composition	630.35	671.73	694.50	718.93	773.3	872	900	950
Drop analysis	631.92	672.9	693.70	718.32	774.18	873	901.34	948.39
Center of sphere	619.00	671.8	693.77	718.13	785.95	881.78	907.31	950.0
Sphere surface, average	633.31	673.75	694.33	718.12	772.95	872.50	898.95	947.70
Maximum difference	15.00	1.95	0.40	0.36	13.15	9.28	8.88	3.08

All compositions are in parts of silver per thousand by weight. The drop analysis represents the composition of the liquid. The formulas are based on erroneous atomic weights: they approximate modern ratios if the number of copper atoms is halved: AgCu is equivalent to Ag_2Cu, Ag_3Cu_4 to Ag_3Cu_2, etc.

been mainly used as the simplest or cheapest means of shaping metal, and the defects of its products accepted or corrected by subsequent working. For the future, can we hope for the understanding of detailed mechanisms to be combined into an understanding of the whole complex system, which would lead to the production of castings accurately shaped, containing no shrinkage or embrittling impurities at grain boundaries, hardened by precipitation, and matching their wrought competitors in every way but cost? One can almost imagine a well-programmed computer running a foundry as it mathematically mimics the movement of heat and matter in a mold.

The apex of casting technology in 1966 is undoubtedly the production of single crystals in complex shapes for engineering service. A turbine bucket made of a hard nickel base alloy is shown in figure 6.38. The Pratt & Whitney Company, which is responsible for this development, reports that parts of a complex shape over five inches long and two inches wide have been made by this process.[89]

Continuous casting, zone melting, single crystal growing are all enticing mechanical and theoretical ingenuity on a scale never previously applied to casting,[90] and from a type of man who would experience no joy in a foundry. In 1540 Biringuccio described the foundryman as "like a chimney sweep, covered with charcoal and distasteful sooty smoke, his clothing dusty and half-burned, . . . his hands and face all plastered with mud," his mind fearful, and "his spirit disturbed and almost continually anxious." But he goes on to say that the metallurgist willingly endures all this because of "a certain expectation of novelty" and, "as if ensnared, he is often unable to leave his place of work" for indeed he finds it "a profitable and skillful art and in large part delightful." This at least will not change.

Notes and References

1
H. H. Coghlan, *Notes on the Prehistoric Metallurgy of Copper and Bronze* . . . (Oxford, 1951).

2
R. F. Tylecote, *Metallurgy in Archaeology . . . in the British Isles* (London, 1962).

3
L. Aitchison, *A History of Metals*, 2 vols. (London, 1960).

4
T. A. Wertime, "Man's first encounters with metallurgy," *Science* 146:1257–1267 (1964).

5
O. Karlbeck, "Anyang moulds," *Bull. Museum of Far Eastern Antiquities (Stockholm)* 7 (1935).

6
N. Barnard, *Bronze Casting and Bronze Alloys in Ancient China* (Canberra and Nagoya, 1961).

7
R. J. Gettens, *The Freer Chinese Bronzes* (Washington, D.C., 1970).

8
H. Drescher, *Der Überfangguss* . . . (Mainz, 1958).

9
One may speculate on a possible connection between the Chinese metalworker's preference for casting over forging and the philosopher's *yin-yang* concept. The relation of rilievo and intaglio, so familiar to the foundryman, has had little mention from scientists. It inverts left and right, but unlike a mirror drawing it also inverts distance from the reference plane. The Dutch artist Maurits Escher has drawn intriguing two-dimensional patterns in which recognizable animals form space-filling units, separated by a *yin-yang* line and often penetrating deeply into each other. Many of the patterns possess rotational and translational symmetry in high degree. There can be much more to space-filling units than the crystallographer's plane-bounded cells. Can it not indeed be argued that such cells are closer to the "real" units of structure than the mathematical

fiction of a unit cell which, to simplify symmetry, divides even atoms into parts? It is significant that replication in biology occurs with essentially linear molecules, although a two-dimensional molecule would carry the same amount of information more compactly: it is a mold-making problem with a high density of undercuts.

10
(*Note added 1980*) See N. Barnard and T. Sato, *Metallurgical Remains of Ancient China* (Tokyo, 1975).

11
W. Gowland, "The art of casting bronze in Japan," *Journ. Soc. Arts* 43:522 (1895).

12
H. Wilson, *Silverwork and Jewelry* (London, 1951).

13
A. Duca, M. C. Flemings, and H. F. Taylor, "Art casting," *Trans. Amer. Foundryman's Society* 70:801–810 (1962).

14
Theophilus, *De Diversis Artibus* (ca. 1125 A.D.; for an English translation see item 121 in the bibliography).

15
V. Biringuccio, *De la Pirotechnia* (Venice, 1540; for the English translation see item 46 in the bibliography).

16
C. ffoulkes, *The Gunfounders of England* (Cambridge, 1937).

17
Diego de Prado y Tovar, *Encyclopaedia de Fundicion de Artilleria y su Platica Manual* (Manuscript, 1603; Cambridge University Library, MS add. 2883).

18
P. Surirey de Saint-Rémy, *Mémoires d'Artillerie* (Paris, 1697).

19
D. Diderot, "Fonderies de canons," *Encyclopédie ou Dictionnaire Raisonné des Sciences . . . Recueil de Planches*, vol. IV (1765).

20
Miriam S. Balmuth, "Forerunners of coinage in Palestine and Phœnicia," *Proc. International Numismatic Convention, Jerusalem, 1963.*

21
Since writing this I have learned of the papers by Karel Castelin in which essentially the same idea is presented ("O' litých' střižcich keltských duhovek," *Numismatické Listy* 9:73ff. [1954]; a German translation appears in *Germania* 38:32–42 [1960]). He reports that the clay body of some Celtic molds had not been heated above 500°C and suggests that piles of metal particles were melted individually in the cavities by the superficial heat of a kind of blowpipe arrangement.

22
See the postscript to item 126 in the bibliography.

23
Prince Rupert, "To make small shot of various sizes," letter quoted by Robert Hooke in his *Micrographia* (London, 1665), pp. 23–24. See also Houghton, Letter No. 264, in the collection cited in note 28 below.

24
T. F. Carter and L. C. Goodrich, *The Invention of Printing in China and Its Spread Westward* (New York, 1955).

25
J. Moxon, *Mechanick Exercises, or The Noble Art of Printing* (London, 1683). Annotated reprint, ed. H. Davis and H. Carter (London, 1958).

26
That most persistent popular belief that type metal expands on solidification is false: see T. Takase, "On the volume changes of . . . some type and fusible alloys during solidification," *Nippon Kinzoka Gakkai-si* 3:117–123 (1939); an English abstract may be found in *Metallurgical Abstracts* 6:304 (1939). The good casting characteristics depend upon other physical properties, not the least being low melting point and low surface tension combined with high density.

27
A. Félibien, *Principes de l'Architecture* (Paris, 1676).

28
J. Houghton, "A patent for thimble-making," Letter No. 260, 23 July 1697, in *A Collection for the Improvement of Husbandry and Trade* (London, 1692–1703).

29
D. Papin, reported in T. Birch, *History of the Royal Society* (London, 1756), vol. I, pp. 343–344.

30
B. Cellini, *Due Trattati, uno intorno alle otto princi-*

pali arti del l'oreficeria. L'altro in materia dell'arte della scultura . . . (Florence 1568). Eng. trans. by C. R. Ashbee (from the Italian edition edited by Milanesi, 1857) (London, 1898).

31
R. A. F. de Réaumur, "Nouvel art d'adoucir le fer fondu . . ." (Paris, 1762). Issued as part of the section on iron of the *Descriptions des Arts et Métiers* of the Académie Royale des Sciences under the general editorship of Duhamel du Monceau.

32
Gaspard de Courtivron and E. J. Bouchu, "Art des forges et fourneaux de fer," in *Descriptions des Arts et Métiers* (Paris, 1761).

33
D. Diderot, "Forges ou art du fer," *Encyclopédie ou Dictionnaire Raisonné des Sciences . . . Recueil de Planches* (Paris, 1765), vol. IV. The plates on loam molding of pots are nos. 3 and 4 in the third section, "Des fourneaux en marchandise." Sand molding is shown for making a sadiron (Pl. 5), pots (6–8), and waterpipes (9–11). Some of these are reproduced in *A Diderot Pictorial Encyclopedia of Trades and Industry*, ed. C. C. Gillispie (New York, 1959). There are nine volumes of plates, published between 1762 and 1772. Others of special interest to the foundryman are as follows: "Fonderies de canons," vol. V (1767); "Fontes de cloches," vol. V (1767); "Fontes de la dragée," vol. V (1767); "Fonte de l'or," vol. V (1767); "Poterie d'étain," vol. VIII (1771); and "Sculpture, fonte de statues," vol. VIII (1771).

34
G. Monge, *Description de l'Art de Fabriquer les Canons* (Paris, An. 2 [1794]).

35
J. Boizard, *Traité des Monoyes de Leurs Circonstances et Dependances* (Paris, 1696).

36
Tsuna Masuda, *Kodô Zuroku* [Picture Book on the Smelting of Copper] (Osaka, 1801).

37
D. R. Hull, *Casting of Brass and Bronze* (Cleveland, 1950). A book of delightful personal reminiscences and practical advice on brass ingot casting. For a more scientific account, see R. Genders and G. L. Bailey, *The Casting of Brass Ingots* (London, 1934).

38
T. Tredgold, *Practical Essay on the Strength of Cast Iron* (London, 1822).

39
C. S. Smith, *A History of Metallography* (Chicago, 1960).

40
R. Descartes, *Principia Philosophia* (Paris, 1644); *Oeuvres*, ed. P. Tannery (1904), vol. VIII.

41
R. Boyle, *Essay about the Origine and Virtue of Gems* (London, 1672). The quotation is from vol. III of the collected works, ed. T. Birch (London, 1744).

42
R. A. F. de Réaumur, "De l'arrangement que prennent les parties des matières métalliques & minérales, lorsqu'après avoir été mises en fusion, elles viennent à se figer," *Mém. Acad. Sci.* (Paris, 1724), pp. 307–316.

43
J. H. Zannichelli, *De Ferro Ejusque Nivis Preparatione* . . . (Venice, 1713).

44
P. C. Grignon, "Mémoire sur les métamorphoses du fer, ou réflexions chymiques et physiques, sur les différentes situations du fer dans la terre dans son traitement jusqu'à sa perfection et sa destruction . . .," *Mémoires de Physique sur l'Art de Fabriquer le Fer* (Paris, 1775), pp. 56–90. For an English translation see item 150 in the bibliography.

45
L. B. Guyton de Morveau, *Journal de Physique* 9:303–305 (1777).

46
J. A. Mongez, "Réponse à M. de Cissay (Sur les cristallisations métalliques)," *Journal de Physique* 18:74–76 (1781).

47
D. Mushet, "On the crystallisation of cast iron," *Phil. Mag.* 61:22–24, 83–87 (1823).

48
F. Savart, "Recherches sur la structure des métaux," *Ann. Chim. Phys.* 41:61–75 (1829). Abridged English translation in *Edinburgh Jour. Science* 2:104–111 (1830).

49
R. Mallet, *On the Physical Conditions Involved in the Construction of Artillery* (London, 1856).

50
D. K. Tschernoff, "Izsledovanie, otnosiashchiesia po struktury litikh stalnykh bolvanok" [Investigations on the structure of cast steel ingots], *Zapiski Imperatorskago Russkago Tekhnicheskago Obshestva* (1879), pp. 1–24. An English translation appeared in *Proc. Inst. Mech. Eng.* (1880), pp. 152–183.

51
D. K. Tschernoff, "A remarkable steel crystal," *The Metallographist* 2:74–75 (1899). A better photograph of the crystal is given by N. T. Belaiew, *Kristallisatsia, Struktura i Svoistra Stali* [St. Petersburg, 1909].

52
O. Lehmann, *Molekular physik, mit besonderer Berücksichtigung mikroscopischer Untersuchungen und Anleitung zu solchen*, 2 vols. (Leipzig, 1888–1889).

53
H. A. Miers, "Growth of crystals in supersaturated liquids," *J. Inst. Metals* 37:331–350 (1927).

54
G. Quincke, "The transition from the liquid to the solid state, and the foam structure of matter," *Proc. Royal Soc.* 78A:60–67 (1907). The application of the theory to metals is summarized in a later paper, *Zeit. Metallographie* 3:23–36, 79–101 (1913).

55
J. Plateau, *Statique Expérimentale et Théorique des Liquides . . .*, vol. I (Paris, 1863).

56
C. H. Desch, "First report to the Beilby Prize Committee . . . on the solidification of metals from the liquid state," *J. Inst. Metals* 11:57–118 (1914); "Second report . . .," *J. Inst. Metals* 22:241–276 (1919).

57
W. Rosenhain and D. Ewen, "Intercrystalline cohesion in metals," *J. Inst. Metals* 8:149–182 (1912).

58
A. W. Brearley and H. Brearley, *Ingots and Ingot Moulds* (London, 1918).

59
R. W. Ruddle, *The Solidification of Castings*, Institute of Metals Monograph No. 7 (London, 1950). Second edition greatly enlarged (London, 1957).

60
G. Montefiori-Levi and C. Kunzel, *Essais sur l'Emploi . . . du Bronze Phosphoreux pour la Coulée des Bouches à Feu* (Brussels, 1870).

61
See J. Percy, *Metallurgy: Iron and Steel* (London, 1864), pp. 840–849.

62
The early use of aluminum in deoxidizing iron and steel is well treated in J. W. Richards, *Aluminum, Its History . . . Metallurgy and Applications . . .*, second edition (Philadelphia, 1890).

63
L. Ercker, *Beschreibung Allerfürnemsten Mineralischen Erzt und Berckwerksarten* (Prague, 1574; for an English translation see item 70 in the bibliography).

64
A biographical note on H. L. Pattinson, giving a history and detailed technical account of his desilverizing process, appears in J. Percy, *The Metallurgy of Lead* (London, 1870), pp. 121–148.

65
J. A. Cramer, *Elementa Artis Docimasticae* (Leyden, 1739; English translation: London, 1741).

66
G. Jars, "Observations métallurgiques sur la séparation des métaux . . ." (1769), in his *Voyages Métalliques*, vol. III (1781), pp. 260–286. In a comprehensive study of gold alloys carried out for the British mint in 1803, Charles Hatchett uncovered some alarming density differences in ingots of gold-copper alloys, but he soon found that this was attributable entirely to inhomogeneous liquid in the crucible. See C. Hatchett, "Experiments and observations on the various alloys, on the specific gravity, and on the comparative wear of gold . . .," *Proc. Royal Soc.*, pp. 43–194 (1803).

67
A. Levol, "Mémoire sur les alliages . . .," *Ann. Chim. Phys.* 36:193–224 (1852); 39:163–184 (1853).

68
A. Matthiesen, "Report on the chemical nature of alloys," *Report of the Thirty-Third Meeting British Assn. Adv. Sci.*, 1863, pp. 37–47; "On alloys," *Journ. Chem. Soc.* 20:201–220 (1867).

69
W. C. Roberts [-Austen], "On the liquation, fusibility and density of certain alloys of copper and

silver," *Proc. Royal Soc.* 23:481–495 (1875).
70
F. Guthrie, "On Eutexia," *Phil. Mag.* 17:462–482 (1884).
71
F. Stubbs, Discussion on paper by J. Parry, *J. Iron & Steel Inst.* (i):199–200 (1881).
72
G. J. Snelus, "On the distribution of elements in steel ingots," *J. Iron & Steel Inst.* (ii): 379–396 (1881).
73
E. Maitland, "The treatment of gun steel," *Proc. Inst. Civil Eng.* 89:114–154 (1887).
74
H. M. Howe, *The Metallurgy of Steel* (New York, 1890), pp. 202–209.
75
W. H. Hatfield, *The Work of the Heterogeneity of Steel Ingots Committee*, Iron and Steel Institute Special Report No. 12 (London, 1936). This is a short summary and bibliography of the work of the committee from its beginning in 1924 to the end of 1935. Subseqently followed the *Seventh Report*, I.S.I. Special Report No. 16, 1937, and *Eighth Report,* I.S.I. Special Report No. 25, 1939.
76
N. B. Vaughan, "Inverse segregation: a review," *J. Inst. Metals* 61:35–60 (1937).
77
S. W. Smith, "Surface tension and cohesion in metals and alloys," *J. Inst. Metals* 17:65–103 (1917).
78
O. Bauer and H. Arndt, "Seigerung-serscheinungen," *Zeit. Metallkunde* 13:497–506, 559–564 (1921).
79
S. W. Smith, "Liquation in molten alloys and its possible geological significance," *Trans. Inst. Mining and Metallurgy* (London) 35:248–300 (1926).
80
J. Phelps, Discussion on paper by S. W. Smith, *Trans. Inst. Mining and Metallurgy* 35:304–306 (1926).
81
G. H. Gulliver, Discussion on paper by S. W. Smith, *Trans. Inst. Mining and Metallurgy* 35 (1926).
82
R. Genders, "Mechanism of inverse segregation in alloys," *J. Inst. Metals* 37:241–285 (1927).
83
R. Kühnel, "Umgekehrte seigerung," *Zeit. Metallkunde* 14:462–464 (1922). See also W. B. Price and A. J. Phillips, "Exudations on brass and bronze," *Trans. AIME (Proc. Institute of Metals Division)*, pp. 80–89 (1927).
84
C. Benedicks, "Action of a hot wall . . . related to segregation in hot metals," *Trans. AIME* 71:597–626 (1925).
85
D. Hanson, Discussion on paper by S. W. Smith, *J. Inst. Metals* 17:112–114 (1917).
86
F. Johnson, Discussion on paper by R. T. Rolfe, *J. Inst. Metals* 20:274–278 (1918). G. Masing, "Zur Erklarung der umgekehrten Blochseigerung," *Zeit. Metallkunde* 14:204–206 (1922).
87
J. H. Watson, "Liquation or 'inverse segregation' in the silver-copper alloys," *J. Inst. Metals* 49:347–358 (1932).
88
B. Chalmers, "Melting and freezing," *Trans. AIME* 200:519–532 (1954).
89
B. J. Piearcey and F. L. VerSnyder, *Monocrystaloys: A New Concept in Gas Turbine Materials*, Report No. 66-007, Advanced Materials Research and Development Laboratory, Pratt & Whitney Aircraft Company (Hartford, CT, February 1966).

7
Porcelain and Plutonism

[This article begins rather abruptly, for it was written for presentation at a conference of historians of geology who were fully familiar with the background of the events referred to. Although the earliest steps toward modern scientific geology were taken mainly in relation to ore bodies and mineral formation, the more comprehensive views of the earth's history that developed in the eighteenth century were at first based on the record of sedimentary rocks, which suggested a catastrophic flood and supported the belief of the neptunists, following A. G. Werner, that all rocks had been formed by sedimentation or crystallization from a once all-encompassing turbid ocean. In opposition to this, the Scot Sir James Hutton advanced the view that the geological processes that had formed the world in the past were similar to those still operating. He also emphasized the role of internal heat in driving geological processes, whence he and his followers were dubbed plutonists. Both schools were in some degree right, but it was Hutton's insistence that geological processes had not discontinuously changed their nature and so could be studied in operation that led directly to the fertile theories of Sir Charles Lyell early in the nineteenth century and to the possibility of obtaining validation for them by measurement in the field and laboratory.

This article is not to be read as a history of these events: it is simply a tracing of a single thread in a very complicated fabric. It is intended only to show how an idea can connect with seemingly unrelated ones to contribute to an environment that encourages without mechanistically determining the appearance of new thoughts and new events.]

Although Sir James Hutton, the "father of modern geology," was himself scornful of those who "judge of the great operations of the mineral kingdom from having kindled a fire and looked into the bottom of a little crucible," the experiments of Sir James Hall, beginning with the crystallization of glass and lava into stony masses, played a very important role in convincing geologists of the general correctness of Hutton's theory of the earth.

The purpose of this note, and the excuse for bringing as exotic a material as porcelain into down-to-earth geology, is to examine some of the ideas on the crystallization of matter that had themselves begun to crystallize during the eighteenth century. It was difficult for men to see the identity between the matter in well-formed crystals and that in confused aggregates of crystal grains, and it was hard to abandon the early belief that crystals grew only from aqueous solutions; both stumbling blocks favored the neptunists.

The beauty of crystals in their geometric perfection resulted in an overemphasis on external characteristics and tended to obscure the significance of their internal order. Crystals seemed to be a special form of matter, rare in nature, and produced in the laboratory only from a few substances, under unusual and carefully controlled conditions. The fact that virtually all solid inorganic matter is crystalline was not commonly appreciated even at the beginning of the present century. In the late eighteenth and early nineteenth centuries, it was a question for debate among leading scientists. Other aspects of solids that were slow to come out were the relationships between the simple compounds required by Daltonian chemistry and the prevalence of nonstoichiometric compositions, and, conversely, the facts that crystals could be extremely small and that matter was often structurally heterogeneous when it was visibly uniform. In both the eighteenth and nineteenth centuries, the idea of the molecule provided all the flexibility that chemists needed. Mineralogists could classify by composition and by external shape; mathematicians were bemused by space groups; and it was not until after the introduction of x-ray diffraction, when crystallography could move from external symmetry and effectively grapple with internal structure and imperfections, that the irregular shapes of crystal grains could be properly appreciated by anyone but an empirical observer. If any early scientist saw that a random aggregate of well-shaped crystals could not fill space, he was more likely to abandon the crystallinity than he was to accept as a crystal anything that did not have regular polyhedral shape.[1]

Hall's important experiment, so controversial to geologists at the time, was, I believe, a natural outcome of what had been one of the main activities of eighteenth-century chemists. Back of the spectacular theoretical achievements of Lavoisier and Dalton were centuries of refinement in analytical techniques that established the identity of things and the conservation of some chemical units through numerous changes of state. Much, perhaps most, of this knowledge had been developed by assayers who had established multitudinous separatory reactions for metals and metallic minerals. Some discoveries came under medicinal stimulus. But parallel to all this was the study of the chemical and thermal behavior of earths and rocks that was to a large degree incited by the desire to duplicate Oriental ceramics.

Some Chinese porcelain came to Europe in the Middle Ages, and it began to have a strong influence on European taste from the end of the fifteenth century on. During the sixteenth and seventeenth centuries, some beautiful soft-paste wares in Oriental styles were made in Italy and France, but they were not true porcelain and were fired at relatively low temperatures. Successful factories were established at St. Cloud near Paris in 1693 and at Meissen in Saxony in 1713 to exploit the discoveries of François Morin and J. F. Böttger (who used kaolin with marble or alabaster, respectively) and at Chantilly, which, after 1725, produced some aesthetically superb work in the Chinese manner but in soft ware with a tin-oxide glaze. A new wave of research was precipitated by the receipt in France of letters written in 1712 and 1725 by François Xavier d'Entrecolles, a Jesuit missionary in China. He gave a fairly complete description of the process as carried out on a mass-production basis in 3000 kilns at Ching-te-chen (Kiangsi Province), and he gave the Chinese names, *petuntse* and *kaolin*, for the two essential materials, though he could not identify them in European terminology. (*Petuntse* is actually feldspar.)

The great versatile scientist René Antoine Ferchault de Réaumur studied samples that d'Entrecolles had sent from China, and in papers published in 1727 and 1729 reported that *kaolin* was infusible but *petuntse* was relatively easily melted. He incorrectly identified the former as talc or a talclike stone and the latter as a fusible kind of flint, but he correctly saw that the essential nature of Chinese porcelain lay in its being a mixture of a very fine-grained infusible earth and a glass. Similar, indeed, were the porcelains then being made in France and Germany, but these could be melted in a laboratory furnace, while the Oriental product was quite infusible.

There followed a frantic search for superior potters' materials and for means of characterizing them. Böttger's successor at Meissen, J. H. Pott (one of Stahl's students), is said to have carried out more than 30,000 experiments in trying to discover the secret of true porcelain; he published many of the results in 1746 in his *Lithogeognosia*. He used a high-temperature wind furnace with a tall chimney and, on the basis of the effect of fire on various earths, classified them as calcareous, gypsumlike, argillaceous, or vitrifiable. The last—siliceous materials—he found would act as a flux for most of the others when mixed with them, rendering them more easily fused and giving a glass on cooling. Many natural rocks fused by themselves. Pott reported the color and appearance of the fracture of the resulting material and usually the hardness, but made no chemical analyses, for techniques in this field did not then exist.

Twenty years later, Jean Darcet, or D'Arcet, published a book *Mémoire sur l'Action d'Un Feu Égal, Violent et Continu . . . sur un Grand Nombre de Terres . . .*, in which he reported a large number of experiments similar to those of Pott, to whom he obviously owed much. He did his heating in "porcelain" crucibles in closed saggers, heated in the large porcelain furnace at the factory of the Count de Lauragais. Darcet showed that mixtures were commonly more easily melted than were pure minerals, and that a siliceous mixture would usually result in a glass. Most pertinent in the present connection are his experiments on the melting of a blackish stone coming from Auvergne that had been given to him as a product of ancient volcanoes and that had a prismatic crystallization (i.e., faulting) like those at the Giant's Causeway and elsewhere. This melted to give a coffee-colored enamel like the glaze on one kind of Oriental porcelain, with a perfectly glassy fracture. Several other rocks acted similarly, and he also found that he could easily melt lava, which gave a glass like common bottle glass. He thought that stones of volcanic origin (which, like spars and granites, had been originally at least partially subject to attack by fire) could well have been formed by the coming together and general melting of several materials of a different nature, contrary to lava, which had been more completely liquefied. "However this may be, it is natural enough to conclude that the fire which is used in baking a true porcelain is by far superior to that which works in the terrible overturnings in the terrestrial globe."

Neither Pott nor Darcet suggested that their melted and solidified materials were crystalline, and they ignored the possible effect of rate of cooling. Their studies showed systematically what many potters and metallurgists knew from experience, namely, that many rocks could be melted and that quartz or flint would flux calcareous rocks, clays, earths, and many metallic minerals to give a melted mass with a vitreous look. Since the effect of rate of cooling was at this stage ignored, the experiments led to the belief that rocks that had been molten inevitably became glassy. When the plutonic origin of rocks was suggested, it therefore appeared to be necessary to invoke some other agency in addition to heat.

By 1750 chemists had, of course, become familiar with the formation of crystals in the laboratory, though they generally regarded them as a product of a solution in water or some other liquid: free growth from solution allowed a material to take its natural geometric form. Virtually no attention was paid to the structure of either natural or artificial minerals that were not well shaped. Whether based on aesthetics or on science, this preoccupation with well-formed polyhedra certainly delayed inquiry into internal constitution. Even Haüy's great contribution to crystallography was external, and his structural units were essentially homomorphic little crystals.

After the middle of the eighteenth century, the idea slowly spread that crystallization could follow the removal of heat as well as that of an aqueous solvent. Grignon (1761) complained of chemists' overemphasis on aqueous systems and pleaded with men in industry who used fire on a large scale to counterbalance this by themselves adopting an observant and analytical attitude toward their processes. He extracted crystals from many glassy slags and remarked that art most closely approaches the operations of nature in the furnaces of the ironmaster, which imitate the products of volcanoes. He believed that if all the forms of matter that precipitate in volcanic craters were introduced into ironmaking furnaces, they would give the same results. He then described the crystallization of various glassy slags and showed their similarities to lava, chalk, etc.: "Both fire and water can modify matter, but the important operations are reserved to [the former] which is the most powerful and most active agent in nature, achieving in an instant results for which water would require centuries."

Though Romé de Lisle in 1772 still believed that water was a principal and perhaps the only agent in nature for forming crystals, the popular Macquer (1766), in a remarkably clear article, says that crystallization resulted whenever the integrant parts of a body that had been separated by the interposition of any fluid became disposed to unite again and form solid regular and uniform masses. (The primitive integrant molecules were of unknown shape, but he believed that all those of any one body were of a "constant uniform and peculiar figure.") Thereafter, chemists produced crystals by the partial solidification of many molten materials. The beautiful crystals of metals produced by Grignon (1775) and Guyton de Morveau (1776) made it quite clear that "fire is to metals what water is to salts," and the old objections subsided.[2]

The parallel with the neptunists' views on the origin of rocks is clear. If it was hard to visualize the crystallization of a molten metal, it was even harder to think of crystals forming in solid glass. But as it turned out, it was observations on glass that first suggested to geologists that basalt might be a crystallization of lava.

The devitrification of glass (not, of course, at first understood as crystallization) has a long history. It must have been observed long before it was recorded, for it occurs whenever common glass is heated for an hour or so at a low red heat. Manufacturers of glass could hardly have avoided some crystallization of their product whenever it was allowed to cool in large pots or bulky furnaces. The earliest reference that I have found is in a twelfth-century manuscript of the *Mappae Clavicula*, which gives a garbled recipe for turning "fragile glass into the nature of a stronger metal" for use as a mortar, by cooking it after it has been coated with a lute of ashes and blood. The use of lute suggests that the process had been discovered by the heating of a retort or other laboratory vessels, which were habitually coated with blood-cemented ashes or clay to protect them from thermal shock and to provide support during prolonged heating at high temperatures. The suggestion that the glass will become like a stronger "metal" follows glassmakers' terminology, though it may also be an echo of the legendary malleable glass of the Roman artisan reported by Pliny and other ancient writers.

Caspar Neumann, who died in 1737, is reported by Lewis (1763) to have observed that the bottom of a glass retort used for distilling milk acquired the appearance of porcelain, which he attributed to the fine white earthy matter of the milk being forced by heat into the glass. In his 1756 English translation of Neumann's work, Lewis refers to experiments of his own and promises a full account of them, a promise that he made good seven years later.

Before this, however, Réaumur's work on the devitrification of glass had been published and had aroused considerable interest. In his 1727 and 1729 papers on porcelain, Réaumur had concluded that Oriental ware was a semivitrified material composed of a refractory earth and a glass. He suggested that there might be two general ways of achieving the right structure, one being to catch a vitrifiable material in an imperfectly vitrified state during its transition from terra cotta to glass on heating, and the other being to fire a paste composed of a highly refractory material mixed with one that is easily vitrified. In 1740 he announced a third method (Réaumur, 1739). He had found that common glass, when annealed for a long time, lost its vitreous character and became internally white and fine-grained, like porcelain, and at the same time more refractory as well as more resistant to thermal shock and to chemical action. He says that he had made the first discovery of the porcelain process 20 years before publication, i.e., about 1720, in the midst of his research on malleable cast iron, which constitutes Réaumur's other famous contribution to the practical arts (Réaumur, 1722). He was stimulated to publish when a man named Montamis, in the service of the Duke of Chartres, showed him a porcelain-like object that Montamis had made by annealing glass after he had observed the change in character of the bottom of a flask that had been exposed to fire. Réaumur read his paper before a public meeting of the Académie des Sciences at Easter 1740, where it created so much excitement that it was published in the delayed *Mémoires* of the Academy for 1739, which actually appeared in 1741.

Réaumur's process for making malleable cast iron involved the annealing of a brittle white cast iron, packed in bone ash mixed with a little charcoal. His stated motive for making it was to cheapen works of art so that everyone could possess works as elaborately decorated as those produced by the best smiths. He was equally proud of the cheapness of his porcelain. "The methods of attaining it are so simple that there is no one who cannot transform all the bottles in his cellar into bottles of porcelain." He wanted no fine vases but only utilitarian ware.

He was sensitive to the importance of the structural changes that were deducible from the appearance of the textures of the fractured surfaces of test pieces. He was the last scientist for well over a century who had a serious concern with the structure of materials on this level. The essentially Cartesian corpuscular views that had guided his work on steel and iron served equally for porcelain:

> Glass . . . has a polish, a lustre, that is never seen in the fracture of true porcelain. Porcelain is granular, and it is partly by its fine grain that the fracture of porcelain differs from that of terra cotta, and finally it is by the size and arrangement of the grains in them that the kinds of porcelain differ from each other and become closer or less close in nature to glass. (1739, p. 375)

Just as steel could be made either by adding something to wrought iron (Réaumur supposed "sulphurs and salts," for carbon was not suspected until 1774) or by removing something from cast iron, so, also, porcelain could be approached in different ways: to his two earlier ways of achieving the mixed structure, he now added the devitrification of glass. If metals can be returned to the metallic state after being converted to their calces or dissolved in glass, why should not the sand and stones that gave rise to glass be restorable?

Réaumur made his porcelain by heating glass packed in a refractory powder in saggers in an ordinary potter's kiln. After experiments with many cements, he came to prefer a mixture of sand and gypsum, and he thought that the physical change was accompanied by a chemical one, as in the cementation of iron. He observed that the change in nature of the glass started at the surface (a fact supporting the supposed chemical change) and grew inward in the form of silklike fibers that were composed of extremely fine grains. It could be made entirely granular, like ordinary porcelain, under some heat treatments. Any kind of glass except the best crystal (lead glass) could be devitrified to give porcelain under proper conditions, but the ordinary colored glass used for bottles and chemical matrasses worked best.

Réaumur calls his product "porcelain by transmutation, by revivification, or porcelain from glass." It would have been an anachronism for Réaumur to have used the word "crystallization," but he did see his devitrification as a return to the state of stoniness characterizing the materials that went into the glassmaker's pot.

Réaumur complained that chemists, with their meltings, calcinings, and distillings, had neglected cementation as a means of studying bodies, for he thought this to be

> perhaps the method of operation that comes closest to that of nature, which makes its mixtures only slowly and imperceptibly, and, similarly, decomposes bodies only little by little and only very slowly. Everything is mixed together too suddenly by fusion, and often the materials, before being mixed, have undergone too much change. . . . The heat that a solid body endures during a long period dilates its parts, separates them; it opens thousands of passages through which can seep volatile particles that continuously detach themselves from the materials that touch the body on all sides; or particles from the body itself escape: its composition is altered, is perceptibly changed. (pp. 380–381)

The most detailed scientific study of Reaumur porcelain ever published is that of William Lewis (1763). He confirmed Réaumur's observations in all respects but gave far more experimental details on the rate at which the reaction occurred in several different kinds of glass. He found that if the material was overannealed, the elongated fibers began to break up into grains that grew larger until eventually the whole mass became friable, like a mass of loosely adhering white sand. The fibrous state was the hardest, striking sparks with steel and cutting glass, and was resistant to corrosion by acid or alkaline liquors. If the outside of a piece of glass was transformed by a short heat treatment and the temperature then raised, the inside would run out, for the fusibility of the transformed material was much less than that of the original glass. Lewis showed that the action of many different cementing materials was identical except for superficial coloration. Glass simply heated with no surrounding powder devitrified in the same way, although slowly, and it slumped out of shape before the reaction had given it stiffness. He saw that the loss of alkali was not significant, although some alkali did diffuse from the glass into the surrounding sand.

Lewis's work was experimental, and he had little to say on theory beyond stating that the change to the fusible fibrous material formed at the beginning "may depend in part on an alteration produced by the heat in the glass itself considered as a compound or in the nature of its alkaline ingredient"; and at the end, "Thus glass whose production has been commonly supposed to be the utmost limits of the power of fire, has its earth and its salt, which one degree of fire had so firmly united, almost disjoined by another."

For over a century, Réaumur's porcelain was regarded as an interesting material. It is discussed in most books on pottery and porcelain in the nineteenth century. Typical are Lardner (1832) and Knapp (1849), who mention Réaumur porcelain in a laudatory fashion, but they add nothing to Lewis's understanding of it and provide no indication that it was in current production. Knapp mentions that "the idea is well worthy of more attention than has really been expended on it." Even in 1883, Weeks thought that annealed glass should replace porcelain in most of its uses.

For all this, nothing seems to have happened industrially to this supposedly fine invention. The principal collections of art porcelain contain no examples of it, which is perhaps not surprising, since Réaumur's porcelain was aesthetically unimpressive.[3] Although internally it matched the whiteness of the best Chinese ware, its surface was dark and rough. Its physical properties however—its high melting point (St. Cloud "porcelain" could be melted in a crucible of Réaumur's devitrified glass), its resistance to thermal shock and to the action of both acids and alkalies—suggested its use for chemical apparatus, and both Lewis and Macquer extol its virtues for this purpose; but even here it failed. Liphardt (1785) discusses its use for laboratory utensils at some length. Ehrmann (1786), studying the effect of the newly discovered oxygen torch on materials, "sacrificed" a retort made of Réaumur porcelain that had come to him from the Strasbourg chemist, J. R. Spielman. If historic samples exist, they are more likely to be in museums of science than of art. Despite the praises of its properties in the literature, it must have been an inferior material (probably because of the plane of weakness existing in the region where the surface-nucleated crystallization met in the center), and there is no indication that it was ever produced commercially in appreciable quantities. It was forgotten in the twentieth century until it came back in a vastly improved form, indeed as a totally new material based on extremely fine controlled nucleation, in the Pyroceram of Donald Stookey of the Corning Glass Company (Stookey, 1959).

The words *crystal* or *crystallization* appear nowhere in either Réaumur's or Lewis's paper on devitrification. The change was regarded as a loss of glassiness, not a gain of crystallinity.

With James Keir's paper (1776), glass, crystals, and geology merged for the first time. In 1771 Keir had published his English translation of Macquer's *Dictionary of Chemistry*, which contains the suggestion that the transformation in Réaumur porcelain was due to the diffusion of vitriolic acid from the surrounding gypsum to form a kind of salt within the glass. Shortly afterward Keir established a glass factory of his own in Stourbridge, Worcestershire. In a mass of glass that had been allowed to cool slowly, he observed a white crust and numerous hexagonal crystals. In reporting on these to the Royal Society, he begins with as good a discussion of the structure of polycrystalline bodies as any that I have found in eighteenth-century literature:

In many substances, when broken, the parts appear to have some determinate figure. This determination of figure, or grain, as it is called, is obvious in bismuth, regulus of antimony, zinc, and all other metallic bodies which may be broken without extension of parts; and although the ductility of gold, silver, lead, and tin, prevents the appearance of the peculiar grains when pieces of these metals are broken, yet we have reason to believe, that, by exposing them to proper circumstances, they also would shew a disposition to this species of crystallization, as it may be called, by a further extension of that term; for Mr. Homberg has observed that when lead is broken while hot, in which state it is not ductile, a granulated texture appears. Perhaps all homogeneous bodies, in their transition from a fluid to a solid state, would, if this transition were not effected too hastily, concrete into crystals or bodies simularly figured. Instances of such crystallization have occurred to me in glass which had passed very slowly from a fluid to a solid state; and the form, regularity, and size of these vitreous crystals have varied according to the circumstances with which their concretion had been accompanied. I send along with this paper a few specimens of this crystallized glass, together with a drawing of some of the most remarkable crystals.

Some of these regular crystals occurred in window glass, composed of "sand, kelp, calcareous earth and lixiviated vegetable ashes," and also in common bottle glass of similar composition except for the incorporation of slag from an iron blast furnace. These glasses form the best crystals, Keir believed, because they are more fluid and less tenacious, and the minute particles of which crystals consist can more easily "apply themselves to each other with less resistance from the medium: Perhaps also the greater portion of calcareous and other earthy particles may dispose these glasses to crystallize more than others, which contain a larger quantity of saline and metallic fluxes." He saw that, despite the loss of weight (nearly 2 percent in one experiment), the alteration was a change of texture, not basically chemical. "The properties of bodies depend merely on the different arrangements of their integrant parts or on their modes of crystallization." There are many other instances of such nonchemical property changes, for example, the hardening of steel.

Many of Keir's glassy crystallizations were spherulitic aggregates of fibrous crystals. Some, however, were well-formed hexagonal crystals, and it was these that led him to speculate as follows:

Does not this discovery of a property in glass to crystallize reflect a high degree of probability on the opinion that the great native crystals of *basaltes*, such as those which form the Giant's Causeway, or the pillars of Staffa, have been produced by the crystallization of a vitreous *lava,* rendered fluid by the fire of volcanoes?

This opinion is further confirmed by the following considerations. The prismatic and other regularly-shaped *basaltes* have been almost always found to be accompanied with *lava*, pummice-stones, and other vestiges of the fire of the volcanoes whenever they have been carefully examined by intelligent naturalists, as has been shewn by M. Desmarets, in his Memoir on the *Basaltes* of the province of Auvergne, in France; *Mém. de l'Acad. des Sciences*, 1771. Basaltic columns have even been discovered, according to the same author, among the productions of volcanoes now existing, as of those of Mount Etna and of the Isle of Bourbon.

He was, of course, quite wrong in attributing the hexagonality of basaltic faulting to crystallization, but its geometry had misled many others.[4]

The devitrification of glass continued to interest chemists—especially French ones—for some time. Dartigues (1804), the proprietor of a glassworks, carried out extensive tests on different kinds of glass and came out very firmly for the theory that devitrification is crystallization. He ended up by referring to lava and its devitrification, and thought that the laboratory facts could provide explanations of the phenomena of geology on which there had not been agreement because "nothing can make people believe that stones had previously been glass." Other important papers were those of Guyton de Morveau (1810) and Pelouze (1855).

F. Leydolt (1852) described and illustrated several examples of true crystallization in glass. He then went on to apply to the structure of glass a technique that he had found useful in studying meteorites. After etching glass with hydrofluoric acid, he found large crystals. These were actually superficial growths of reaction product; his paper is of more interest for the fact that it contains a beautiful plate printed from an electrotype made directly over a crystallized sample—an early scientific use of the electrotype process that had been developed a decade earlier. Similar misleading studies of crystals forming on etched glass were done by Wetherill (1866).

Nineteenth-century chemists, so impressed with the simple atomic ratios that compounds had to have according to Daltonian theory, were extraordinarily resistant to the idea that there could be crystalline solid solutions, and they were even reluctant to accept the idea of microscopic heterogeneity or of diffusion in the solid state that was necessary to give rise to it. Pelouze, for example, affirmed that crystals arising in a vitreous mass must have a composition essentially the same as the medium in which they grew, although Dumas (1855) argued against this. Studies of the growth of crystals in glasses continued, however, and their formation in borax beads became a useful branch of blowpipe analysis. Nevertheless, the question of the devitrification of glass remained mainly a practical rather than a scientific problem until Stookey's studies of nucleation in the 1950s precipitated a new flurry of interest on a different structural level. (But this has little to do with eighteenth-century geology, to which we return.)

Concern with the nature of basalt and lava remained active after Keir's paper. In 1781 Joseph Priestley wrote to Josiah Wedgwood to thank him for a sample of basalt and to report experiments on the amount of air that could be extracted from it and from lava. After melting in a retort and cooling, basalt had become a perfectly black glass. Priestley also asked Wedgwood's opinion on Keir's view that basalt "though a perfect glass at first . . . might become the substance it now is by length of time," and wondered if Wedgwood's pottery kilns would be convenient for an experiment on the subject.

In 1785 James Hutton presented the first outline of his theory of the earth, which had such great influence after its publication in fuller book form (1795) and elaboration and championing by Playfair (1802). Although Hutton does not mention Thomas Beddoes, his mature ideas were probably stimulated by the latter's paper "On the affinity between basaltes and granite," presented before the Royal Society in 1791. This was an astute discussion of crystallization. Beddoes expressed the interesting view that the results of chemical analysis (then properly so much in vogue) would often serve to perplex mineralogists rather than to reduce their science to order, and he discussed the whole question of how the transition from a state of fusion or solution to a solid may give rise to various homogeneous or heterogeneous internal structures:

> Nor is it perhaps difficult to assign highly probable reasons why a mixture of different earths with more or less of metallic matter, in returning from a state of fusion to a solid consistence, may assume sometimes the homogeneous basaltic, and sometimes the heterogeneous granitic internal structure. No fact is more familiar than that it depends altogether on the management of the fire, and the time of cooling, whether a mass shall have the uniform vitreous fracture or an earthy broken grain arising from a confused crystallization. The art of making Réaumur's porcelain consists entirely in allowing the black glass time to crystallize by a slow refrigeration; and the very same mass, according as the heat is conducted, may, without any alteration of its chemical constitution, be successively exhibited any number of times as glass or as a stony matter with a broken grain.

Beddoes's disregard of chemistry led him to the erroneous conclusion embodied in his title, but his emphasis on crystallization was badly needed in geological thinking at the time. He described the joint crystallization of minerals in a two-phase eutectic (not, of course, so named).[5] Beddoes was careful to warn that his ideas did not test the igneous hypothesis any more than the opposite one, for regardless of whether the disunion of the parts in forming initial solution "had been effected by the repulsive power of fire or the intervention of water, it is just as easy to conceive of heterogeneous earthy crystals shooting from different points of a uniform liquid" according to either supposition. However, he did believe that both granite and basalt arose from the injection of "a fused mass raising, rending and shivering the incumbent strata, while its heat hardened them into laminated stone." This, and not the infiltration of water as proposed by de Saussure, appeared to be the true origin of granite veins.

We come, at last, to the main point of this paper. As is well known, Hutton's theory of the earth derived much of its support from the series of laboratory experiments duplicating geological phenomena that began with the work of Sir James Hall; and Hall's experiments were a direct outgrowth of the earlier work on glass. One of Hutton's main contentions was that basalt represented intrusions of igneous material essentially identical with lava. It would seem that there would not have been much occasion for argument had not Guettard's original misconception of the nature of basalt been fortified by the dogmatism of Werner and his followers, but there was certainly something in plutonism that cut across the grain of conventional thinking, and we will therefore first consider the scientific basis of the objections to it.

The versatile Irish chemist and geologist Richard Kirwan (1733–1812) was inspired to much of his best work by criticism of others. His objections to Lavoisier's antiphlogistonic chemistry and revised chemical nomenclature are well known to historians of chemistry. In 1793 Kirwan had advanced many arguments of a chemical nature against Hutton's preliminary statement of his theory, and he could not let Hall's later experimental support of it pass unchallenged. William Nicholson, editor of the *Journal of Natural Philosophy, Chemistry and the Arts*, reprinted, or rather preprinted, in his April 1800 issue Hall's paper that had been read before the Edinburgh Royal Society, and followed this in the June issue with a rather bitter attack on it by Richard Kirwan. Kirwan, conceding that "fanciful and groundless as the Huttonian theory seems to me to be, it may, like the researches for the philosopher's stone, be highly useful for suggesting new experiments," based much of his argument on assumptions as to the nature of crystals and crystallization. He placed great emphasis on the association of perfection with crystallization. In the case of granite, he thought that crystallization would have to result in regular crystals of quartz, feldspar, and mica, formed and stratified in the order of their melting points, a structure never found in nature. He objected to Hall's use of the word "crystallization," and even to the term "crystallites" for imperfect crystals:

To the vague term of crystalization I must however object, for as those stones [basalts] in their original state present no regular crystals, but are at most internally and imperfectly crystalized, so they must be when reduced from a glassy state to one resembling the original, and thus discover [i.e., reveal] rather a *nisus* toward crystalization than perfect crystals, which latter the term crystalization [as] generally applied would lead us to expect.

The consolidation should not be called crystallizations,

for according to every sense in which this term [crystallization] has ever been employed, whether that operation was perfect or confused, it denotes at least an union of particles previously dispersed through a liquid medium, they must, therefore, be at liberty to move through this medium in order to coalesce and reunite to each other . . . but in Sir James' experiments we find the consolidation to take place in a fragment of glass which still retained its solid state, and consequently the particles were not at liberty to move toward each other. This consolidation must, therefore, evidently have arisen from some internal change in the constitution of the glasses in which it was observed. (pp. 154–155)

The change, he thought, was the volatilization of soda, which several investigators of Réaumur porcelain had discussed. He did not perceive that volatilization could have nothing but a superficial effect if the particles had no internal mobility, and he thought he had entirely destroyed the validity of the analogy between glass and basalts. Kirwan did, however, accept the fact that chemical combination is affected by temperature and that separation can occur on cooling, and he was therefore not surprised by the loss of the close vitreous texture that occurred on slowly cooling melted whinstone, or at the appearance of the looser texture of a more stony substance. Except for the emotional overtones of the word "crystallization," this is essentially Hall's theory; but in Kirwan's mind, it did not support the Huttonian theory. At best, to an unprejudiced mind it might render the origin of whins ambiguous

by making them assume the appearance of Neptunian origin when in fact they owe it to fusion; but it is only an appearance, for natural whins are accompanied with circumstances, and contain substances which contradict that appearance and prove it to be deceitful. (p. 156)

After listing various aspects of the external characteristics of whins, Kirwan concluded that there could be no doubt that the production of even the basaltic pillars of Staffa and the Giant's Causeway had occurred in the moist way. As a parting shot, he reiterated his belief that lavas must carry with them a substance capable of maintaining their heat and fluidity, a substance that burns in contact with the atmosphere.

His opponent, Sir James Hall (1761–1832), was a man of very wide interests. Hall was active in the Royal Society of Edinburgh from its beginning. His concern with structure led him not only to geology but also to studies of details of Gothic buildings, which he believed developed from structural antecedents in simpler materials. He introduced Lavoisier's chemistry to the members of the Society and debated phlogiston with Hutton, whom he seems to have admired greatly, despite constant disagreements on scientific questions. In 1788 he discussed the origin of pumice. On January 4 and March 1, 1790, he presented a paper supporting the plutonic origin of granite, written "a few weeks" after he had seen a large pot of green bottle glass at the Leith glassworks that had been allowed to cool slowly and had assumed a stony structure.[6] Hall reported experiments analogous to Réaumur's, reversibly converting bottle glass by slow cooling into a stony substance that would revert to a glass if it were rapidly cooled after being more strongly heated. On March 5 and June 18, 1798, just a year after Hutton's death, Hall presented a detailed paper to the Society. In a preliminary report on this (Hall, 1798), he specifically related his ideas on crystallization to Réaumur's porcelain, which clearly was his inspiration, although reference to Réaumur is omitted from the fuller publication (Hall, 1800, 1805).

The advice of a geologist friend had caused Hall to drop the granite of his early work and concentrate on basalt, choosing for experiment several samples of whinstone, a Scottish term for a variety of stones elsewhere known as "basaltes, trap, wacken, *grünstein* and porphyry." By heating these in a crucible in "the great reverberating furnace at Mr. Barker's iron foundery," he easily converted whinstone from a quarry in the neighborhood of Edinburgh into a black glass, but if the crucible were very slowly cooled it yielded "a substance different in all respects from glass, and in texture completely resembling whinstone. Its fracture was rough, stony and crystalline; and a number of shining facettes were interspersed through the whole mass." Seven different varieties of whin behaved similarly.[7]

Hall made the important observation (related to Lewis's partial melting of partly devitrified glass) that the glass made by melting and cooling whinstone became soft enough to yield readily to the pressure of an iron rod at a temperature far below that at which the original rocks softened. Glass that had been recrystallized in the laboratory was intermediate. The temperatures, measured with a Wedgwood pyrometer, are detailed in table I. Crystallization of a typical whinstone took place in the range 21 to 28 degrees Wedgwood. A minute was time enough to start crystallization in the range of 21 to 23 degrees, regardless of whether temperature was rising or falling. At higher temperatures, up to 28 degrees, the reaction was slower but gave more perfect crystallization; at lower temperatures, down to 21 degrees, the reaction became faster and gave imperfect crystallites. Hall clearly understood the nature of isothermal transformation.

Having established that whinstone would turn to a lavalike glass, and that this could be converted to a stony material either on slow cooling or on prolonged sojourn at intermediate temperatures, Hall then proceeded to examine lava itself. He selected six kinds of

Table I Table of softening temperatures of crystalline basalt and glass (Hall, 1800)

Substances	Original softened	Glass softened	Crystallite softened
1. Whin of Bell's Mills Quarry	40	15	32
2. Whin of Castle Rock	45	22	35
3. Whin of Basaltic Column, Arthur's Seat	55	18	35
4. Whin near Duddingstone Lock	43	24	38
5. Whin of Salisbury Craig	55	24	38
6. Whin from the Water of Leith	55	16	37
7. Whin of Staffa	38	14½	35
1. Lava of Catania	33	18	38
2. Lava of Sta. Venere, Piedimonte	32	18	36
3. Lava of La Motta	36	18	36
4. Lava of Iceland	35	15	43
5. Lava of Torre del Greco	40	18	10
6. Lava of Vesuvius, 1785	18	18	35

The numbers represent the temperatures at which the samples were soft enough to yield to the touch of an iron rod. They are expressed in degrees Wedgwood, one degree of which corresponded to a $\frac{1}{240}$ change in length of a little test cylinder of a standard clay. The pyrometer had a horribly non-linear scale, although it was historically important as the first experimental means of measuring high temperatures and was useful in firing pottery, for it combined time and temperature in a manner analogous to the ware itself. Hall's temperatures range from about 800° to 1200°C. The glass samples were made by melting the whin or lava "by the application of a strong heat and subsequent rapid cooling." The "crystallite" samples are the whins or lavas first turned to glass, subsequently annealed at a lower temperature or slowly cooled to make them crystalline.

The samples were analyzed chemically by Robert Kennedy (1800, 1805). Except for the presence of about 5 percent volatile matter (mainly water and CO_2) in the whinstones, they were essentially all the same in composition, 46 to 51 percent "silex" (i.e., SiO_2), 16 to 19 percent "argil" (i.e., Al_2O_3), 14 to 17 percent "oxyde of iron" (i.e., Fe_2O_3), and lesser amounts of lime, soda, and HCl.

lava, five of which were already partially crystalline, and converted them all into homogeneous glasses by heating and rapid cooling, then caused them to become crystalline by treatment at a lower temperature. On a visit to Vesuvius in 1785, Hall had taken samples of lava directly from the molten stream, and these, because of their rapid cooling, were completely glassy; he made them crystalline by annealing. He measured the temperature of softening in each case (see table I). With his data, and by invoking the effect of time and temperature on crystallization, he was able to explain all of the glassy and stony characteristics of lava and basalts. Since the glass was more fusible than crystallized material of the same composition, it was no longer necessary to invoke, as Kirwan and Dolomieu had done, some foreign substance or combustible to keep it molten. Geology had become, at least in part, a laboratory subject.

Kirwan's view on the nature of crystals and his overemphasis on perfect shapes was quite typical of mineralogists and chemists at the time; indeed, something like it continued among crystallographers well into the twentieth century. Concern with imperfect reality was slow to come. Its roots were in the work of Réaumur, who knew of the irregular grains visible in fractured surfaces, and whose work on the structure of iron and steel (Réaumur, 1722) later bore fruit in modern metallography and the ability to control the structure of materials in line with the desired properties. Similarly, Réaumur's observation on the structure of his glass-porcelain precipitated thoughts about the crystallization of lava, although he himself had no inkling that the structural parts he observed were related to the striking geometric crystals of the mineralogist. The similarity was in internal structure, and only practical observation could keep the messy facts alive until theory had arrived at the state at which it could properly disentangle them. Réaumur and his porcelain, Hall and his whins, serve beautifully to illustrate the interaction between scales of structure and scales of thought which gives so much life to science.

Notes

1
Even today the dominant role of the interfaces between crystals in determining their shape is slow of acceptance among geologists. Contacts between crystals in metamorphic or metasomatic rocks are more frequently a result of interfacial energy balances than they are of the sequence in which the various crystal species were nucleated. See, for example, R. L. Stanton, "Mineral interfaces in stratiform ores," *Trans. Instn. Mining & Metallurgy* 74:45–79 (1964). In sedimentary rocks, which have not yet been studied in detail from this viewpoint, it appears that interfacial energy excludes aqueous solutions from between calcite-calcite and calcite-silicate intercrystalline contacts, but maintains open capillary channels between crystals of quartz and many silicates, as well as pushing aside mineral particles to prevent their entrapment in pyrite and garnet crystals growing in shales and schists.

2
For a discussion of the history of understanding crystallization in metals, see Smith (1960). The work of the experimental geologists in the nineteenth century provided the background for microscopic studies of engineering metals and alloys, in which the crystals are generally invisible to the naked eye and never well shaped. The great metallographer Osmond frequently cites Vogelsang, Daubrée, Fouqué, and Lévy, who had been making artificial mineral crystals; and much of the early metallographic terminology was adapted from that of the mineralogist.

3
(*Note added in proof, 1969*) See, however, the recent article describing a box made in Austria around 1750 that seems to be of Réaumur porcelain: R. von Strasser, "A box of Schackert porcelain," *Journal of Glass Studies* 9:118 (1967). Von Strasser cites a book by Robert Schmidt, *Die Brandenburgische Gläser* (Berlin, 1914), describing attempts in Potsdam to make porcelain from glass in the years 1750 to 1783.

4
The hexagonality of basalt prisms results, of course, from the unidirectional thermal gradient, which produces a network of contraction cracks that adjust to the most economical geometry as they advance through material that is so finely crystallized as to be uniform on the scale of the crack. (The prisms are not uniquely hexagonal, as the pentagons and septagons in the reject pile of Staffa blocks at the end of the Zuider Zee dam show.) The crystalline appearance of the prisms nevertheless provides a nice illustration of structural principles, of the scale-free relationships that occur in aggregates or segregates of anything.

5
That the idea was not a simple one to grasp can be seen from the fact that Hutton advanced the obvious co-crystallization in graphic granite as a conclusive argument against its igneous origin.

6
Hall's ideas of 1790 were published only in summary form, for he had no wish to offend Hutton, whom he later said "was impressed with the idea that the heat to which the mineral kingdom has been exposed was of such intensity as to lie beyond the reach of our imitation, and that the operations of nature were performed on so great a scale, compared to that of our experiments, that no inference could properly be drawn from one to the other." Although Hutton believed that a congelation or crystallization had taken place with more or less regularity to produce the stony and crystallized structure common to all unstratified substances from large-grained granite to the fine-grained and almost homogeneous basalt, he thought that the texture of basalt arose entirely from the pressure to which the parental lava had been subjected. Hall saw that slow cooling was responsible, not pressure.

7
Some of Hall's experimental samples still survive. A fragment of a crucible containing a piece of Rowley Rag (basalt) melted by him that once belonged to Michael Faraday is now in the Science Museum, London. In an excellent article on Hall's scientific contributions, V. A. Eyles (1961) gives photographs of melted whinstone samples that are preserved in the Geological Museum in London and photomicrographs of three of them that, though crystalline enough, are not strikingly similar to the original whinstone. Hall's experiments continued to be shown to students even in the present century—Professor Thomas Turner displayed a crucible containing glass made from the local Lickey Hills basalt when discussing slags before the author's class in metallurgy at the University of Birmingham in 1922. Another cross connection with metallurgy is provided by H. C. Sorby's Hall-like experiments on syenite that were done in the same year, 1863, in which Sorby first revealed the true microscopic structure of iron by careful polishing and etching.

References

Beddoes, Thomas (1791). "On the affinity between basaltes and granite," *Phil. Trans. Royal Soc.* 81:48; abridged in the *Transactions* 17:8–18.

Blake, J. F. (1892). *Catalogue of Collection of Metallurgical Specimens Formed by the Late John Percy . . . Now in the South Kensington Museum* (London).

Darcet, Jean (1766). *Mémoire sur l'Action d'Un Feu Égal, Violent et Continu . . . sur un Grand Nombre de Terres, de Pierres & de Chaux Métalliques . . .* (Paris).

Dartigues (1804). "Mémoire sur la dévitrification du verre et les phénomènes qui arrivent pendant sa crystallisation," *Annales de Chimie* 50:325–342.

Dumas, J. B. A. (1855). Discussion on paper by M. Pelouze, *Comptes Rendus* 40:1327–1329.

Ehrmann, Friedrich Ludwig (1786). *Versuch einer Schmelzkunst mit Beihülfe der Feuerluft* (Strasbourg). French trans. with added papers by Lavoisier and Laplace: *Essai d'un Art de Fusion à l'Aide de l'Air du Feu, ou Air Vital . . .* (Paris, 1787).

Eyles, V. A. (1961). "Sir James Hall, Bart.," *Endeavour* 20: 210–216.

Gerhard, Carl Abraham (1773–76). *Beiträge zur Chymie und Geschichte des Mineralreichs* (2 vols.; Berlin).

——— (1788). *Abhandlung über die Umwandlung und den Ubergang einer Erd und Stein-Art in die Andere* (Berlin).

Grignon, Pierre Clément (1761). "Mémoire sur des crystallisations métalliques, pyriteuses et vitreuses artificielles, formées par le moyen du feu" (written in 1761). In his *Mémoires de Physique, sur le Fer* . . . (Paris, 1775), pp. 476–481.

Guyton de Morveau, L. B. (1776). "Observation de la cristallisation du fer," *Journal de Physique* 8:348–353.

——— (1810). "Diverses observations relatives à l'art de la verrerie," *Annales de Chimie* 73:113–146.

Hall, James (1790). "Observations on the formation of granite," in report of meeting for January 4 and March 1, 1790, summarized in pp. 8–12 in the History of the Society, *Trans. Royal Soc. Edinburgh* 3.

——— (1798). "Curious circumstances upon which the vitreous or stony characteristics of whinstone and lava respectively depend . . . ," *[Nicholson's] Journal of Natural Philosophy* . . . 2:285–288. (Abstract of paper read before Royal Soc. Edinburgh on March 5 and June 18, 1798. The full text appeared in 1800 and 1805—see next two items.)

——— (1800). "Experiments on whinstone and lava," *[Nicholson's] Journal of Natural Philosophy* . . . 4:8–18, 56–65.

——— (1805). "Experiments on whinstone and lava," *Trans. Royal Soc. Edinburgh* 5:43–75.

Hutton, James (1785). "Theory of the earth, or an investigation of the laws observable in the composition, dissolution, and restoration of land upon the Globe" (paper read March 7 and April 4, 1785), *Trans. Royal Soc. Edinburgh* 1:209–304.

——— (1795). *Theory of the Earth* (2 vols.; Edinburgh).

Keir, James (1776). "On the crystallizations observed in glass," *Phil. Trans. Royal Soc.* 66:530–542.

Kennedy, Robert (1800). "A chemical analysis of three species of whinstone and two of lava," *[Nicholson's] Journal of Natural Philosophy* . . . 4:407–415, 438–442.

——— (1805). "A chemical analysis of three species of whinstone and lava," *Trans. Royal Soc. Edinburgh* 5:76–98.

Kirwan, Richard (1793). "Examination of the supposed igneous origin of stony substances," *Trans. Royal Irish Acad.* 5:51–81.

——— (1800). "Observations on the proofs of the Huttonian Theory of the Earth, adduced by Sir James Hall, Bart.," *[Nicholson's] Journal of Natural Philosophy* . . . 4:97–102, 153–158.

Knapp, Friedrich Ludwig (1847–1853). *Lehrbuch der Chemischen Technologie* . . . (2 vols.; Braunschweig). Eng. trans., *Chemical Technology* (Philadelphia, 1849).

Lardner, D., ed. (1832). *A Treatise on . . . the Manufacture of Porcelain and Glass* (London).

Leroux, F.-P. (1867). "Sur la trempe de quelques borates," *Comptes Rendus* 64:126–128.

Lewis, William (1756). "Vegetable Earth." Note appended to chapter on artificial vegetable productions, Sect. IV in William Lewis, ed., *The Chemical Works of Caspar Neumann* (London).

——— (1763). "Experiments on the conversion of glass vessels into porcelain, and for establishing the principles of the art," in his *Commercium Philosophico-Technicum or Philosophical Commerce of Arts* (London), pp. 230–255.

Leydolt, F. (1852). "Ueber die Krystall-Bildung in gewöhnlichen Gläse . . . ," *Sitzungberichte der K. Akademie der Wissenschaften, Vienna* (Math.-Naturwiss. Klasse) 8:261–275.

Liphardt (1785). "Einige Versuche über das Reaumurische Porcellan," *Crell's Chemische Annalen* (ii):132–138.

[Macquer, Pierre Joseph] (1766). *Dictionnaire de Chymie, contenant la Théorie & la Pratique de cette Science* . . . (2 vols.; Paris). Eng. trans. by James Keir, *Dictionary of Chemistry* . . . (2 vols.; London, 1771).

Mappae Clavicula. The text of a twelfth-century version of this Latin compilation of recipes is given by Sir Thomas Phillipps in *Archaeologia* 32:183–244 (1846). For an English translation see item 172 in the bibliography.

Neumann, Caspar (1756). *The Chemical Works of Caspar Neumann,* edited by William Lewis (London; 2nd ed., 2 vols., 1773).

Pelouze, J. (1855). "Mémoire sur la dévitrification du verre," *Comptes Rendus* 40:1321–1329.

Percy, John (1875). *Metallurgy . . . Introduction, Refractory Materials and Fuel* (Rev. ed.; London), pp. 51–54.

Playfair, John (1802). *Illustrations of the Huttonian Theory of the Earth* (Edinburgh).

Pott, J. (1746). *Chymische Untersuchungen welche fürnehmlich von der Lithogeognosia . . .* (Potsdam). French trans. *Lithogéognosie ou Examen chymique des pierres et des terres en général . . .* (Paris, 1753).

Priestley, Joseph (1781). Letter from Joseph Priestley to Josiah Wedgwood, August 8, 1781. In Robert E. Schofield, ed., *A Scientific Autobiography of Joseph Priestley* (Cambridge, MA: The MIT Press, 1966), pp. 204–205.

Réaumur, R. A. F. de (1722). *L'Art de Convertir le Fer Forgé en Acier et l'Art d'Adoucir le Fer Fondu . . .* (Paris).

——— (1727). "Idée générale des différentes manières dont on peut faire la porcelaine, et quelles sont les véritables matières de celles de Chine," *Mém. Acad. Sci.*, pp. 185–203.

——— (1729). "Second mémoire sur la porcelaine, ou suite des principes qui doivent conduire dans la composition des porcelaines de différens genres . . . ," *Mém. Acad. Sci.*, pp. 325–344.

——— (1739). "L'Art de faire une nouvelle sorte de porcelaine . . . ou de transformer le verre en porcelaine," *Mém. Acad. Sci.* (published 1741), pp. 370–388.

Romé de Lisle, J. B. (1772). *Essai de Cristallographie* (Paris), p. 321.

Saussure, Horace Bénédict de (1779–1796). *Voyages dans les Alpes; précédés d'un essai sur l'histoire naturelle des environs de Genève* (4 vols.; Neuchatel).

Smith, Cyril Stanley (1960). *A History of Metallography* (Chicago: University of Chicago Press).

——— (1967). "The texture of matter as viewed by artisan, philosopher, and scientist in the seventeenth and eighteenth centuries," in *Atoms, Blacksmiths, and Crystals* (Los Angeles: William Andrews Clark Memorial Library, UCLA).

Sorby, Henry Clifton (1863). "On the microscopical structure of Mount Sorrel Syenite, artificially fused and cooled slowly," *Geol. and Polytechnic Soc. West Riding of Yorkshire* 4:301–304.

Stookey, S. D. (1959). "Catalysed crystallisation of glass, in theory and practice," *Industrial and Engineering Chemistry* 59:805–808.

Thy, Nicholas Christiern de, Le Comte de Milly (1771). *L'Art de la Porcelaine* (Paris, 1771 [colophon 1772]). Part of the *Descriptions des Arts et Métiers* of the Académie des Sciences.

Watt, Gregory (1804). "Observations on basalt, and the transition from the vitreous to the stony texture which occurs in the gradual refrigeration of melted basalt . . ." *Phil. Trans.*, pp. 279–314.

Weeks, Joseph D. (1883). *Report on the Manufacture of Glass* (New York).

Wetherill, C. M. (1866). "On the crystalline nature of glass," *Amer. Journ. Sci. and Arts* [*Silliman*] 91:16–27.

Note: The punctuation of the quotations from the various sources in English has been modernized. Quotations from works in other languages are from published translations where listed; otherwise they are new translations by P. Boucher and the author.

8
Art, Technology, and Science: Notes on Their Historical Interaction

Introduction

It is misleading to divide human actions into "art," "science," or "technology," for the artist has something of the scientist in him, and the engineer of both, and the very meaning of these terms varies with time so that analysis can easily degenerate into semantics. Nevertheless, one man may be mainly motivated by a desire to promote utility, while others may seek intellectual understanding or aesthetic experience. The study of interplay among these is not only interesting but is necessary for suggesting routes out of our present social confusion.

Humanists have shown a widespread disregard for technology's role in human affairs, but if they had seen technology as an eminently human experience, they could have better guided society's choice of objectives and controls. Civilization has been an ecological process with interacting contributions coming from an infinite diversity of individual human characteristics and social institutions. As historians have turned away from their older concern with the great movements headed by kings, generals, or businessmen, they have naturally emphasized the role of people like themselves (scientists and other intellectuals), and they have, until recently, largely disregarded the rather messy technology that has been associated with virtually every important historical change and which continually impinges directly upon Everyman in his daily life. Neither religious conviction nor institutional conservatism has, until today, sensed in technology a peril sufficient to prompt an examination of its nature and its growth. Certainly, at the extremes, the concepts of the cosmos and of the ultimate nature of matter developed by philosophers and scientists are of overriding importance, for they have basically influenced man's opinion of himself: Men have gone to the stake for their ideas on the nature of the universe, and all men know of it. Ideas on ultimate atomism have aroused bitter philosophic debate. Conversely, however, anyone who considers the nature of materials, advocates a new way of making pottery, or advances a new theory of the hardening of steel meets with both intellectual and popular indifference. Yet the voyage to the moon depends on men making metal as well as on computations based on the theories of Newton and Einstein.

Art, Techniques, and Materials

The present paper is an outcome of my realization, some years ago, that many of the primary sources I had selected for a study of the history of metallurgy were objects in art museums. Though materials are not all of technology, they have been intimately related to man's activities throughout all of history and much of prehistory, and they therefore provide an excellent basis for a study of some of man's most interesting characteristics under greatly different social and cultural conditions. A materials-oriented view of history may overemphasize the association of technology with art; yet it was precisely the artist's search for a continued diversity of materials that gave this branch of technology its early start and continued liveliness despite an inner complexity which precluded scientific scrutiny until very recently.

Several writers have discussed the manifest interactions between artistic expression and the basic view of the world embodied in contemporary scientific or religious concepts. Such interactions certainly exist at the highest level of insight, but artists have had far more intimate and continuing association with technology than they have had with science. In turn, the attitudes, needs, and achievements of artists have provided a continuing stimulus to technological discovery and, via technology, have served to bring to a reluctant scientific attention many aspects of the complex structure and nature of matter that simplistic science would have liked to ignore. The antecedents of today's flourishing solid-state physics lie in the decorative arts. One must conclude that creative discovery in any field is a matter for the whole man, not his intellect alone. Though it occurs in an individual mind, it is strongly interactive with society and tends to seek out the least rigid parts of a community structure.

Leonardo da Vinci said in his treatise on painting: "Those who are in love with practice without science are like a sailor who gets into a ship without rudder or compass and who never can be certain where he is going."[1] At the same time, Leonardo strongly opposed the view that knowledge that is both born and consummated in the mind is enough: "It seems to me that all sciences are vague and full of errors that are not born of experience . . . , that do not at their origin, middle or end pass through any of the five senses."[2] And, of course, all his extant works reflect continual interplay between sensual experience and intellectual analysis. The same view is to be found in the writings of many scientists, though for most of the last three centuries science has rightly been more concerned with the unreliability of the senses than with their essential contribution to whatever knowledge human beings can acquire.

When discussing the new routes to the understanding of nature in the preface to his *Micrographia* (1665), Robert Hooke remarks: "So many are the links upon which the true Philosophy depends, of which, if any one be loose, or weak, the whole chain is in danger of being dissolv'd; it is to begin with the Hands and Eyes, and to proceed on through the Memory, to be continued by the Reason; nor is it to stop there, but to come about to the Hands and Eyes again, and so, by a continual passage round from one Faculty to another, it is to be maintained in life and strength."[3] Hooke believed that the advancement of knowledge depended upon both the senses and the intellect—upon the mind, the hand, and the eye in cooperation. His writings repeatedly reflect his obvious enjoyment of natural phenomena and his intuitive understanding of them. However, Hooke's slightly younger contemporary Isaac Newton was engaged in demonstrating the great power of mathematical science and setting the stage for three centuries of superbly unfolding knowledge based on the belief that the senses are unreliable and that science advances best if, at any one time, it is limited to those small areas in which rigorous methods can be applied. Though the domain accessible to such science is steadily expanding, there remain many important aspects of natural and man-made systems that are too complicated for complete analysis. The present-day political and intellectual unrest reflects increasing awareness that the scientist's understanding of things "in principle" is not enough. The more holistic view of the Renaissance artist may be returning—though whether it will be put into practice by people who allow themselves to be called artists is another question.

Just as the meaning of the words "art," "science," and "technology" have varied greatly throughout history, so has the role in society of the various practitioners. Perhaps technology has been the most constant in its aims. Science has encompassed many different approaches to the collection and analysis of data, just as art, in different places and periods, has combined in vastly different degrees the functions of decoration, symbolism, illustration for didactic purposes, the projection of feeling, and (by no means the least important) pure enjoyment. In what follows the "art" may sometimes be of a kind beneath the notice of an art historian, but it will always be concerned with a man's doing something that is not strictly necessary for the performance of a function, something extra done to give enjoyment to the producer himself and usually also to others who subsequently come in contact with his work.

Not all peoples have regarded "art" as a separable human activity, and the self-conscious production of paintings, sculpture, and *objets d'art,* like the organized commerce in them, has by no means always occupied the privileged place that it has had in Europe since the Renaissance. Most of what follows is concerned with the decorative arts—those arts relegated to the minor category in most museums today—although it might be remarked that the best of today's nonobjective paintings have more in common with sensitively wrought useful objects of ceramic and metal than they have with many of the "fine" arts displayed on museum walls.

There is some analogy between the exploration and exploitation of the materials of nature in chalcolithic times and earlier, the detailed exploration of the forms of nature that followed increased representational skill in the thirteenth and fourteenth centuries, and the experiments with perspective, light, and shadow in the Italian Renaissance. The driving force in all three was an essentially scientific curiosity directed to the discovery of some fairly practical means of achieving an aesthetic end.

The relation between art and the artist's materials was well discussed by Henri Focillon.[4] Remarking that art is bound to weight, density, light, and color, he says that it is borne along by the very matter it has sworn to repudiate: matter in its raw state "evokes, suggests and propagates other forms according to its own laws." The ceramics of the Far East appear to be "less the work of a potter than a marvellous conglomerate created by subterranean fire or accident." The raw stuff of [Chinese ink brush painting] partakes of both water and smoke . . . yet . . . such a painting possesses the extraordinary secret of being able to stabilize these elements and at the same time leave them fluid and imponderable." Though overemphasis on technique is clearly dangerous, Focillon believed that "the observation of technical phenomena not only guarantees a certain objectivity to [the studies of] a historian but affords an entrance in the very heart of the problem by presenting it in the same terms and from the same point of view as it is presented to the artist." In discussing the artist's various techniques to get different qualities of line, shadings, and gradations, "such alchemy does not, as is commonly supposed, merely develop the stereotyped form of an inner vision: it constructs the vision itself; gives it body and enlarges its perspectives."

Technique, of course, mainly gives details of form, not the gross outlines and balances. Nevertheless, much of the refinement of an artist's vision as he works toward its realization comes from his interaction with his materials. The whole quality of a line and surface depends upon both the material and the tool as well as upon the artist's hand, whose movements they subtly control. Compare the same pictorial concept as it is realized in different media—with a brush in oils, watercolor, or tempera on canvas, wood, or paper; by printing from a metal plate with intaglio lines made by etching or engraving or from surfaces left in relief on a chiseled wood block; or by repoussé work, tracing or otherwise working directly on the final metal surface. It is understandable that those students who must work from reproductions of works of art are usually more interested in iconography than in the more subtle questions of technique and quality, but it is regrettable that technical ignorance should so frequently prevent art historians from considering the whole experience of the artist. In much the same way, science historians have tended to overlook the less logical side of science.

The Discovery of the Properties of Matter

In studying man's earliest history, when the evocative qualities of certain forms and the power of symbolism in nonrandom shapes and sounds was being discovered, it is difficult to separate things done for "pure" aesthetic enjoyment from those done for some real or imagined "practical" purpose. The man who selected for admiration a beautifully shaped and textured stone was yielding to a purely aesthetic motivation, but the man who molded clay into a fertility figurine was simultaneously an artist, a scientist learning to understand the properties of matter, and a technologist using these properties to achieve a definite purpose. Supposedly most of the innumerable fertility figures recovered by the archaeologist's spade from periods even before 20,000 B.C. were made as a kind of industry, acquired for reasons of fashion, and employed practically to make more probable some desired result. This does not, however, destroy their fundamental aesthetic quality.

More important is the fact that in the earlier stage of discovery, first of form and later of materials that, once shaped, would retain desirable form, the motive can hardly have been other than simply curiosity, a desire to discover some of the properties of matter for the purpose of internal satisfaction. Paradoxically man's capacity for aesthetic enjoyment may have been his most practical characteristic, for it is at the root of his discovery of the world about him, and it makes him want to live. It may even have made man himself, for, to elaborate a remark by the poet Nabokov, it seems likely that verbal language (to which anthropologists now assign vast evolutionary advantage) was simply a refined use of the form-appreciating capabilities first made manifest in singing and dancing.

A natural step after the collection and admiration of unusual natural stones and animal or vegetable debris would have been the use of the properties of some natural materials to produce unnatural shapes and textures in others. This supposedly began by matching the hard cutting edge of stone to softer wood, hide, sinew, and bone, and was followed by the discovery and exploitation of the special properties of a host of substances. The last were mainly minerals that could be ground and used as pigments, undoubtedly far more for the decoration of the body and other long-perished surfaces than for the incredibly preserved cave paintings that we admire so greatly today. It is not only the nature of the record that makes one feel the joy that early man took in the discovery of the properties of materials. The cracking propensity of different stones, the plasticity of moist clay, the fine granular color of pigments were all used for what they are and appreciated directly by the senses in shaping or in use.

Aesthetically satisfactory forms have repeatedly developed from interaction between cultural requirements and the real properties of a new material or technique: the forms are not just superimposed. A returning sensitivity to this is at least partially behind the present passion for primitive art, for a simpler technology makes the properties of materials more evident.

Over and over again scientifically important properties of matter and technologically important ways of making and using them have been discovered or developed in an environment which suggests the dominance of aesthetic motivation. The presence of flowers in Neanderthal graves[5] suggests that the transplanting of flowers for enjoyment preceded the development of agricultural technology for food supply. The first use of both ceramics and metals occurs in decorative objects. Fire-hardened figurines of clay precede fired pots in many Middle Eastern archaeological sites. The seventh millennium B.C. copper dress ornaments and beads at Chatal Huyuk in Anatolia and at Ali Kosh in Iran considerably precede the use of copper for weapons, though the useful needle appears early. Although there is some evidence for earlier pyrotechnological experiments with ores, the replacement of simply hammered native copper by smelted metal did not occur until about the time that copper oxide was being used in blue glazes on ceramics, though probably only after high temperatures had become available for firing useful ceramic sickle blades.[6]

The modern metallurgist uses alloying elements to strengthen metals and to lower their melting point; he cold-works to harden and anneals to soften. He uses their differing chemical reactivities, immiscibilities, and surface energies in refining and joining processes. The discovery of all these effects is very old. To take a single point in history, an examination of the jewelry and other metal objects from the famous Royal graves at Ur,[7] dated about 2600 B.C. (figures 8.1 and 8.2), reveals knowledge of virtually every type of metallurgical phenomenon except the hardening of steel that was exploited by technologists in the entire period up to the end of the nineteenth century A.D. One must not, of course, overlook the fact that royal burial objects are far from being representative samples of contemporary use of any material. The court would appropriate the best work to its own ends, but just for this reason it provides the best index of both the most novel and the most sophisticated techniques.

The transition from copper ornaments to axes and swords of bronze in the fourth millennium B.C. was paralleled in the fifteenth and sixteenth centuries A.D. by the transition from the casting of monumental bronze doors, statuary, and especially bells, to the casting of cannon. If the objects themselves are not sufficient evidence, a comparison of the vivid circumstantial account of bell founding given by the early twelfth-century artist-craftsman Theophilus[8] with the discussion of the casting of cannon by the eminently practical Biringuccio[9] some four centuries later will show how much the warrior depended upon the churchmen's technique. To be sure, existing technology is applied to whatever need may be seen by a government or people: my point is only that the *invention* of a technique has, until re-

Figure 8.1
Gold beaker and cup made by raising from sheet metal, decorated by repoussé work and tracing. Cup height, 15.5 cm. From the Royal Graves at Ur, ca. 2600 B.C. Photo courtesy of University Museum, Philadelphia.

Figure 8.2
Gold rings made from square and round wire almost invisibly soldered, 2.1 and 1.7 cm diameter. From the Royal Graves at Ur, ca. 2600 B.C. Photo courtesy of University Museum, Philadelphia.

cently, been far more likely to occur in an aesthetically sensitive environment than in a practical one. We will see later that even the development of efficient quantity-production methods owed much to the art industries, if not directly to the artist.

Technology's debt to the artist is inseparable from the converse. Though both the most exquisite and the most ugly objects can be made with the same technique, technique is essential to beauty. The technique of the artist merges by invisible stages into the technology of his materials. Among the best examples of this are superb cast bronzes of Shang dynasty China.[10] The earliest ones reflect an advanced ceramic technology, so necessary in making the molds, and have linear decoration obviously cut into the mold (figure 8.3). The almost brutal strength that characterizes the later bronzes of the Shang and early Chou periods, the flanges, and almost every aspect of their form arise in a direct interplay between design and the practical details of the foundry. The molds were divided into a number of sections that would have produced unpleasant breaks in the surface decor had this not been designed for division, and leaky mold joints would have produced ugly fins if these had not been exaggerated into flanges the edges of which could easily be dressed (figure 8.4). The attractive difference in quality of the fine and the bold intaglio lines probably arises in the technical difference between carving the former directly into the mold surface and making the latter by applying convex lines of clay to a molded concavity. Still later comes the building up of designs from a few units by the use of some method of three-dimensional replication (figure 8.5)—a clear forerunner of the printing process.

In view of the centrally important role that welding plays in today's space-age structures, it is interesting to note the facility with which Greek and Roman founders welded together the parts of their statuary. Almost any classical bronzes, when closely studied, reveal some patching of foundry defects, but recent studies[11] have uncovered the widespread use of a welding process involving the running-in of superheated molten metal for making joins between precast parts, with an accuracy and permanence that would challenge a modern welder. The beautiful and famous statue of Poseidon (figure 8.6) made in 475 B.C. and now in the National Museum at Athens would not exist were it not for the welds securing its many parts together (figure 8.7).

The technique of casting-on preceded this joining of long seams and was widely exploited in the Middle East, in Europe,[12] and in the Far East[13] because of the freedom that it gave to the designer (see figure 6.7). It also enabled different metals to be combined, as in the application of an elaborately detailed cast bronze handle to a serviceable steel blade—a literal welding of beauty and utility—and in the combination of cold-worked and cast bronze.

In later times the little mouth blowpipe used by the goldsmith in his soldering operations was adapted to laboratory use, first for the examination of ores and later as the basis of the first comprehensive scheme of qualitative chemical analysis.[14] The larger blast lamps worked with foot bellows that were used for making glass beads and for decorating them with colored enamels (figure 8.8) led directly to the oxy-gas blowpipe. This was first used in high-temperature research about 1782, became fully commercial with the melting of platinum in the 1850s, and finally became the modern welding torch.

Figure 8.3
Chinese cast bronze ceremonial vessel, type *tsun*, early Shang dynasty. Height, 27 cm. Photo courtesy of the Arthur M. Sackler Collection, New York.

Figure 8.4
Chinese cast bronze ceremonial vessel, type *ting*, late Shang dynasty or early Chou (11th–10th century B.C.). Photo courtesy of the Fogg Art Museum, Harvard University.

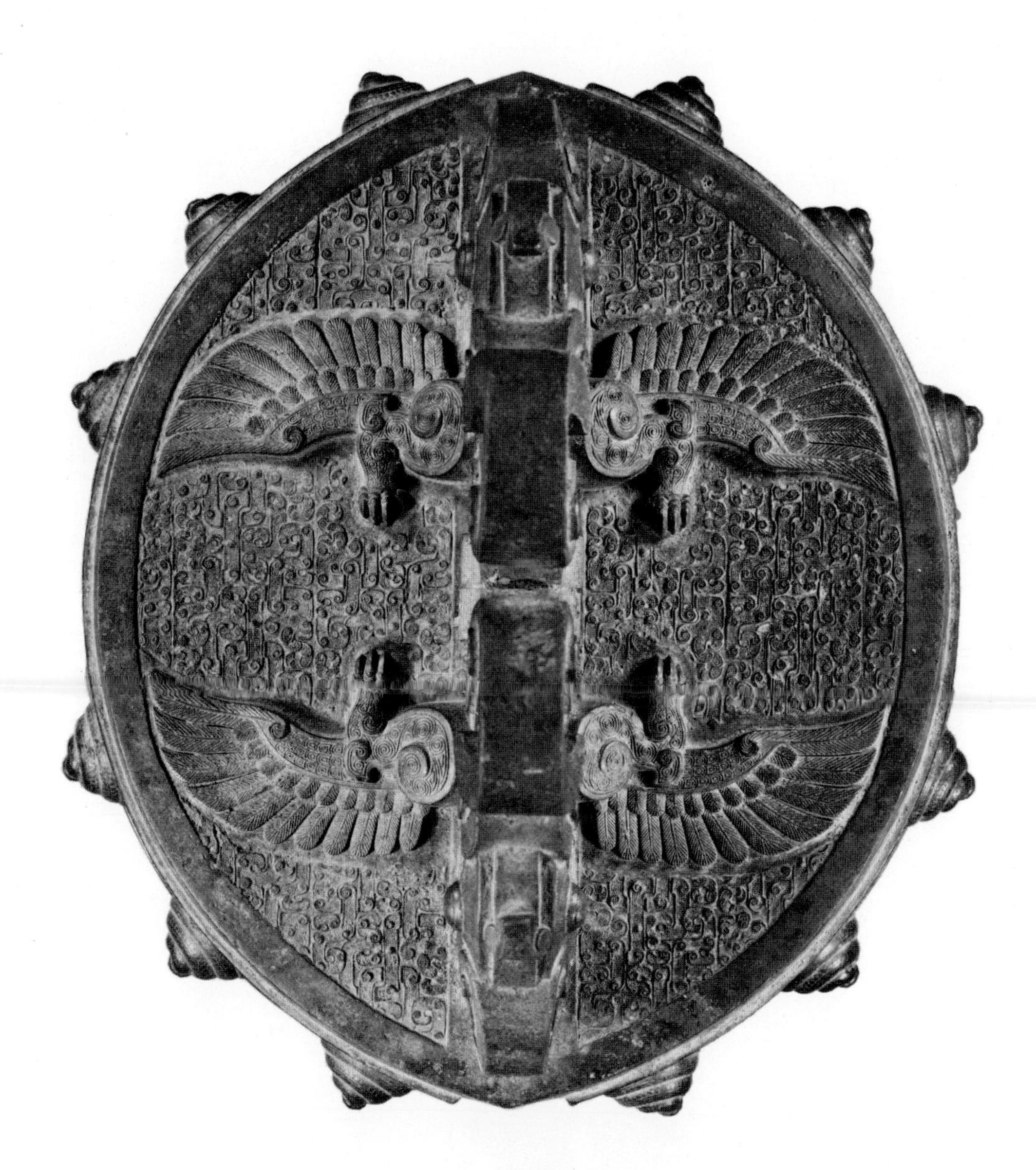

Figure 8.5
Chinese cast bronze bell, Chou dynasty. The detail in the ground is built up in three successive stages of replication. Photo courtesy of the Smithsonian Institution, Freer Gallery of Art, Washington, D.C.

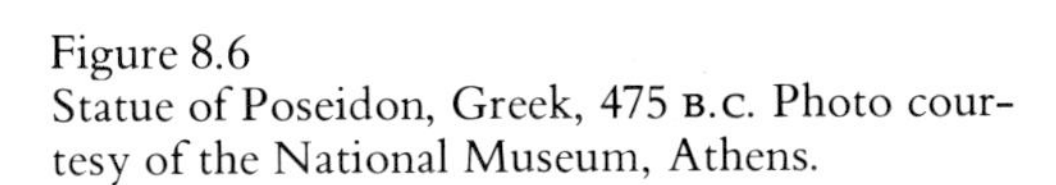

Figure 8.6
Statue of Poseidon, Greek, 475 B.C. Photo courtesy of the National Museum, Athens.

Figure 8.7 (*right*)
Welded ankle joint in the Poseidon statue, natural size. Photo by Arthur Steinberg.

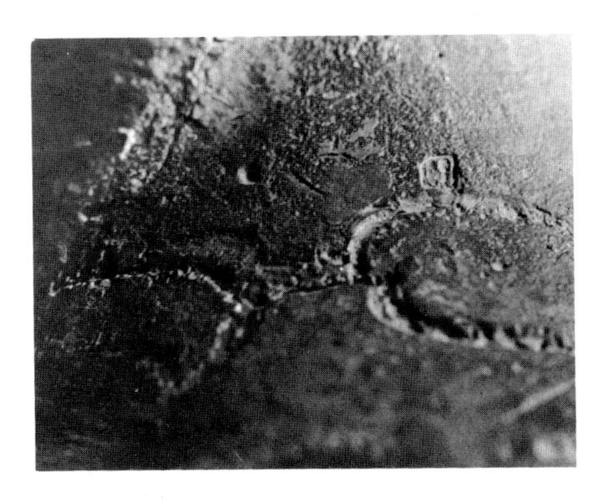

Figure 8.8
Blast lamps being used in making colored enameled glass beads. Johan Kunckel, *Ars Vitraria Experimentalis* (1679). Photo courtesy of the Corning Museum of Glass.

The decoration of pottery with colored pigments and later with glazes repeatedly brought to man's attention the chemical diversity of natural minerals and led to new techniques. The cementation process that was probably used in the fourth millennium B.C. to make Egyptian blue frit (faience) involves very subtle behavior of alkalies and silicates in differential contact with lime and silica surfaces.[15] It is highly likely that it gave rise directly to the manufacture of the first "sand-cored" glass vessels (which probably had a calcareous, not a siliceous, core). Though its relationship to early metallurgy has not yet been explored, this cementation process gives hints of the way in which the first alloys may have been made. It may relate to the smelting of complex sulphide ores by the use of highly alkaline fluxes, to say nothing of its later use in the soldering, parting, and coloring of gold and eventually in the making of brass and steel.

In the eighteenth century, European desire to duplicate beautiful porcelain from the Orient inspired not only geological search but also experiments in high-temperature chemistry and the development of the first realistic methods of chemical analysis for anything but the precious metals. Reports of the large-scale operations at Ching-te-chen may have inspired the integration of mass-production operations at Wedgwood's factory in Staffordshire, and, at the other end of the spectrum of knowledge, it was an interest in porcelain that led to Réaumur's studies of the devitrification of glass, which later played a role in the understanding of lava and the development of Hutton's plutonic theory of the earth.[16]

Chinese fireworks for pleasurable celebration inspired more diverse chemical experimentation than did military explosives. Today's rocket ships and missiles are an outgrowth of fun-fireworks, and their guidance systems depend on knowledge first acquired from that ubiquitous toy, the top. All optical devices have their roots in the polishing of ancient mirrors and the cutting of accurate facets on gems for a more decorative glitter. The chemist's borax-bead test, now alas passé, arose from the use of metal oxides in making stained glass windows and colored enamels (as well as fake gem stones).

Colors and chemistry are inseparable. The earliest pigments were naturally occurring minerals, but the preparation of artificial ones, such as red and white lead, verdigris, and marvelous sublimed vermilion, mark a chemical industry in classical times. The subtleties of surface tension on which the modern flotation process for the beneficiation of ores depends were first used in the purification of lapis lazuli to give fine ultramarine.[17] The important metal powder industry of today began with gold ink.[18] Art historians rarely go behind the blue and gold splendor and the iconography of a medieval illuminated manuscript to see the ingenious technology that made it possible and that reflects men's lives on another, no less necessary, level. It is the same with organic dyes: think of the chemical knowledge behind an oriental rug or an emperor's robes! The chemist's indicators and his eventual awareness of pH came directly from the chameleon colors of the miniature-painter's turnsole. . . . The list is endless.

The Development of Mechanical Technology

The relation between design, structural engineering, and knowledge of materials in architecture is a well-known example of the inseparability of aesthetic and technological factors. Here it must suffice to make only the passing comment that it has usually been nonutilitarian structures such as temples and monuments that have stretched the limits of existing techniques and led to the development of new ones.[19]

The popular belief that technology is recent is partly based on the fact that intricate machines were, in fact, slow to develop. The advanced knowledge of materials in the ancient world was not paralleled by mechanical devices of seemingly comparable ease of discovery. The ancient military devices (which have usually followed not far behind aesthetic needs in promoting discovery) and hoisting machines of importance to the builder are all relatively simple. Mechanical devices of any intricacy appear only as toys, as aids to priestly deceptions, or as theatrical machinery. It was not utility in the usual sense—though it may have been a search for the public's money—that prompted the mildly ingenious devices described by Hero of Alexandria.[20] It may be, as has often been suggested, that the availability of cheap labor rendered the Persians, Greeks, and Romans unable to appreciate the advantages of mechanical power; but their failure to develop other types of intricate mechanisms is, I believe, attributable to the fact that the aesthetic rewards to beginning experimentation by the curious in this area are not large. Indeed, for simple mechanical experiments to be intriguing, they require a kind of overlay of intellectual analysis: they are too easily reproducible to provide a rich and varied sensual experience of the kind that comes directly from play with minerals, fire, and colors. Not until the mid-twentieth century have artists shown much desire to experiment with machinery, and their efforts sometimes seem to be more directed toward

Figure 8.9
Cast bronze funerary bucket with lathe-tool marks on bottom, Roman, ca. A.D. 200. 25 cm diameter. Photo courtesy of W. J. Young, Museum of Fine Arts, Boston.

Figure 8.10 (*right*)
Three products of the sixteenth-century ornamental turning lathe, showing some of the complicated shapes made possible by the mechanical combination of simple motions. Made of ivory by Jacob Zeller, shortly before 1600. National Museum, Dresden. Photo courtesy of Deutsche Fotothek.

catching up with and exploiting the technologists' world than toward leading it.

The association of the earliest clocks with mechanical automata was a natural one, for, with the possible exception of organ makers, only the makers of automata had the necessary skill and sense of mechanism.[21]

Machine tools, like materials and mechanisms, had a period of prehistory within the decorative arts. The earliest is probably the rotary drill, which, though it was perhaps developed for hafting axes, found wider use in making beads, seals, stone pots, and sculpture.[22] The inverse geometric motion of material against a fixed tool begins with the potter's wheel and progresses to the simple lathes that supposedly produced the soft-stone products of Glastonbury and the Roman bronze objects such as mirrors and pots having decorative bottoms with deep, heat-catching circular grooves (figure 8.9). Then followed Theophilus's twelfth-century description of lathes for turning bell molds as well as for the molds of pewter pots and the metal pots themselves. By this time rotary motion was commonly used in the grindstone. The first machine with intermittent motion after the Oriental rice-pounding mill is Theophilus's little device for cutting the criss-cross ground for decorative overlay of precious metal on iron. The cam- and template-guided lathes of Jacques Besson (1578) not only cut screws but also turned decorative work of great variety. They were followed by the ornamental turning lathes of the seventeenth through the nineteenth centuries, used mainly by gentlemen hobbyists and for decorating gold snuff boxes. These were devices of great mechanical ingenuity applied to a mechanically trivial purpose (figure 8.10); nevertheless, they provided the experimental environment in which definable compound motions were generated, and

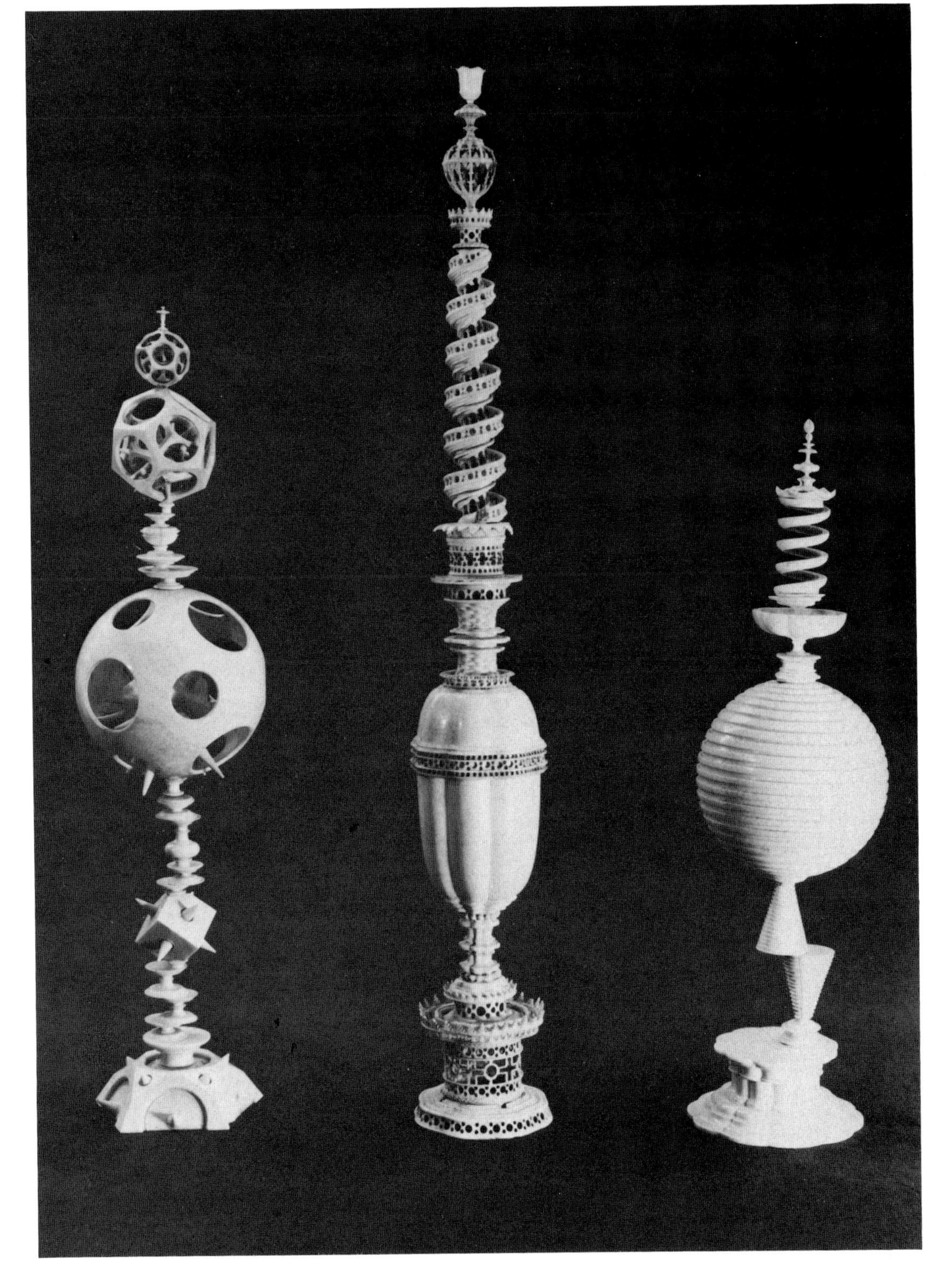

they served not only as a basis for instrument making and later industrial machinery but also to disseminate a feeling for the composition of mathematical curves.[23] The toylike nature of these lathes resulted in their being rather briefly dismissed in the standard machine-tool histories,[24] but it is easy to see how the desire to produce a decorative effect was once more the motivation for the discovery of phenomena that would later be applied to more serious purposes.

Decorative fountains—for example, Versailles with its magnificent pumps and pipes—stretched the capacities of hydraulic engineers more than did plebeian water supply. Savery's fire engine was pumping water for a garden in Kensington in 1712 not long after its use in mine drainage.

The Graphic Arts

The introduction of printing illustrates the same point, though here the art is even less separable from the technology. The obvious advantage of transmitting information in written form kept thousands of scribes busy for millennia, but the functional business of recording the commands of the government or the information needed by merchants did not lead to printing—this came from the desire to reproduce images and patterns. The ceramic decorative stamps at Chatal Huyuk,[25] the cylinder seals made in such profusion throughout the Middle East, the tools for the impression of decorative details in ceramic vessels and tiles as well as in molds for casting, the punches for repetitive stamping of metal, the dies for striking metal coins, and the block printing of textiles—all these precede "useful" typographic printing and lay the groundwork for it. The sequence from rubbing to woodblock print to movable type in the Far East is a direct one.[26] The first true printing was for the dissemination of a Buddhist sutra—utility and aesthetics united in the service of religion. In Europe, although the precise stages of the invention are hard to trace, the sequence is similar. The reproduction of pictures with text from woodblocks was a popular art early in the fifteenth century, though for the step to reusable type Gutenberg's solution involved the transfer of technique from a humbler craft, that of the pewterer, whose permanent molds with replaceable parts for decorative detail and whose alloy needed little change to make type.[27] The earliest type seems to have been cast from a tin-base alloy perhaps containing bismuth, but cheaper, harder lead alloys were common in the sixteenth century and thereafter.

A strong aesthetic motivation is visible in the works of the early typographers. Much of it obviously derived from the desire, or perhaps the necessity, of duplicating the quality of the manuscripts with which they were initially in competition. But art and technology are even more inextricably interwoven in the reproduction of pictures, which began before typography but received an enormous impulse from their use to illustrate printed books. Though to some extent the mere possibility of making multiple copies is the enemy of art, limited reproduction brings an artist's works to a greater audience, and the techniques themselves give rise to aesthetic qualities not otherwise obtainable. Woodcuts, etchings, lithographs—especially if the artist's hand prepares the printing surface—are often preferable to unique works executed in the traditional media of the painter.

Print-making from intaglio lines in metal plates was late in appearing, but its roots are deep. Decorative engraving on the surfaces of bone or soft stone objects, of course, precedes the use of metal, and it was widely used pictorially on three-dimensional objects of bronze, gold, and silver.

The earliest date on a print from an engraved plate is 1446. Some playing cards printed about four years later have attractive animal designs that are similar to some of the marginalia in the great manuscript Bible of Mainz (dated 1452–1453, now in the Library of Congress), and Lehmann-Haupt has suggested that the plates may have originated in abortive experiments by an engraver working in collaboration with Gutenberg, who at the very time and in the same city was at work on his famous Bible and would naturally have liked marginal embellishment matching the best contemporary manuscripts to appear alongside his typographic text.[28] Plausible and attractive though this hypothesis is, there is no intaglio printing that can be definitely associated with Gutenberg. In any case, for hints as to possible technical steps behind the invention, we must move to Italy, where the first engraved prints—those of Maso di Finiguerra, 1452–1455—were made slightly later than in northern Europe. Sulphur casts associated with the Italian prints are preserved in both the British Museum and the Louvre.[29] Goldsmiths were accustomed to make such replicas of engraved objects, both to check the designs before filling them with niello and to provide a record for future use. It was a simple matter to make a mold (perhaps of plaster) from the engraving and to obtain an exact replica of the original intaglio lines by casting sulphur in it; smearing this with soot and oil would make the design clear and produce a general effect of black lines on a yellow background much like the final niello on gold. Transfer to paper would follow naturally and soon render the cast copies obsolete. Northern engravers may have been more ingenious: the casting of the mold material on a dirty engraving might have suggested direct transfer to paper without the need for double molding or a sulphur intermediary. In any case, fine prints could not have been made in the fifteenth century had not centuries of earlier work with niello developed both the technique of using the graver as well as the sense of design appropriate to it, had not the caster of art bronze had experience in the replication of models with fine detail, and had not the new oil-based inks and presses become available for the printer. Once the process of transferring to paper had been invented, it spread rapidly, and artists throughout Europe produced prints which used to the full the possibilities of rendering fine detail and controlled shading that were implicit in the technique. A few years later, shortly after 1500, engraved plates began to meet competition from those in which some or all of the lines were bitten with acid, giving them a special quality that many artists prefer to engraved ones. Here, too, an old technique was ready to be adapted to a new purpose, for armor makers had been using etching in the decoration of the more elaborate of their products for at least a century, probably much longer.[30]

The beginnings of etching—the removal of metal by localized chemical attack—are obscure. Supposedly jewelers and coppersmiths had long used vegetable acids or minerals such as copperas to remove the oxide scale produced by annealing their ware, but there is no early record of this practice. Chemical attack was certainly used in the cementation process to remove silver from solid gold at least as early as the sixth century B.C.[31] Etching with a design pro-

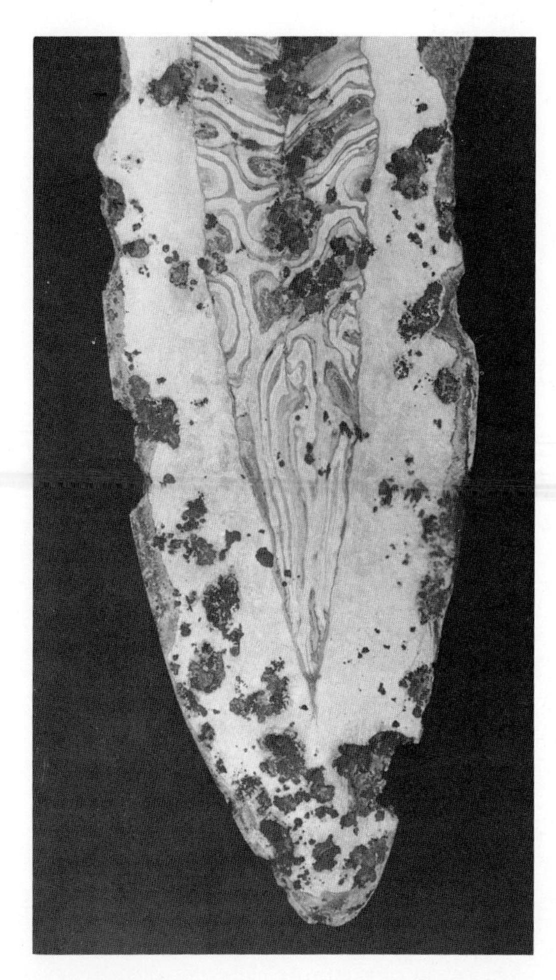

Figure 8.11
Tip of a pattern-welded iron sword, Merovingian, sixth century A.D. Width, 3.9 cm.

duced with the aid of a resist was done on calcareous shells at Shaketown Pueblo in Arizona, dated between 900 and 1200 A.D.[32] A related effect is seen in the reserved areas of white electrum in depletion-gilded ornaments of sheet copper-silver-gold alloys sometimes found in pre-Columbian South America.[33]

In the Old World, swords made of different kinds of iron and steel welded together into a consciously decorative pattern appear in La Tène sites, and supposedly some chemical attack would have been used to reveal the pattern. Quite apart from its decorative function, the visibility of the pattern in the welded composite would serve simultaneously to control the work in the smithy and to provide a kind of index of quality to the customer. The patterns were sometimes faked—an early use of a resist. The patterns on the swords of the Franks and Vikings (figure 8.11) are referred to in Viking sagas in terms that leave no doubt as to their visibility.[34] The beautiful textures of Damascus swords were also acclaimed by poets long before the technique of forging and etching them was described.[35] These were certainly etched to bring out the pattern, and etching was probably done on the European blades, though polishing alone can leave a just-visible texture on the surface if there is enough slag mixed in the metal. Japanese swords owe both their effectiveness and their beauty to the distribution of intensely hard areas left by an intricate control of the forging and heat-treating operations. These, with finely dispersed slag particles, are subtly revealed in the final polishing operation without the use of acids. There is no better symbiosis of the highest aesthetic and technical standards than in these swords.[36]

Some paragraphs in Pliny may refer to chemical attack on iron, but the first clear reference to etching in European literature is in the eighth-century chemical manuscript at Lucca, *Compositiones Variae,* which contains a recipe for the treatment of an iron surface with a mixture of corrosive salts containing copper as a preliminary to gilding it. A similar technique appears in the ninth-century *Mappae Clavicula.*[37] With the omission of copper and the use of a stop-off to localize the effect, decorative etching was born (figure 8.12). Although the earliest extant etched decoration is on late fifteenth-century iron armor, there is earlier evidence for its use. Conrad Kyeser's 1405 manuscript, *Bellifortis,*[38] describes the preparation of distilled nitric acid for this purpose, and he even calls it *aqua martis,* in clear reference to its use on iron. It seems highly probable that the discovery of this first mineral acid about a century earlier had come directly from the experimental distillation of an etching mixture containing saltpeter and acid sulphates. Parenthetically, hydrochloric acid, distilled from a mixture of chlorides and sulphates, also appears first in connection with decorative embellishment—in a work on dyeing, *Plictho,* published in 1548—and in 1589 decorative etching with it is described, but on marble, not metal.[39]

Figure 8.12
Etched design on Italian helmet, Milan, sixteenth century (detail). Photo courtesy of the John Woodman Higgins Armory, Worcester, Massachusetts.

The technique of etching passed directly from arms to the production of etched iron plates for printing, which was at first a part-time activity of armorers. But, having begun as art, etching eventually began also to influence science. As the Damascus and Merovingian swords showed, etching is a sensitive means of revealing heterogeneity in steel, but metallurgists did not begin to use it consciously for this purpose until 1762. In the period between 1773 and 1786, observations on the etching of Damascus gun bar-

Figure 8.13
"Instantaneous light box" with case made of green *moiré métallique.* Made in London about 1820 by "J. Watts and Co., Chymists No. 478 Strand." Height, 8.0 cm. This device made fire by bringing a wooden match tipped with potassium chlorate and sulphur into contact with concentrated sulphuric acid. *Moiré métallique* was tin-plated iron that was given a special treatment to develop a fancy crystallization, subsequently etched and covered with colored lacquer. Photo courtesy of Bryant and May Ltd. and the Science Museum, London.

rels, which were then being made in Europe, led to the first identification of carbon as the material responsible for the differences between wrought iron, steel, and cast iron.[40] The investigation of an essentially decorative phenomenon, and an oriental one at that, thus led directly to the most important single scientific discovery in metallurgical history!

Soon thereafter etching gave rise to a new decorative technique known as *moiré métallique*.[41] This was invented in 1814 and aroused considerable excitement for a few decades. (Figure 8.13 shows a fire lighter made by this technique.) It was no more than etched and lacquered tin plate, but the plate was sometimes treated by local heating and cooling to give very fancy crystallization patterns, even semblances of flowers and landscapes!

New methods of printing illustrated books repeatedly redounded to the advantage of both science and technology. An interesting printing technique—first published in 1555,[42] though Leonardo had described it earlier in his notebooks—was to make direct impressions of objects such as leaves by coating them with printer's ink and impressing them directly on paper. The process (which is not unrelated to the much earlier and more versatile oriental method of producing rubbings on paper laid over objects with details in relief) was later called "nature self-printing." In the eighteenth century a number of botanical books were published with illustrations printed this way, the first being J. H. Kniphof's *Botanica in Originali,* published in 1733.[43] The same technique was used by Schreibers and Widmanstätten in 1813 for recording the etched structure of a section of the Elbogen meteorite. Their print was a spectacular improvement in clarity and accuracy over the lithographs of other meteorites that accompanied it in their published book[44] or the engravings by Gillet de Laumont in the *Annales des Mines* of 1815. For a time thereafter many methods of obtaining relief or intaglio impressions of an object directly on a printing surface were experimented with for both scientific and other purposes.[45] Nature printing from a collage of textured surfaces is the basis of a flourishing school of printmakers today.

The early history of photography itself is a classic example of the symbiosis of art and invention. Della Porta in 1558 recommends the *camera obscura* as a device to lighten artists' labors and help them with perspective. Niepce's famed photochemical etchings on glass (1826) were done to reproduce art, not reality. The processes of Daguerre and Talbot were of both worlds, as photography has been ever since. When the invention of photomechanical methods displaced most other methods in the printer's shop, etching had become a common laboratory technique The science of metallography—indeed, practically the whole structural side of modern materials science—stems from the work of Henry Clifton Sorby in 1863–1864 in the famous steelmaking center of Sheffield, which was the world center of supply for engravers' steel plates. By applying to the preparation of laboratory specimens the methods used for giving these plates their fine finish, and by using etching, which he had heard discussed at a meeting of the local Literary and Philosophical Society, Sorby was able to reveal for the first time in history the true microstructure of steel without disfigurement by fracture or deformation.[46] In the present connection it is interesting to note that the next paper on metallography, by the German railroad engineer Adolf Martens in 1878, was directly inspired by some work on the quality control of metal for use in the exquisite art castings of iron for which Germany was rightly famous at the time.

Electroplating and Electrical Engineering

Electrochemistry is another area in which the interest of the artist or the art industry accelerated scientific knowledge and technological development. An old and pretty parlor trick was the *Arbor Dianae,* mentioned with other "metallic vegetations" in most chemical textbooks of the seventeenth and eighteenth centuries. Eighteenth-century assayers knew of the electrochemical series (though they did not call it that) in the form of sequential replacement of silver in solution by copper, then of copper by iron, and of iron by zinc.

The medieval use of an acid cupriferous solution to give a coating of copper on iron was mentioned above. Such electrolytic replacement remained a common observation and was sometimes used for recovering copper from waste mine waters as well as to confuse people with the semblance of transmutation. It gave rise to a minor art in the seventeenth century in the form of a very pleasant ware made from cement copper in the town of Herrengrund in the Bohemian Erzgebirge.[47] As in figure 8.14, these objects commonly bear inscriptions reflecting their polymetallic origin. [For further discussion of this ware see chapter 14 in this volume.] It has been reported that this ware was shaped in iron and then plated by immersion in cement water. Some folk objects were certainly so made, but the real Herrengrund ware is actually nonmagnetic and was probably made from cement copper powder that was melted, refined, cast, hammered into sheets, and shaped as any other copper would have been. The role of electricity in

Figure 8.14
Copper dish (Herrengrund ware) containing model of minehead equipment with working miners and mineral specimens. Length, 33 cm. Heavily gilded. The inscription, "Eisen war ich, Kupfer bin ich, Silber trag ich, Gold bedecket mich," refers to the recovery of the copper from mine waters by displacement with scrap iron. Made in Herrengrund, Bohemia, early eighteenth century. Photo courtesy of Abegg-Stiftung, Bern.

these operations was not, of course, suspected, any more than it was in the mysterious decay of rudder irons on the English ship Phoenix that in 1670 had been sheathed with sheet lead, which had just then become available in wide sheets from the new rolling mill at Deptford.[48]

If any of these effects had been looked at by a sufficiently curious mind, Galvani's discovery could easily have been made a century or more earlier and without the intervention of a frog. However, even after Galvani and Volta, even after Wollaston's and Cruickshank's demonstrations of the cathodic deposition of copper and other metals, and even after Michael Faraday's elucidation of the laws of electrolysis, no use was made of the phenomenon until 1838. In that year Jacobi, Spencer, and Palmer, in somewhat confused priority, all began the art of electrodeposition for the duplication of coins and other small art objects as well as for the reproduction of printing surfaces—at first for illustrations (figure 8.15) and later for letterpress.[49] Within twenty years large sculpture was being made—for example, the ten-foot-high statue of Prince Albert (figure 8.16) and pieces twice this size for the Paris Opera House.

Henry Bessemer later claimed that in 1832, when a young man of nineteen, he had reproduced plaques by electrolysis, but he did not publish. After 1838 this quickly became a very popular hobby and resulted in widespread knowledge of electricity. Smee, writing in 1842, remarked that "there is not a town in England that I have happened to visit, and scarcely a street of this metropolis [London] where prepared plasters are not exposed to view for the purpose of alluring persons to follow the delightful recreation af-

Figure 8.15
Print from one of the earliest American electrotype plates. This formed the frontispiece of Daniel Davis, Jr., *Manual of Magnetism* (Boston, 1842), alongside a print from the original engraved plate, indistinguishable from the copy.

forded by the practice of electrometallurgy." The new metallurgy quickly spread from copper to other metals. Gold was naturally one of the first, but the most commercially significant was the electrodeposition of silver upon the beautiful white copper alloy now known as nickel silver which rapidly displaced the more expensive Sheffield plate (figure 8.17). The base alloy itself had been imported from China for about two centuries and was used for fireplace equipment, candlesticks, and other domestic objects of some elegance,[50] but it was not analyzed until 1776 (twenty-five years after the discovery of nickel itself), and it was not available commercially until 1833, just in time for its wedding with the new plating method.

Electroplating soon became an important art industry. It provided economic support for the beginning of the nickel industry, it gave birth to the first commercial electric generator (figure 8.18), and, with telegraphy, it provided training and experience for innumerable men who were soon to combine their empirical knowledge with a growing science to give birth to electrical engineering.

Art for the Masses

The story of electroplating is only one example among many in which a desire to simulate a precious material in a cheaper form has stimulated technical advance. All students of the historical literature on iron and steel know Réaumur's classic *L'Art de Convertir le Fer Forgé en Acier et l'Art d'Adoucir le Fer Fondu* (Paris, 1722), with its curious subtitle *faire des ouvrages de fer fondu aussi finis que de fer forgé*. Réaumur specifically states that his incentive was to provide cheaply for the masses decorative objects of the kind expensively made for and previously available only to the rich. It was not engineering devices but elaborately chiseled door-knockers (figure 8.19) at which he first aimed—but he ended by extolling the virtues of mass production of interchangeable parts in industry generally.[51] The motive of cheap art was also behind Réaumur's development of his "porcelain," a devitrified glass of the type that has recently been revived in superior form.[52] He was also the first to suggest the use of wood pulp in the making of paper.

Today's steel rails, I-beams, and other structural shapes also have their origin in decorative needs—the rolling of H-shaped lead cames for stained glass windows (figure 8.20). Around 1750, fancily profiled sections of iron for use in balcony railings, window moldings, and the like (figure 8.21) were being made in grooved rolls, three decades before Henry Cort applied the process to the large-scale consolidation of wrought iron bars.[53] An antecedent to this was the slitting mill (figure 8.22). Another even more portentous mechanical invention was Jacquard's loom (1801) with its punched-card control: this was not needed for plain fustian but for the fanciest of lace.

Parcel gilding can be justified on purely aesthetic grounds as producing an agreeable contrast in color, as in inlay, but most gilding operations have been done simply to

Figure 8.16
Memorial to the 1851 Exhibition and its patron, Prince Albert. The central statue, 10.5 feet high, and the flanking figures are all of electrodeposited copper, not cast in the foundry. From an electrotype in the *Illustrated London News*, June 1863. The memorial is still in place in South Kensington, now appropriately flanked by Albert Hall and the chemistry laboratories of Imperial College.

Figure 8.17
Lithograph in Elkington, Mason and Co.'s catalog of 1847, showing some of the electroplated silver hollow-ware offered for sale.

Figure 8.18
The first commercial electric generator. Height, 160 cm. The patent, issued to John S. Woolrich in August 1842, also specified solutions for the electrodeposition of silver. From the Museum of Science and Industry, Birmingham. Photo courtesy of N. W. Bertenshaw.

save money and to make expensive-looking objects available to others than the rich. The preparation of thin gold leaf, the most extreme utilization of the malleability of any metal, is similarly inspired. The fact that composition gradients could be produced in solid metals was made quite clear, long before diffusion became a subject of scientific inquiry, by the common use of gilding via gold amalgam, as in Europe, or by chemical methods of surface enrichment, as in pre-Columbian South America and in Japan.

The many changes of properties and surface coloration of metals produced by goldsmiths could hardly have failed to support the belief that transmutation is possible—as indeed it is, if "transmutation" is not limited to nuclear changes but is applied to major changes of physical properties.[54] Hopkins,[55] in particular, has argued that alchemy was an outgrowth of the joining of Greek philosophy with a knowledge of workshop practices. Yet the value of empirical knowledge naturally fades as a field advances, and the replacement of alchemy by modern chemical theory is attributable more to the logical than to the practical approach.

The above examples show that the art industries have contributed greatly to the development of techniques and to the knowledge of reactions on which today's science and technology are based. Perhaps, indeed, the mixture of aesthetic and commercial motivation involved in such developments was quantitatively the most powerful stimulus of all, for basic discovery of new effects inspired only by curiosity is by its very nature rare, as rare, perhaps, as any great individual work of art.

Sources in Art for the History of Technology

Little reliance can be placed on any of the written sources relating to technological history prior to about A.D. 1500 unless they are confirmed by contemporary nonverbal evidence. Even today technologists are not noted for literacy, and men like the Benedictine monk Theophilus (early twelfth century), whose hands were accustomed to both the hammer and the pen, have always been rare. In books, ideas (both true and false) naturally fare better than technology. Moreover, the chances of survival of written technological information in medieval libraries were not high. For all this, there are many records that must be studied in the absence of anything better, and in these a strong bias toward the decorative arts is evident. Following the Roman Vitruvius's *De Architectura*, the best pre-Renaissance technological sources are the Leyden papyri, the Lucca manuscript entitled *Compositiones Variae*, the *Mappae Clavicula*, Theophilus's *De Diversis Artibus*, and Eraclius's *De Coloribus et Artibus Romanorum*. Every one of these deals with the artist's materials and techniques, to the exclusion of almost everything else. Manuscripts presenting primary information on machinery for warfare, mining, and other industrial occupations do not exist before the fourteenth century. The famed Theophilus's manuscript of about A.D. 1125 is an outstanding source of pure technology, though he confines himself to giving intimate details on painting, stained glass, and metalwork for the embellishment of the church.

Although it was far from the artist's conscious intent, many paintings of religious subjects, especially those of the thirteenth to

Figure 8.19
A door knocker—the first use of malleable cast iron. From Réaumur's *L'Art de Convertir le Fer Forgé en Acier . . .* (Paris, 1722), plate 16.

Figure 8.20
The glazier's shop. Woodcut by Jost Amman, 1568. The small rolling mill at the left has grooved rolls for shaping lead cames; it is the forerunner of the mills producing rails and steel shapes for building construction today.

Figure 8.21
Mill for hot-rolling iron bars with decorative profiles. From Diderot's *Encyclopédie*.

Figure 8.22
Slitting mill for making iron rods. From Jean Errard, *Le Premier Livre des Instruments Mathématiques* [*et*] *Physiques* (Nancy, 1584). Errard says that this is not his own invention but that of Charles Desrué (Desruet). The slitting mill was a most important link in the development of large power-driven machinery for metal fabrication. This engraving is the earliest evidence for its use. Photo courtesy of Albert France-Lanord.

sixteenth centuries, convey information on agricultural and building techniques, and they are particularly important in reflecting current attitudes toward labor and machinery. Lynn White, Jr.,[56] has studied from this viewpoint the changing depiction of the seven Virtues. As late as the twelfth century Temperance was in little esteem, but during the thirteenth century she became identified with measure and subtly associated with internal and external control. By 1350 she is depicted with the newly invented sandglass; by 1450 (in a manuscript now in Rouen) all the seven Virtues are depicted with technological appurtenances, but Temperance displays eyeglasses, rowell spurs, a mechanical clock, and a tower windmill. The showing in such a scene of these objects—all recent inventions—expresses "a reverence for advancing technology, a sense of its spiritual value, which is peculiar to the West and which has been essential for the building of industrial society."

There are innumerable illustrations known to historians of art but almost untapped by their technical confreres in which an artist interested in human activity (either for its own sake or to satisfy an ecclesiastical or princely command) used the decorative aspects of tools and mechanical devices, quite commonly with disregard of mechanical details but nevertheless providing useful information to the historian.[57] And, of course, the very materials of the artist are themselves a superb record of the technology that produced them, a record that can be read in intimate detail by modern laboratory techniques. The output of the spectroscope, microscope, and x-ray spectrometer will soon become as important to the technological historian as his older verbal sources, to which we now return.

The two earliest printed works on the prosaic subject of steel both have artistic overtones. The first was a little pamphlet on etching, the *Stahel und Eysen* (1532), and the second was on ornamental ironwork and locks—Mathurin Jousse, *La Fidelle Ouverture de l'Art de Serrurier,* published at La Flèche in 1627 (figure 8.23). By this time the artist was aiding the technologist in substantial ways, for the new techniques of producing more accurate representation of visual appearances served increasingly to convey precise technical information. The accurate, detailed drawings of the liquation process for the desilverization of copper, and those of lathe details, pile drivers, etc., contained in the fifteenth-century "Hausbuch" of the Mendel brothers[58] and in the Nuremberg "Hausbuch,"[59] are a far cry from earlier illustrations in which technology is only incidentally reflected. One of the leading German illustrators of the sixteenth century, Jost Amman, sought inspiration in the technical crafts for 86 of the 118 woodcuts in his popular Book of Trades.[60]

By the mid-sixteenth century many carefully written books on both science and technology were being printed with woodcut illustrations (figure 8.24). Both the biological sciences and technology required and inspired some of the best efforts of the artist in rendering realistic details without confusion. The woodcuts in the well-known treatises of Vesalius, Agricola, and Ercker are about as attractive as any book illustrations of the sixteenth century, and their instructional value was correspondingly high. The mystical side of alchemy, though scientifically sterile, appealed to the artist's imaginative approach and gave rise to some attractive books.[61]

In general, physics, with its abstract concepts and simple diagrams, neither attracted

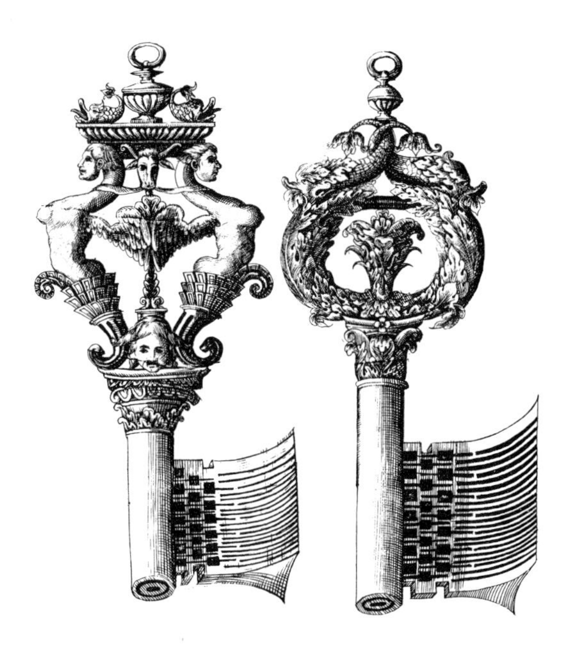

Figure 8.23
Design for a key of forged and chiseled iron. From M. Jousse, *La Fidelle Ouverture de l'Art de Serrurier* (1627), plate 1.

Figure 8.24
Woodcut view of an assay laboratory. From Lazarus Ercker, *Beschreibung allerfurnemisten mineralischen Ertzt und Berckwercksarten* (Prague, 1574).

nor needed the artist. If a physicist used illustrations at all, they were likely to be in the form of colorless linear diagrams making visible the geometry implied by his equations. Galileo in 1638 depicts a weed-encrusted stone wall supporting his elastically deflected beam (figure 8.25), but later elasticians eschewed such realism. The terrellae in William Gilbert's *De Magnete* (London, 1600) have a pleasant look, perhaps contributed by the man who cut the block; and to illustrate his observations on the magnetization of cooling iron, Gilbert allowed himself the luxury of including a woodcut view of a blacksmith's shop that is in the direct tradition of the series of such views in the medieval *Speculum Humanae Salvationis*, where they illustrate (amid changing hearth and anvil design and with occasional detachability of the horse's leg to simplify the smith's work) metallurgy's first contribution to the fine arts—Tubal Cain's rhythmic clangor giving rise to the idea of melody in a listener's mind. Gilbert's other illustrations, however, are purely linear diagrams.

Figure 8.25
Woodcut diagram to illustrate the bending stresses in a beam. Galileo, *Discorci e Dimostrazioni Matematiche Intorno a Due Nuove Scienze* (Leiden, 1638).

The engravings in Robert Hooke's *Micrographia* (1665) reflect both approaches. Most of these are well-shaded sensitive representations of exciting vistas in the New Landscape that his microscope was exposing for the first time (figure 8.26), but Hooke's diagrams of the paths of rays of light (figure 8.27) have a sharp austerity which matches the abstraction of the idea and which came to characterize most scientific diagrams thereafter.

The engraved copper plates that seventeenth- and eighteenth-century book publishers preferred over wood blocks permitted accurate delineation of apparatus and were excellent for showing machinery (figure 8.28). The enormous growth of the graphic

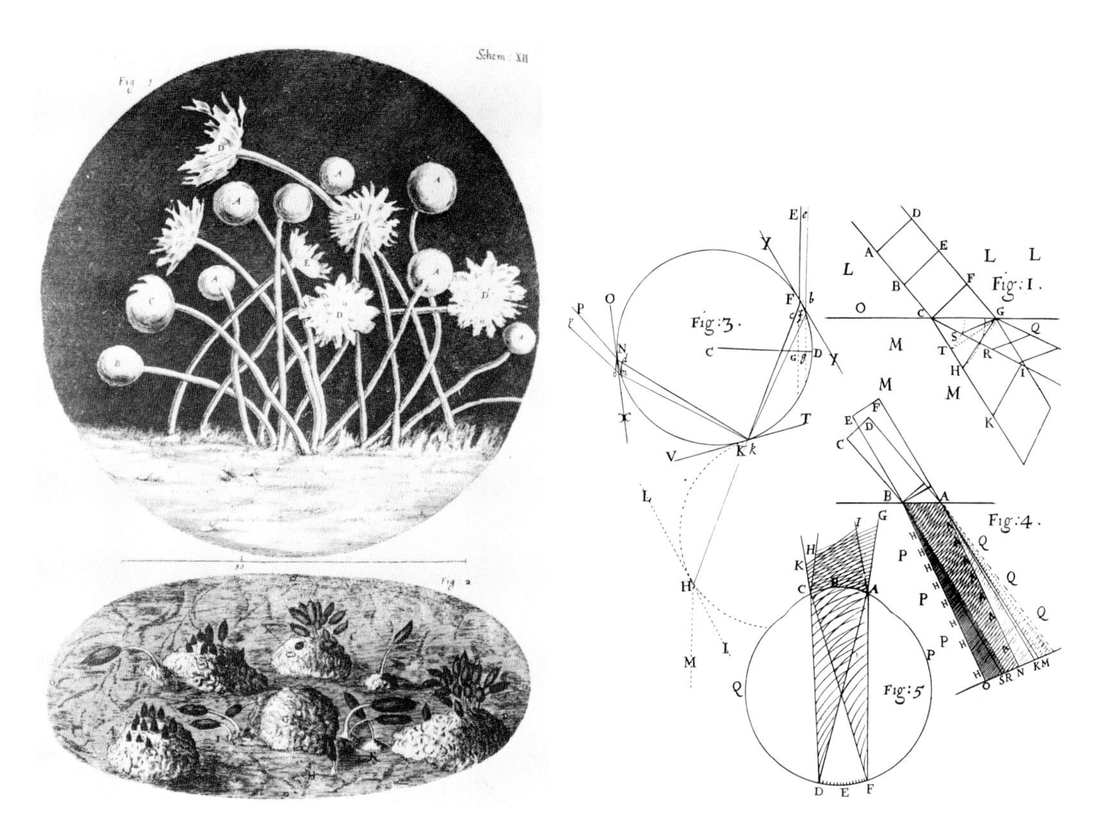

Figure 8.26
Engraving showing "nothing else but the appearance of a small white spot of hairy mould, multitudes of which I found to bespeck and whiten over the red covers of a small book" (Robert Hooke, *Micrographia* [1665], plate 12). The scale line is 1/32 inch, corresponding to an original magnification of about fifty.

Figure 8.27
Engraved diagrams showing paths of rays of light in the eye and in other media. From R. Hooke, *Micrographia* (1665), plate 6.

Figure 8.28
Engraving showing machinery for blanking and striking coins. From André Félibien, *Principes de l'Architecture* (Paris, 1676).

arts in eighteenth-century France coincided with a rationalist viewpoint to result in the publication of massive collections of engravings of technical subjects. The hundreds of folio-size plates in the series of *Descriptions des Arts et Métiers* published by the French Academy of Sciences (figure 8.29) and the seven volumes of plates accompanying Diderot's famed *Encyclopédie* provide a profuse record of technical crafts and industry. Our knowledge of the technology of that time is probably more complete than that of any other period in history, for before this there was scant interest in making records and after it the profusion of technology both outran the possibility of fully recording it and stifled an interest in the details of its minor variations.

On the Segregation of Disciplines

The conscious separation and classification of an activity or viewpoint as science, technology, or art is recent and came about rather slowly. It is misleading to apply modern classifications to earlier periods in which distinguishable professions did not exist and a desired end result dominated over conscious particularities of method. Nevertheless, it is obvious from the above that I regard the somewhat less fully intellectualized activities of the technologist as having much in common with those of the artist and, until recently, interacting rather less with those of the scientist.

The Renaissance marks a natural interaction between a rejuvenated art and a beginning science.[62] In the fourteenth century many artists delighted in using their newly awakened powers of observation and their increased skill in representation to embellish the margins of manuscripts with precisely limned naturalistic living forms,[63] while at the same time they came to observe and emphasize the essential aspects of their subjects in a manner that later became appropriate for scientific illustration. Conscious studies of the interaction of light with matter and almost mathematical considerations of perspective are reflected in the mid-fifteenth-century paintings and sculpture by Ghiberti, Brunelleschi, and others, who both set the tone of the new times and absorbed its spirit. Fifteenth-century writing about art is very different in tone and intent from the earlier collections of pigment recipes or the practical how-to-do-it treatise of Theophilus. Stillman Drake[64] shows how the conflict of theory and experiment in sixteenth-century music contributed directly to the development of the style of Galileo and other great seventeenth-century scientists. Yet, as the different viewpoints that had been combined in the artists' activities came to be consciously realized, an inevitable result was that each of them should become a separate field of specialization.

The artist became an important individual, highly visible in society, while craftsmanship sank to a markedly lower status. Logical thought had always aided the artist in making his materials conform to his vision, but when the critical interplay between logic and experiment was consciously separated as a method of learning about the world, it became the new science, growing and changing beyond all recognition of its origins. For four centuries now it has outrun the other aspects of the artists' approach and has done so by exploiting the power of partial isola-

Figure 8.29
Interior view of workshop producing hammered copper vessels. From Duhamel du Monceau, "Description de la manufacture du cuivre de M. Raffaneau établie près d'Essone," appended to Galon, *L'Art de Convertir le Cuivre Rouge . . . en Laiton* . . . ([Paris], 1764).

tion. If mathematics could deal with music and perspective, it could also deal with falling bodies—but it could handle the planets better than a terrestrial feather, for it only applies to ideal isolated systems of simplified forces and bodies, one or two or at most a very few things at a time. Science in its very essence is simple. The new physics could deal with ideally elastic bodies, but it could do nothing with plasticity or with the host of other structure-sensitive properties on which the arts depend.

The geometry of perspective could be well handled by mathematicians, but the perspective of color could not be. The artist's intuitive knowledge of the psychology of perception has interacted strongly with science in the twentieth century, but for the most part science developed without art, and art was affected by science only through the changing world view that science promulgated or indirectly through the effect of science on technology.

The experience at moments of insight must be much the same among creative men of all kinds. However, the communication of new ideas, and especially their validation in terms that others will accept, is vastly different in different fields. As science and technology have become simultaneously broader in scope and more precise in individual purpose, their connection with art has become less and less apparent. Despite the austere and magnificent beauty of the order that is being uncovered by science, art has remained closer to technology than it has to science. As science has discovered the strength of simplicity, technology has become more complex. There is even a kind of aesthetic quality displayed by the interdependent relationships between the parts of an intricate machine, a complex process, or a large organization. Order per se is not art, and neither is complexity, but the finding of order in complexity is.

Looking back from the twentieth century, it is obvious that engineers, if not exactly aesthetes, have always had a rich and valid aesthetic experience in building their structures and devising their machines. A Newcomen engine at work with its massive rocking beam of oak mounted in a simple stone structure, with clanking chains and resonant iron bars, with its fiery furnace and jets of steam and its slow and irregular oscillation, *was* a work of art even if it was not consciously built as one or so appreciated in its own time. A modern artist, Garry Rieveschl, has proposed building a full-scale working Newcomen engine as a public monument, providing at once a reconstruction of a forgotten experience, a glance at a critical moment in technological history, and a reminder of the beauty and portentous quality of new contrivances.

In the theater, documentary drama is a similar art form based on a selective reconstruction of the past. Less conventional is the work of another Boston artist, Harris Barron, who has devised a performance which by effectively and unforgettably evoking the emotional experience of early aviators exemplifies the way in which art can extend human experience. Perhaps a poet on the first lunar landing would have done more for technology than an astronaut; certainly it will be poetic interpretations of space travel that will remain most in men's minds.

Symbolism in Art and Science

Both art and science are basically symbol-making activities, and both have the quality of yielding metaphors that match far more than their creators intended. The scientist's equations and the conceptual models on which they are based often relate to other parts of nature which are mathematically similar but physically unrelated. This relationship is matched by that fundamental evocative quality of art, in which relationships developed by the artist with one aspect of form in mind turn out to suggest many other things to the eye of a viewer who has had different experiences. The artist consciously exploits the similarities in shape, color, texture, orientation, or other qualities of things of quite different natures; in fact, if there were not some such resonance, the viewer of a picture would find little to hold his attention. The scientist finds that a few basic patterns reappear at different levels and in different systems, but this is mainly because the types of interaction between the few units with which alone he can deal are, after all, quite limited: simplicity and symmetry do not allow many alternative arrangements.

Historically, it is interesting to note occasions on which the decorative artist has developed designs that later were reinvented to represent important scientific concepts. One of the best examples of this is the use of circular mosaic tiles to build up two-dimensional polygonal patterns having all the characteristics of order, symmetry, and angular relationship between planes that are the basis of crystallography. This occurs in the Sumerian palace at Uruk, built in the middle of the fourth millennium B.C. (figure 8.30). Mosaicists ever since have been displaying examples of the combinatorial possibilities of simple geometric forms, none more magnificently than the Islamic tile workers of the fifteenth to seventeenth centuries A.D.[65] The qualities of the quincunx were appreciated and used in the design of ancient Persian, Greek, and Roman gardens. The crystal lattice dislocation, which was conceived in 1926 and has become extremely important in solid-state physics, was modeled much earlier in the fitting of medieval suits of mail armor, in the studded decoration of Japanese cast-iron tea kettles, and with slight distortion in innumerable other repetitive designs.

The best three-dimensional models of the close-packed-sphere arrangement of atoms on a crystal lattice occur in the famed granulation work of Etruscan goldsmiths in the sixth century B.C., though the technique was already 2000 years old at that time. Figure 8.31 shows an octahedral ear ornament composed of tiny gold spheres, made in Persia in the ninth or tenth century B.C. Curiously, none of the Greek atomists hit on the basic principle that these things illustrate, namely, that the mere stacking of equal isotropic spheres would give rise to the directional anisotropy of crystalline matter; and it was left to Johannes Kepler in 1611 first to publish this principle in a scientific treatise—if that is the proper term for his playful essay inspired by the hexagonality of the snowflake.[66] Though the symmetry (not always sixfold!) of the snowflake appears commonly enough on today's Christmas cards, its decorative qualities do not seem to appear in art until after its depiction in scientific works. In the Far East, window lattice patterns representing interfering ice crystals on a frozen pond are common,[67] but the earliest oriental use of true snowflake symmetry appears to be that on a Japanese swordguard made by Harukiro Hirata in 1828, obviously related to the drawings (figure 8.32) that were made by Toshitsura Doi under the influence of Dutch science and published five years later.[68] Figure 8.33 shows an elegant iron guard with snowflake design made

Figure 8.30
Inlaid mosaic decoration on columns at the palace at Uruk, ca. 3500 B.C. From the National Museum, Berlin. Photo courtesy of Bildarchiv Foto-Marburg.

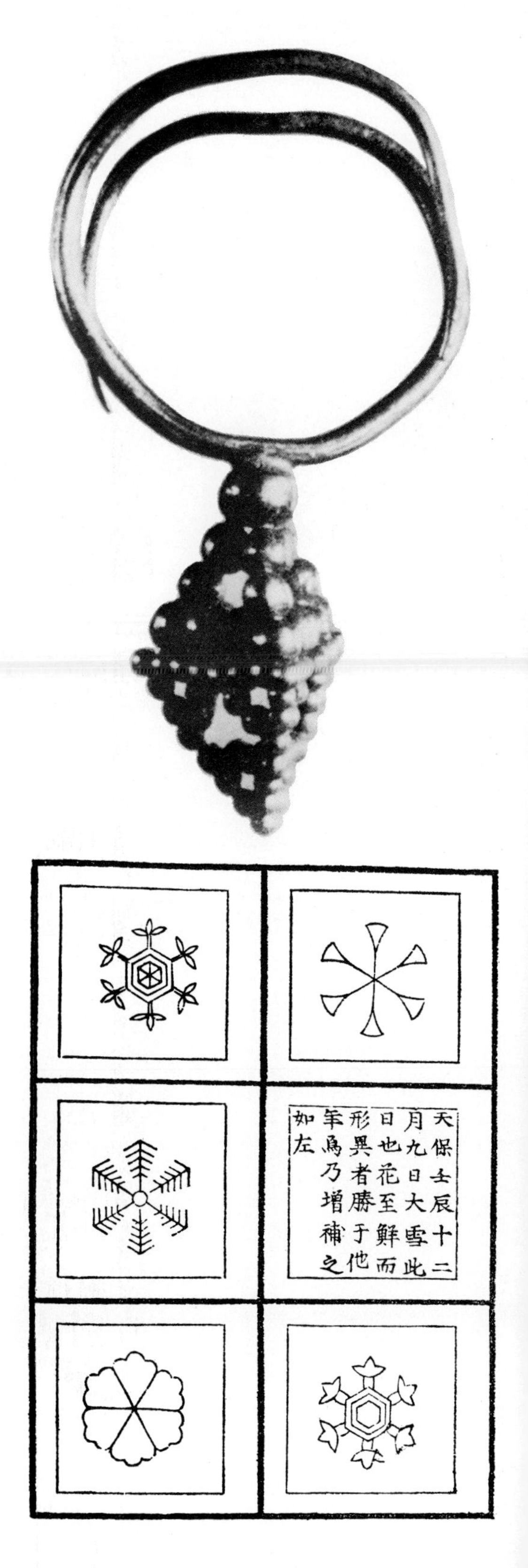

Figure 8.31
Gold earring in polyhedral form composed of gold granules accurately soldered together and unconsciously illustrating the concept of the crystal lattice. From Marlik, ca. 700 B.C. The granules in the top and bottom polyhedra are of different sizes, and their junction in the central plane illustrates an intercrystalline boundary. Photo courtesy of Iran Bastan, Tehran.

Figure 8.32
Drawings of snowflakes made with the "Dutch glass." From Toshitsura Doi, *Sekka zusetsu* (Tokyo, 1833).

Figure 8.33
Japanese swordguard with snowflake design. Iron with inlay. Goto School, ca. 1850. Photo courtesy of the Toledo Museum of Art.

somewhat later by a member of the famed Goto family.

Doi's drawings have a symmetry that is quite un-Japanese. Virtually every drawing of a snowflake that has been published, whether in a work of art or of science, depicts almost exact symmetry, reflecting the unwarranted but firm belief in the basic order of nature and the inability of the eye to see the unexpected. A glance at any photograph or, better, the flakes themselves will show many small differences between the six dendritic branches of even the best flake. And, of course, most snow falls as irregular aggregates displaying no symmetry whatever.

The stacked-ball model of the crystal lattice is in every elementary textbook today; yet it proved difficult to accept, and despite Hooke's elaboration of the idea and Huygens's very effective use of it in explaining the properties of calcite crystals, it virtually disappeared for two and a half centuries as scientists preferred the concept of elementary polyhedra and, later, more elegant mathematical abstractions.[69]

Among the innumerable geometric patterns painted on early pottery in most cultures, there are many reminiscent of the magnetic-domain patterns of today's solid-state physicist. A more recent example of an artist's prescience lies in the work of the Dutch artist Maurits Escher, whose experiments with space-filling and repetitive patterns later provided the illustrations for an introductory book on symmetry for students of crystallography[70] and who in 1942 illustrated color-group symmetries "well before official crystallography even thought about them" and quite independently of their mathematical treatment by Shubnikov. Islamic tile workers had used them earlier, however, notably in the Alhambra (figure 8.34).

In recent years there have been many exhibitions and books relating scientific photographs to abstract art.[71] One of my photomicrographs of a copper silicon alloy once hung in the Museum of Modern Art! A particularly interesting coincidence is in some of the paintings of Piet Mondrian which were later found to have almost exactly matched the microstructures of some cubic crystals containing randomly nucleated plates of a precipitated phase growing at right angles to each other until interference.[72] Such correspondence, of course, is only possible in a period in which artists are unconcerned with representation of the human world and are, for whatever reason, seeking a simplicity commensurate with that of the physicist. Perhaps, however, sculpture and paintings with human symbolism will some day be found to have played a similar role in connection with psychological science. Op art certainly belongs in laboratories studying the simpler aspects of the neurophysiology of perception.

Let us return to history. If the crystal lattice was slow of conception, the idea of the atom, of course, was not. Is there perhaps a connection between the use of pebble mosaics to depict human and animal figures in fourth century B.C. Greece and Leucippus's and Democritus's theories of matter? The concept that the distinguishing characteristics of matter arose in the shape, order, and orientation of parts in aggregation was certainly illustrated by the new mosaic forms, even if it was not suggested by them. At the present day, the printer's halftone is useful to illustrate information theory and discussions on structural hierarchy in matter.[73] And, again, in the same vein, is it absurd to suspect some connection between the revived receptivity to atomism at the end of the sixteenth century and the concurrrent interest in the fine structure of a work of art that accompanied the new graphic methods? A rapid improvement in the quality of metal engraving accompanied the making of niello prints. Shading in both a woodcut and an engraving depends upon the control of discrete, nearly invisible lines which build up to a recognizable body. The painter, with continuous gradation of darkening and lightening even within a single brush stroke, does not need this kind of analysis; neither does the goldsmith with his repoussé bas-relief. Woodcuts in the nonatomistic Orient, exploiting mainly a variable quality of line and texture, are basically different from Western ones.

Today and Tomorrow

After this excursion into some of the past interaction between art and technology, it is tempting to speculate on their joint futures. It is fashionable today to note the similarity between the artist's creative insight and that of the scientist, but for some reason the technical side of art has been downgraded as "mere" technique. Yet the handling of matter will always be necessary to give reality to the artist's all-important vision. Without it he cannot influence the minds and feelings of other people. Moreover, since technique relates more closely to the everyday experience of most men and women, especially when they are young, an interest in it can provide a path to the deeper meaning of art and lead to an understanding of things that the intellectual has never been able to communicate. The artist, if not every art historian, has al-

Figure 8.34
Mosaic tile work in the Alhambra at Granada, ca. A.D. 1325. Photo by Phylis Morrison.

ways known that technology is a basically important human activity.

The recent trend away from representational art in the Western world has, however, been accompanied by a perceptibly increased interest in techniques and materials and consequently by more widespread appreciation of the "minor" arts that make up such a major part of the archaeological record if not of art-historical writings.[74] This has inevitably been accompanied by an increased interest in Oriental art, for, while Western art tells mainly about individuals, ideas, and institutions, Oriental art (in accord with Oriental philosophy) tells more about nature—and it sensitively uses the properties of matter to do this. The subtle representations of the Chinese landscape painter arise from the properties of colloidal carbon and water interacting with the capillarity of various surfaces. The ceramist of the Far East can sometimes reveal the essence of things even better than can a painter because the ceramist's product constitutes in itself a direct example of the balance of natural principles, not merely a representation of them, selected and controlled by the potter just enough to invite appreciation by the human eye and hand.

The natural forces operating on matter are at last being consciously utilized in Western art: for example, Jēkabs Zvilna of Toronto has been producing two-dimensional patterns (visible only with optical enlargement) that result from the interaction of surface tension, shrinkage stresses, volatility, viscosity, and other physical forces working on selected substances under boundary conditions that he deliberately sets up. Though both the method and the results bear some relation to what ceramists have long done, especially in Japan, the nonnatural scale and the freer choice of materials gives his patterns a highly contemporary look. The recent "Kalliroscope" of Paul Matisse is a simple but perpetually fascinating device revealing shear gradients in a liquid containing floating micron-size reflecting platelets: patterns of endless variety and subtlety can be produced by controlling the conditions of turbulence or thermal convection, and close inspection even reveals to the naked eye the effects of molecular agitation (Brownian movement). Many other artists are experimenting with the aesthetic possibility inherent in systems in which the details are fully determined by physical forces but the boundary conditions are set by human intervention.

Many artists are currently exploring the properties of polarized light, of kinetic and balanced motion and flow, of simple magnetic interaction (figure 8.35), and of other phenomena which in the nineteenth century were used as rudimentary lecture demonstrations and laboratory experiments to evoke the interest of students in science.[75] It is high time that scientists admit that their experience in the laboratory is an aesthetic one, at times acutely so: the arid form of presenting their results has disguised this, and their respectable logical front often makes it invisible even to a student. The artist's interest in this aspect of science is very valuable. The introduction of scientific toys, under whatever name, to the general public and the opportunity to experience natural phenomena can only be applauded. The modern sculptor's skill in invoking viewer participation can aid enormously in the teaching of science, but his devices are often of such simplicity that the initial feeling of pleasure cannot deepen by repetition into a rich aesthetic experience.

Figure 8.35
Ballet magnêtique by Greek-American sculptor Takis (1968). The solenoid, intermittently energized, causes irregular movements of the two pendulum bobs, which are suspended on steel wires that strike the taut transverse wire, producing musical notes and dancing of the suspended rod at the left.

The visual excitement of the structures revealed by the microscope and electron microscope, of ion tracks in cloud chambers and interference patterns, has given rise to many fine exhibitions which have enriched the artist's vocabulary at the same time that they have heightened the scientist's sensibility. Yet it seems to me that in most of this the artist is just following others and is not fulfilling his particular role of revealing new significances in large, complex, perhaps social, patterns. Science is proliferating into more and more precise studies of more and more details. Higher energies beckon always away from the understanding of things on a human level to the smaller and simpler units of matter. So much knowledge has been acquired in this way that some scientists have claimed that no valid meaning can be established except by physical science. The most exciting frontier of biology has been on the molecular level, not life itself, which requires higher organization. After decades of neglect, however, something like old-fashioned natural history seems to be coming back into its own: the cell, biological form, and especially that comprehensive subject known as ecology, which is almost the art of science. Can the same thing happen in other areas? And can the artist, when he has learned some of the rudiments of science and technology, help?

Throughout history there has been a slow separation of art from the arts, and of science from both. As science became more definite, it became increasingly useful to technology, and it has given facility and precision to both the design and the control of processes. With art, however, the very utility of its contributions to industrial design and advertising seems rather to have forced out the one component of it that is most needed, and today we are faced with the curious phenomenon of art being mainly a comment and a much-needed protest rather than a constructive suggestion of a way toward deeper understanding. Artists have found much to interest them in both the scientific and technological world, and they have shown that there is much beauty even in things such as galvanized iron roofing and the intricacies of stairs and piping in a chemical processing plant, to say nothing of the elegant patterns of electronic gadgetry. The strength of steel and concrete and the beauty of a streamlined surface are proper aesthetic experiences in today's world, and they become more so as artists explore their meaning. Many sculptors have learned to enjoy the properties of steel and to exploit the cutting and welding torch in producing sculpture. The role of the artist in pointing to common things and making one pause to look at them has always been important. He now plays a similar role in relation to science, not only in finding the visual delights of the New Landscape[76] but also in calling attention to experiences of the other senses that are possible in a scientific or technological environment.

Technology is by its very nature complex and thus is incapable of being completely understood. There are two kinds of simplification that can make this complexity handleable. The first is the scientists' recognition of the units and their interaction on a small precise scale and the other is the recognition of the connectivity of units—which sometimes is systems analysis but more constructively is art. As technology has passed from the individual work of craftsmen to an aggregate of integrated systems, the significance of individual processes has been lost precisely at the moment that they become most efficient. The discovery of new techniques depends less than it once did on artistic curiosity but now occurs in well-financed research laboratories and is increasingly dependent on science. Yet does not the transition from craftsman to technologist itself suggest a new area in which the artist should play a role? The new level of complexity in technology requires a new level of art, perhaps almost a social one. Indeed, the artist is needed now as never before, and only by an introduction of the artist's general sense of relationships will it be possible to restore the balance between social and individual needs. At least some of the artist's work will be devising schemes in which the pleasure of an intensely individual experience can interact with that of others to produce a more viable society than at present. The artist can highlight discrepancies and point up problems that should be solved before they become generally obvious, while it is the job of the technologist to say how to solve them, and to do so.

Here it should be noted that there are more possibilities of diversity above human scale than below it, and the dangers of oversimplification in social matters are correspondingly greater than in the realm of physics and chemistry. The more that individuals are able to enhance their differences without loss of contact, the richer their lives will be. Technology at last makes real diversity possible, but democratic egalitarianism is in danger of eliminating it. Part of the artist's job will be to oppose oversimplification in this world of immensely diverse possibilities. But needed beyond all else is a more aesthetic feeling in the hearts and minds of technologists, who are so rapidly, at other peoples' behest, despoiling the Old Landscape.

Notes and References

1
Jean Paul Richter and Irma Richter, eds., *The Literary Works of Leonardo da Vinci . . .* (London, 1936), vol. 1, p. 119.

2
Ibid., vol. 1, pp. 25–26.

3
Robert Hooke, *Micrographia* (London, 1665), Preface (unpaginated).

4
Henri Focillon, *Vie des Formes* (Paris, 1947); English translation, *The Life of Forms in Art* (New York, 1948), passim, especially pp. 31–41, 76. The illustrations in the English edition are poorly selected to reinforce the author's points.

5
Arlette Leroi-Gourhan, "Le Neanderthalien IV de Shanidar," *Bulletin de la Société Préhistorique Française (Comptes rendus séances mensuel)* 65:79–83 (1968). See also *New York Times,* June 13, 1968.

6
Though blue frit is characteristically an Egyptian product, the earliest examples of it are two frit vessels and some seals and amulets from Mesopotamia, in the Tall Halaf levels at Tall Arpachiyah near Nineveh, a period which lasted from roughly 4900 to 4300 B.C. (M. E. L. Mallowan and J. C. Rose, "Excavations at Tall Arpachiyah 1933," *Iraq* 2:1–178 (1933–1935)). For a later appearance of it see Hans Wulff et al., "Egyptian faience: A possible survival in Iran," *Archaeology* 21:98–107 (1968). See also J. V. Noble, "The technique of Egyptian faience," *American Journal of Archaeology* 73:435–439 (1969), and C. Kieffer and A. Allibert, "Pharaonic blue ceramics," *Archaeology* 24:107–117 (1971).

7
H. J. Plenderleith, "Metals and metal technique," in C. L. Wooley, ed., *Ur Excavations,* Vol. II, *The Royal Cemetery* (London, 1934), pp. 284–310 and Plates 138 and 162. For a discussion of early metallurgy, see T. A. Wertime, "Man's first encounters with metallurgy," *Science* 146:1257–1267. The best comprehensive history of metallurgy is that by Leslie Aitchison, *A History of Metals* (2 vols., London, 1961).

8
Theophilus, *De Diversis Artibus,* manuscript treatise, ca. 1125 A.D. For an English translation see item 121 in the bibliography. Chapters 85–87 deal with bell casting.

9
Vannoccio Biringuccio, *De la Pirotechnia* (Venice, 1540). See the English translation (item 46 in the bibliography), pp. 255–260.

10
Noel Barnard, *Bronze Casting and Bronze Alloys in Ancient China* (Canberra and Nagoya, 1961); R. J. Gettens, *The Freer Chinese Bronzes,* vol. 2, *Technical Studies* (Washington, D.C., 1970). A superb collection of bronzes illustrating stylistic development was assembled for an Asia House exhibition in 1968; every item is described and illustrated in the catalog (Max Loehr, *Ritual Vessels of Bronze Age China* [New York, 1968]).

11
S. Delbourgo, "L'étude au laboratoire d'une statue découverte à Agde," *Bulletin du Laboratoire du Musée du Louvre* (1966), pp. 7–12; H. Lechtman and A. Steinberg, "Bronze joining: A study in ancient technology," in S. Doeringer et al., eds., *Art and Technology: A Symposium on Classical Bronzes* (Cambridge, MA, 1970), pp. 5–35; A. Steinberg, "Joining methods in large bronze statues," in W. J. Young, ed., *Application of Science in Examination of Works of Art* (Boston, 1973).

12
Hans Drescher, *Der Überfangguss* (Mainz, 1958).

13
R. J. Gettens, "Joining methods in . . . ancient Chinese bronze ceremonial vessels," in W. J. Young, ed., *Application of Science in Examination of Works of Art* (Boston, 1967), pp. 205–217. See also the references in note 10.

14
The history of the blowpipe has yet to be written. Blowing through pipes to urge charcoal (?) fires for smelting and melting doubtless preceded the use of bellows. The small mouth blowpipe with a lamp or candle was used by jewelers for local soldering operations and in the seventeenth century was suggested for testing ores. Comprehensive schemes of chemical analysis based on it were developed in Sweden in the last half of the eighteenth century but were slowly displaced, except in the field, by wet methods of analysis. Blowpipe analysis was regarded as an essential part of the training of a young chemist until very recently, and I still trace my feel for the nature of most chemical substances and reactions to my work with the blowpipe as a schoolboy.

15
Wulff et al., "Egyptian faience."

16
See chapter 7 in this volume.

17
Cennino d'A. Cennini, "On the character of ultramarine blue and how to make it," *Il libro dell'arte* [ca. A.D. 1400], ed. and trans. D. V. Thompson, Jr. (New Haven, CT, 1933), chap. 62. A more complete account of the flotation process was recorded slightly later in the Bologna manuscript reported and translated by Mary P. Merrifield, *Original Treatises Dating from the XIIth to XVIIIth Centuries on the Arts of Painting in Oil,* 2 vols. (London, 1849).

18
See item 50 in the bibliography; also Shirley Alexander, "Medieval recipes describing the use of metals in manuscripts," *Marsyas* 12:34–51 (1966), and "Base and noble metals in illumination," *Natural History* 74(10):31–39 (1965).

19
S. Giedion, *Space, Time and Architecture* (Cambridge, MA, 1953); Norman Davey, *History of Building Materials* (London, 1961); L. F. Salzman, *Building in England down to 1540* (Oxford, 1952); Marion E. Blake, *Ancient Roman Construction in Italy* (Washington, D.C., 1947).

20
A. G. Drachmann, *The Mechanical Technology of Greek and Roman Antiquity* (Copenhagen, 1963); B. S. Brumbaugh, *Ancient Greek Gadgets and Machines* (New York, 1966).

21
Alfred Chapuis and Edmond Droz, *Automata: A Historical Account and Technical Study* (Neuchatel, 1958). See also *Les Automates dans les Oeuvres d'Imagination* (Neuchatel, 1947), and several other works by Chapuis.

22
V. Gordon Childe, "Rotary motion," in Charles Singer et al., eds., *A History of Technology* (London, 1954), vol. 1, pp. 187–215.

23
The apogee of ornamental turning and its gadgetry is recorded in John Jacob Holtzapffel, *Turning and Mechanical Manipulation,* vol. 5, *The Principles and Practices of Ornamental or Complex Turnings* (London, 1884). The book by A. K. Snowman, *Eighteenth Century Gold Boxes of Europe* (London, 1966), illustrates innumerable surfaces, both enamelled and plain, whose decorative charm derives directly from engine turning. [In the Green Vault at Dresden there are several ivory pieces of unbelievable intricacy and precision that were turned by Jacob Zeller at the end of the sixteenth century (figure 8.10). It is not difficult to imagine that they were somewhere present in the background of both Kepler's planetary orbits and nested polyhedra and of Descartes's algebraic geometry based upon orthogonal coordinates.]

24
Robert S. Woodbury, *A History of the Lathe to 1850* (Cleveland, 1961); L. T. C. Rolt, *A Short History of Machine Tools* (Cambridge, MA, 1965).

25
James Mellaart, *Çatal Huyuk: A Neolithic Town in Anatolia* (New York, 1967).

26
T. F. Carter and L. C. Goodrich, *The Invention of Printing in China and Its Spread Westward,* 2nd ed. (New York, 1955).

27
The earliest description of typecasting is in Biringuccio's *De la Pirotechnia,* where it appears appropriately in the chapter on the pewterer's art.

28
Hellmut Lehmann-Haupt, *Gutenberg and the Master of the Playing Cards* (New Haven, CT, 1966). An excellent discussion of the beginning of the graphic arts is provided by the works of A. M. Hind, especially *A Short History of Engraving and Etching* (London, 1908), and *An Introduction to a History of Woodcut . . . ,* 2 vols. (London, 1935).

29
A. M. Hind, *Nielli, Chiefly Italian of the XV Century: Plates, Sulphur Casts and Prints Preserved in the British Museum* (London, 1939). For reproductions of other sulphurs and niello prints, see also John G. Phillips, *Early Florentine Engravers and Designers* (Cambridge, MA, 1955).

30
James G. Mann, "The etched decoration of armour," *Proceedings of the British Academy* 28:17–44 (1942).

31
Sidney Goldstein, in a private communication, reports that fragments of hammered gold that had unmistakably been cemented were uncovered in a 550 B.C. workshop site at Sardis. They were found during the 1968 campaign of the Fogg Museum at Sardis. The earlier Lydian coins were of unrefined electrum.

32
Emil W. Haury, "Etched shells," in H. Gladwin et al., eds., *Excavations at Snaketown,* Medallion Papers, no. 25 (Tucson, AR), pp. 148–153.

33
H. Lechtman, "Ancient methods of gilding silver: Examples from the Old and New Worlds," in R. H. Brill, ed., *Science and Archeology* (Cambridge, MA, 1971), pp. 2–30; and "The gilding of metals in pre-Columbian Peru," in W. J. Young, ed., *Application of Science in Examination of Works of Art* (Boston, 1973), pp. 38–52. Pre-Columbian gilding was done cold, not by hot cementation. Basic ferric sulphate, a natural mineral, will remove copper and silver from a dilute gold alloy, leaving a porous layer of pure gold which is easily consolidated by gentle heat or by burnishing. The presence of reserved areas of silver in "gold" objects from Peru has not previously been noted, although there have been speculations on bimetallic construction.

34
H. R. E. Davidson, *The Sword in Anglo-Saxon England* (Oxford, 1962). On the metallurgy of the pattern-welded blades, see chapter 1 in my *History of Metallography* (Chicago, 1960), and the references cited therein.

35
See chapter 3 in my *History of Metallography;* also C. Panseri, "Damascus steel in legend and in reality," *Gladius* 4:5–66 (1965). [A more recent reference is J. Piaskowski, *O Stali Damascenskiej* (Warsaw, 1974).]

36
See item 97 in the bibliography; also B. W. Robinson, *The Arts of the Japanese Sword* (London, 1967). [See also W. A. Compton et al., *Nippon-to, the Art Sword of Japan* (New York, 1976).]

37
H. Hedfors, ed. and trans., *Compositiones ad Tingenda Musiva . . .* (Upsala, 1932); Thomas Phillipps, "A manuscript treatise . . . entitled *Mappae Clavicula*," *Archaeologia* 32:183–244 (1847); Wilhelm Ganzenmuller, "Ein unbekanntes Bruchstrück der *Mappae Clavicula* aus dem Anfang des 9. Jahrhunderts," *Mitteilungen zur Geschichte der Medezin der Naturwissenschaft und der Technik* 40:1–15 (1941). For an English translation of the *Mappae Clavicula* see item 172 in the bibliography.

38
Conrad Kyeser, *Bellifortis,* facsimile, transcript, and German translation by G. Quarg, 2 vols. (Berlin, 1967). Versions of similar recipes, without, however, using distillation to make strong acid, appear in several fifteenth-century sources. The first printed account of etching is the anonymous Dutch *T. Bouch vā Wondre* (Brussels, 1513; reprinted with commentary by H. G. T. Frencken, Roermund, 1934). Next is the important little pamphlet *Von Stahl und Eysen* (Nuremberg, 1532), which was reprinted many times both by itself and with other materials in the series of *Kunstbüchlein* and other books of secrets. For a modern English translation see C. S. Smith, ed., *Sources for the History of the Science of Steel, 1532–1786* (Cambridge, MA, 1968), pp. 2–19.

39
L. Reti, "How old is hydrochloric acid?" *Chymia* 10:11–23 (1965); Sidney Edelstein and H. C. Borghetty, eds., *The Plictho of Gioanventura Rosetti* (Cambridge, MA, 1969); G. B. della Porta, *Magiae Naturalis Libri Viginti* (Naples, 1589). The 1658 anonymous English translation of the section "How to grave porphyr marble without an iron tool" is reproduced in Smith, *Sources,* pp. 37–38. See also Haury, "Etched shells."

40
See chapter 2 in this volume.

41
See Smith, *A History of Metallography,* pp. 63–65. A later description of the process in an American hardware catalog is quoted by H. J. Kaufman, *Early American Ironware, Cast and Wrought* (Rutland, VT, 1966).

42
Alexis [pseud.], *Secreti . . . del Alessio Piemontese* (Venice, 1555). There were innumerable subsequent editions and translations of this book, which is the most complete of all the early books of recipes for artists, craftsmen, and housewives.

43
For a history of nature printing see Ernst Fischer, "Zweihundert Jahre Naturselbstdruck," *Gutenberg Jahrbuch* (1933), pp. 186–213. [E. P. Newman, "Nature printing on colonial and continental currency," *The Numismatist* 77:147–154, 200–305, 457–465, 613–623 (1964), reports that Joseph Breitnall was making prints of botanical specimens of superb quality in Philadelphia from 1731 to 1742, and that Benjamin Franklin used nature prints of leaves on paper currency beginning in 1736.]

44
Carl von Schreibers, *Beyträge zur Geschichte und Kenntniss meteorischer Stein- und Metall-massen* (Vienna, 1820). See also Smith, *A History of Metallography,* pp. 150–156, and "Note on the history of the Widmanstätten structure," *Geochimica et Cosmochimica Acta* 26:271–272 (1962).

45
Alois von Auer, *Der polygraphische Apparat* (Vienna, 1853). This includes a portfolio of fine prints made by all methods of reproducing illustrations then known; several are of scientific subjects.

46
Norman Higham, *A Very Scientific Gentleman: The Major Achievements of Henry Clifton Sorby* (Oxford, 1963). For Sorby's work on steel, see Smith, *A History of Metallography,* pp. 169–185; A. R. Entwisle, "An account of exhibits relating to Henry Clifton Sorby," *Metallography 1963* (London, 1963); and papers by C. S. Smith, D. W. Humphries, and Norman Higham in *The Sorby Centennial Symposium on the History of Metallurgy* (New York, 1965).

47
Gustav Alexander, *Herrengrunder Kupfergefässe* (Vienna, 1927).

48
Thomas Hale, *An Account of Several New Inventions of Improvements Now Necessary for England . . .* (London, 1691).

49
See item 190 in the bibliography for a history of electrochemistry.

50
Alfred Bonnin, *Tutenag and Paktong* (Oxford, 1924).

51
R. A. F. de Réaumur, *L'Art de Convertir le Fer Forgé en Acier . . .* (Paris, 1722); English translation by A. G. Sisco (Chicago, 1956), pp. 340–359.

52
R. A. F. de Réaumur, "L'art de faire une nouvelle sorte de porcelaine, . . . ou de transformer le verre en porcelaine," *Mémoires de l'Académie des Sciences,* 1739 (published 1741), pp. 370–388. Réaumur's porcelain is extremely rare, but a box that seems to be made of it has been described by R. Strasser in the *Journal of Glass Studies* 9:118 (1967). See also chapter 7 in this volume.

53
Hans Sachs, *Eigentliche Beschreibung aller Stände auff Erden . . .* (Frankfurt, 1658), with illustrations by Jost Amman. Plates showing the profiled iron shapes and some new window designs that they made feasible appear in [Bullot] *Mémoire sur les Ouvrages en Fer et en Acier Qui Se Fabriquent dans la Manufacture Royale d'Essonne par le Moyen du Laminage* (Paris, 1753). For more detail see item 167 in the bibliography.

54
See chapter 5 in this volume.

55
A. J. Hopkins, *Alchemy: Child of Greek Philosophy* (New York, 1934).

56
Lynn White, Jr., "The iconography of Temperantia and the virtuousness of technology," in T. K. Rabb and J. E. Seigel, eds., *Action and Conviction in Early Modern Europe* (Princeton, NJ, 1969), pp. 197–219. See also White's important book, *Medieval Religion and Technology* (Berkeley, CA, 1968).

57
Three useful collections of paintings and other works illustrating technological scenes and devices are Heinrich Winkelmann, *Der Bergbau in der Kunst* (Essen, 1958); Vaclav Husa et al., *Traditional Crafts and Skills* (Prague and London, 1967); F. D. Klingender, *Art and the Industrial Revolution* (London, 1947). Many works on the history of technology reproduce artists' works as illustrative material. Emil E. Ploss, *Ein Buch von alten Farben* (Heidelberg and Berlin, 1962), as befits its subject, is an unusually fine mixture of historical and artistic material.

58
Wilhelm Treue et al., eds., *Das Hausbuch der mendelschen Zwölfbrüderstiftung zu Nurnberg . . .*, 2 vols. (Munich, 1965).

59
Helmuth T. Bossert and Willy F. Storck, eds., *Das mittelalterliche Hausbuch* (Leipzig, 1912).

60
See note 53 above. The same blocks were used in printing a Latin edition of Sachs's book in the same year.

61
John Read, *Prelude to Chemistry* (London, 1936), and *Humour and Humanism in Chemistry* (London, 1947). [See also E. E. Ploss et al., *Alchimia, Ideology und Technologie* (Munich, 1970).]

62
Giorgio de Santillana, "The role of art in the Scientific Renaissance," in M. Claggett, ed., *Critical Problems in the History of Science* (Madison, WI, 1959), reprinted in his *Reflections on Men and Ideas* (Cambridge, MA, 1968). See also H. H. Rhys, ed., *Seventeenth Century Science and Arts* (Princeton, NJ, 1961).

63
G. Evelyn Hutchinson, "Psychological and aesthetic factors in the progress toward realism, A.D. 1280–1480," paper presented at the Symposium on Art and Science held as part of the December 1968 meeting of the American Association for the Advancement of Science.

64
Stillman Drake, "Renaissance music and experimental science," *Journal of the History of Ideas* 31:483–500 (1970). [On visual perspective see Samuel Y. Edgerton, *The Renaissance Rediscovery of Linear Perspective* (New York, 1975) and the same author's forthcoming (1980) book discussing the influence of techniques for the accurate representation of objects on medicine, mechanics, and scientific thought. Reproductions of many of the early illustrations of mechanisms are given in Alexander Keller, *Theatre of Machines* (New York, 1956).]

65
Edith Muller, *Gruppentheoretische und strukturanalytische Untersuchungen der maurischen Ornamente aus der Alhambra in Granada* (Ruschlikon, 1944). [See also Keith Critchlow, *Islamic Patterns: An Analytical and Cosmological Approach* (London, 1976).]

66
Johannes Kepler, *Strena seu de Nive Sexangula* (Frankfurt, 1611); English translation by Colin Hardie (Oxford, 1966).

67
Daniel S. Dye, *A Grammar of Chinese Lattices*, 2 vols. (Cambridge, MA, 1937).

68
Toshitsura Doi, *Sekka zusetsu* [Illustrations of snow crystals] (Tokyo, 1833; suppl., Tokyo, 1840). Both sections were reproduced with extended commentary and a summary in English by Teisaku Kobayashi (Tokyo, 1968).

69
John G. Burke, *Origins of the Science of Crystals* (Berkeley, CA, 1966).

70
Caroline H. Macgillavry, *Symmetry Aspects of M. C. Escher's Periodic Drawings* (Utrecht, 1965); M. C. Escher, *Grafiek en tekengen* (Zwolle, 1960); J. L. Locher, ed., *The World of M. C. Escher* (New York, 1971).

71
See, for example, Gyorgy Kepes, ed., *Structure in Art and Science* (New York, 1965); Philip C. Ritterbush, *The Art of Organic Form* (Washington, D.C., 1968); Georg Schmidt, ed., *Kunst und Naturform: Form in Art and Nature* (Basel, 1960); Paul Weiss, "Beauty and the Beast: Life and the rule of order," *Scientific Monthly* 81:286–299 (1955); Lancelot L. Whyte, ed., *Aspects of Form* (London, 1951). [For a somewhat different approach see item 196 in the bibliography.]

72
R. W. Cahn, "Art in science, science in art," *Museum* (UNESCO) 21:16–21 (1968).

73
Paul A. Weiss, "$1 + 1 \neq 2$," in G. C. Quarton, T. Melnechuk, and F. O. Schmitt, eds., *The Neurosciences: A Study Program* (New York, 1967), pp. 801–821.

74
It is interesting to note that the published catalogs of three recent exhibitions have long technical introductions: Herbert Hoffmann and Patricia Davidson, *Greek Gold* (Boston, 1965); Dominique de Menil, *Made of Iron* (Houston, 1966); David G. Mitten and Suzannah H. Doeringer, *Master Bronzes of the Classical World* (Cambridge, MA, 1967).

75
Jack Burnham, *Beyond Modern Sculpture* (New York, 1968).

76
Gyorgy Kepes, *The New Landscape in Art and Science* (Chicago, 1956); see also his edited collection *Structure in Art and Science* (New York, 1965).

9

Metallurgical Footnotes to the History of Art

I do not apologize for the narrowness of my topic, or even for the parochial point of view, though I do acknowledge that my emphasis is not on the most important aspect of art and I doubtless expose ignorance of some of the deeper aesthetic and social significance of the works examined. However, art historians and critics often disregard some things that contribute substantially to the aesthetic appeal of the objects they discuss, and I simply wish to redress the balance, to emphasize for the moment the artist's physical experience in shaping a work of art, without in any way denying the overwhelming importance of what can loosely be called his vision. Even the decorative arts involve far more than technique. Nevertheless technique is an essential aspect of any work of art from a trivial trinket to the greatest painting, and some specialized study of it is essential to full appreciation. Moreover, a study of the relation of parts to wholes on any scale or within any special area of human experience can in some measure suggest a metaphor for all. An appreciation of form is common ground to scientist and artist alike.

The enjoyment of a work of art or an aesthetic environment has to be an internal experience of an individual; yet the copious literature on art history shows how this enjoyment is enriched by ancillary knowledge: the personality of the artist, the sources which suggested his forms and relationships, his iconographic vocabulary, and the constraints applied by his environment, including social factors such as the wealth and wishes of his patron. Except in the field of the graphic arts, however, historians have said astonishingly little on techniques. Though museum labels and catalogs refer to materials and processes—"bronze," "fresco," "parcel gilt," "tempera," "lacquer on wood," and so on—they usually display only superficial attention to the essential details of the artist's technique. It may, therefore, not be entirely inappropriate for a technician of another kind to point out and for the moment to overemphasize the contribution that is made to a work of art by the structure and properties of the materials themselves. Not only does the behavior of matter quite significantly modify the aesthetic quality of the final work but it is a major part of the experience of the artist himself as he shapes it. His fingers physically interacting with matter can no more be neglected than can his emotions or his thoughts on spatial arrangement or symbolic meaning.

Nearly all the industrially useful properties of matter and ways of shaping materials had their origins in the decorative arts. Indeed, prior to the twentieth century, few people except those engaged in aesthetically motivated play were likely to make discoveries; when faced with urgent need for food, shelter, or weapons, a man does not plant flowers, or fool around to see what new colors appear when mixed minerals are heated, or use electricity to copy medals in glittering silver. Yet the cultivation of flowers seems to precede agriculture and the making of ornaments from copper and iron certainly precedes their use in weaponry, just as baked clay figurines come before the useful pot. Alloys come from jewelry and the metal-casting industry began as sculpture. Steel rails are rolled in a mill of the type that was first devised for shaping the lead cames for stained glass windows. Mass production had its origins in the art industries. The first suggestion of anything really new seems always to be in an aesthetic experience, but its development, of course, is purposeful and is determined by interaction with society. But I have discussed this in detail in another place:[1]

the present paper is to examine art rather than industry, and to study some of the ways in which the properties of materials contribute to it.[2]

The work of an artist in getting the details that he wants is greatly facilitated if he selects a material whose inner nature makes it want to take the desired shape. The forces between invisible atoms are strong enough and sufficiently diverse to influence many visible phenomena. In selecting some technique and material to make his concept communicable to others, the artist's past experience with the behavior of matter has already modified what he can envision, and, as he works, the experience of the material under his fingers subtly interacts with the idea in his mind to give the finished work some quality that was rarely fully anticipated. A few artists seem to have such a feeling for their materials that the prevision needs little modification; most say that the idea grows as they work experimentally. The interaction between their ideas and real matter is, indeed, rather like the interchange between scientific theory and experiment: matter in a preset chosen environment acquires a visible form and thereby gives validity to an attractive but at first only tentative idea.

The material is, of course, not determinative, only interactive. It must be compatible with the artist's intent, but it can modify it only in details. The magnificent concepts of, for example, Michelangelo's *Pietà,* El Greco's *Burial of the Count of Orgaz*, or Rembrandt's *Christ Healing the Sick* are so far above any thought of technique that it seems almost sacrilegious to point out that these works did not become visible without down-to-earth chipping and abrasive working of marble, without laborious chemical and mechanical operations on pigments and vehicles, and without the application of unpleasant acids to a stickily coated plate of base metal. The technique rarely controls the major tangible outlines of a work of art, but intangible quality is strongly influenced by the properties of matter, its internal or superficial reaction with light, and its ability to acquire gradients of shape both on a nearly microscopic scale and in gross stability. The artist cannot avoid being intimately and sensually aware of this relationship as he works, even if he had not intellectualized it from the beginning.

The shaping of matter on a large scale can be arbitrarily done by man; on a very small scale he has no influence. On the scale just visible to our senses there is balance, and the very nature of art lies in such balance.

When closely examined, the minute effects due to the material are seen to be just as dependent upon the relations between parts and wholes as is the aesthetic quality of the entire work on the scale at which it is intended to be viewed. Both are ultimately structural, but the relevant structure of the material is on a scale that is not resolved by the naked eye, or only marginally so. It may be a structure on the level of the atom (as in the case of color); on a microcrystalline level (as in the texture of metal, marble, or granite or the reflectivity of pigments); cellular (as in wood); or molecular (as in pigment vehicles or lacquer). Sometimes quality lies in microparticulateness, as in Chinese ink-brush paintings; sometimes, often in fact, in the balance between surface-dependent and volume-dependent properties, most simply surface tension and viscosity. Like the bare content of iconography, the techniques must be transformed and joined by the artist's genius, but they cannot be ignored. It is the main theme of the present paper that man's experience with the properties of matter, his technology, is just as much a part of human history as are the things that are dealt with by historians of politics, society, ideas, and art. Materials are discovered, accepted, or rejected for reasons as complicated as those behind any other facet of culture.

Our examples will be mostly from the decorative arts because in them the qualities of the materials are often central to the artist's aims. Moreover, there has not always been the distinction that we make today between the fine and minor arts: witness medieval European ecclesiastical metalwork or ceramics in Sung-dynasty China. The preoccupation with easel painting is an ethnocentric phenomenon, mainly confined to a post-Renaissance Europe and its derivatives.

If in the following I overemphasize the Orient, this is simply because in the Far East the properties of materials are a little nearer to the surface, a little more consciously a part of what the artist is trying to show. The naturalistic aspects of Oriental philosophy encourage a sensitivity to the quality of materials—or is it the inverse, that an early enjoyment of stone, wood, clay, and fiber gave rise to the philosopher's perception of the soul in all natural things comparable to that in man himself? Westerners tend to override materials, usually in ignorance, but sometimes proudly as a *tour de force*.

Surface Tension Generates Pleasing Forms

Perhaps the most important properties of matter from our point of view (after its basic ability to aggregate) are those that depend upon the nature of the two-dimensional interface between different phases, whether solid or liquid or gas. Everyone knows that oil and water do not mix. The Japanese swordguard-maker's effect known as *gama-hada* (figure 9.1) depends on the fact that metallic iron and silver also do not. Molten silver on a clean iron surface does not spread over it, as would molten copper, but takes the natural form of a segment of a sphere with a definite angle of contact against the underlying surface (figure 9.2). Physically, this simply results from the equilibration of the local surface energies according to the well-known equation:

$$\cos \theta = (\gamma_s - \gamma_{sl})/\gamma_l,$$

θ being the contact angle, γ_s and γ_l the surface energies of the solid and the liquid surface respectively, and γ_{sl} the energy of the interface between solid and liquid.[3] With liquid silver on solid iron under a flux, the angle turns out to be not much below 90 degrees, for the difference in energy between the iron surface and the silver-iron interface is small (i.e., iron and silver neither interact nor reject each other strongly). The silver droplets in the lower part of figure 9.2 contain about 10 percent copper, which lowers the angle perceptibly. If much more copper or a little zinc had been alloyed with the silver, it would have spread avidly over the surface of iron, but the surface tension of the chosen alloy pulls each drop into its individual shape, preserving in the final object the pleasant pattern of randomness in which the silver lumps were first applied.

To the artist, surface or interfacial tensions are properties quite as important as are color or refractive index. In fact, they are more basic, for being interactive, they influence directly the shaping of larger structures and lead to more diverse effects. Surface tension becomes relatively more dominant the smaller the scale, and surfaces mainly shaped

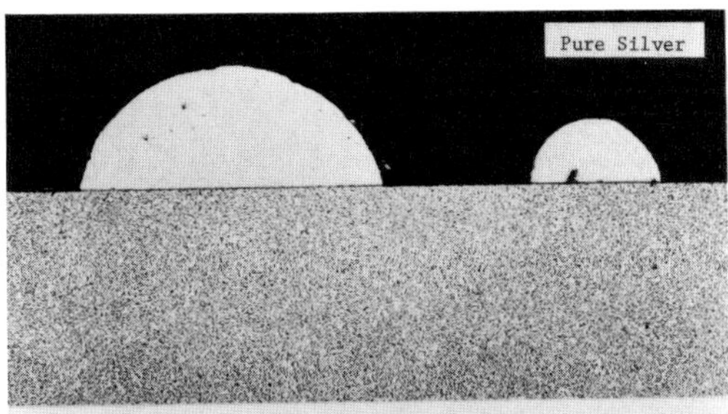

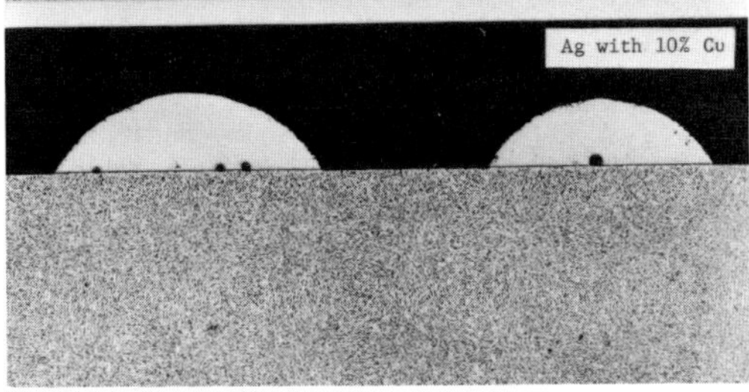

Figure 9.1
Japanese swordguard with *gama-hada* decoration. Signed Toshihiro. Early nineteenth century. Height, 6.8 cm. Photo courtesy of the Ashmolean Museum.

Figure 9.2
Section though silver drops melted on a flat surface of iron (mild steel). × 11. *Top:* Pure silver. Contact angle about 80 degrees. *Bottom:* Silver with 10 percent copper. Contact angle about 60 degrees.

by it invariably display a relationship of parts to wholes that exemplifies art itself.

Nowhere is surface tension used to better effect than in the famed Etruscan gold granulation work of the seventh century B.C. (figures 9.3 and 9.4). Here, tiny spheres are first made by melting gold clippings or filings, dispersed in charcoal powder, in order to let surface tension convert them into perfect little spheres. They are on the verge of invisibility, typically about a quarter of a millimeter in diameter, too small for gravity to flatten them perceptibly. These are then applied to a smooth gold surface in the desired patterns (often highly geometric ones prescient of a modern crystallographer's model, but in the fibula of figure 9.3 the granules are random in size and do not settle into close-packed order), then secured by a nearly invisible solder which is applied in just enough quantity to make the surface wet so that capillarity can carry it into the narrowest crevices where the tiny spheres touch the base. The solder was not applied as granules or clippings of a low-melting alloy as is usual today, but rather it was made in situ by reaction between a very thin layer of a paint containing finely ground copper oxide or carbonate which was reduced in the goldsmith's fire to metallic copper which alloyed with the underlying gold to give a thin layer of liquid.

[The Irish gold lock rings, no. 11 in the catalog of the exhibition *Treasures of Early Irish Art* (New York, 1977), have a superb surface quality produced by a related technique. Examination under a microscope shows that, instead of the usual precisely shaped spherical granules, extremely fine irregular filings of gold were randomly distributed over the surface of the wires and invisibly soldered in place. The rings were made about 700 B.C.]

Figure 9.3
Gold fibula with granulation. Etruscan (Vetulonia?). Ca. 650 B.C. Detail, 3.4 cm wide. Photo courtesy of Tony Hackens and the Rhode Island School of Design.

Figure 9.4
Microscopic view of the granules shown in figure 9.3. They average about 0.18 mm diameter. × 6.

Of course, there are no contemporary written descriptions of this process, but the physical details of surviving objects agree with it and the Greek word *chrysocolla* for the mineral malachite (copper carbonate) clearly implies that it was used to solder gold. Recipes for the process are in the eighth-century *Mappae Clavicula*,[4] and full realistic detail is given by both the marvelous Theophilus[5] (twelfth century) and by Cellini.[6] The last refers to it, not as soldering but as firing in one piece:

> Be ready at hand when the charcoal is beginning to glow and your work is growing fire-coloured, to blow wind over it with your bellows very skillfully and very evenly, so that the flames may play all round it alike. . . . Watching with care you will see the outer skin of gold begin to glow and then to move; as soon as you note this, quickly take a brush and sprinkle a little water on your work, which will there and then be beautifully soldered without any need of special solder being applied to it. And this one might call the first firing.
>
> Indeed, the first soldering ought not to be called soldering at all, but rather firing in one piece, because there is so much virtue in the verdigris when combined with the salts of ammonia and the borax, that it only moves the outer skin of gold, and so amalgamates it together that it all grows to one even strength . . . . (English translation, p. 46)

For some reason the method seems to have been forgotten and it was rediscovered and even awarded an English patent in 1937!

Without surface tension any kind of soldering or welding would be impossible. It also gave shape to the earliest coins (figure 9.5), which are nothing but sessile drops of gold or silver made by melting together fragments of metal selected to give the correct weight, finally authenticated with a punch. Surface tension operating in conflict with a solid oxide skin also gives rise to sophisticated surface effects on metals such as the *samorodok* ("nugget") finish used by the Russian goldsmith Fabergé. It is surface tension which makes oil paints stick to canvas, and surface tension which carries Chinese ink into the finer interstices of the paper against the weaker capillary forces of the brush. Differential wetting by pitch and water of the minerals in crushed lapis lazuli was the basis of the preparation of the superb ultramarine blue of the medieval and Renaissance miniaturist. The fact that the interface between crystals of solid calcite and selected pigments has a lower energy than that of the crystals against water or air ensures both the brilliance of a fresco painting and its permanency. It was also capillarity that in the recent flood drew up salty water through the frescos of Florence and (carrying it to the underside of growing crystals of gypsum or salt) destroyed the frescos, just as it is capillarity that carries corrodants into the interior of decaying bronze sculpture.

Surface Tension in Ceramic Glazes

It is in ceramics that surface tension produces the most interesting effects. The smooth perfection of the glaze on Ming white porcelain is simply the frozen surface of a liquid that could take a minimum-area configuration— though it should be noted that the incompletely smoothed, slightly granular, surfaces of the earliest glazes are more attractive than the later perfect ones. Many potters let surface tension and gravity contend with viscosity to shape their glazes, producing gorgeous gobs of color dripping down the sides of a jar in a naturally defined shape. Much ceramic decoration comes, however, from the pattern in which the glazes or pig-

Figure 9.5
Persian *siglos,* fifth century B.C. Maximum diameter, 1.6 cm.

ments are consciously applied by the potter, with only enough flow or diffusion to modify harsh edges.

Some of the more intriguing effects arise from the local lowering of surface tension in a glaze, allowing the adjacent high-surface-tension material to pull away in the manner of the locally thinned layer of grease over a hot spot in a frying pan, or the alcohol-depleted ring above the main meniscus in a cocktail glass. This, I believe, is the reason for the attractive yellow-green dappled lead glazed ware which is one of the finest products of Tang China (figure 9.6). Occasionally, one can see some residual material unfused in the center of a spot (figure 9.7). The effect can be approximately duplicated by the use of a little soda applied locally to a uniform layer of a lead glaze containing iron oxide to color it.

The opalescent jadelike quality of Sung celadon and the superb early Koryu celadons comes from myriads of tiny bubbles, each shaped by surface tension to scatter light incident from any angle (figure 9.8). Contrast this, on the one hand, with the uniform white of a marzacotta glaze with invisibly small tin oxide particles or, in another medium, the star sapphire which contains tiny crystals shaped by an anisotropic interface energy to give its oriented spatial glow.

Another use of surface tension in ceramics is of great historical interest, for it is the basis of the blue-glazed silica frit ware (often called faience) of Egypt. This first appeared in Sumeria shortly after the middle of the fifth millennium B.C. It was the first fused glaze to be applied to a ceramic body and the direct antecedent of glass. This ware has puzzled generations of archaeologists and art historians, for there is no trace in the glaze of support marks as there is in all other glazed ware: it seems to have been suspended in the

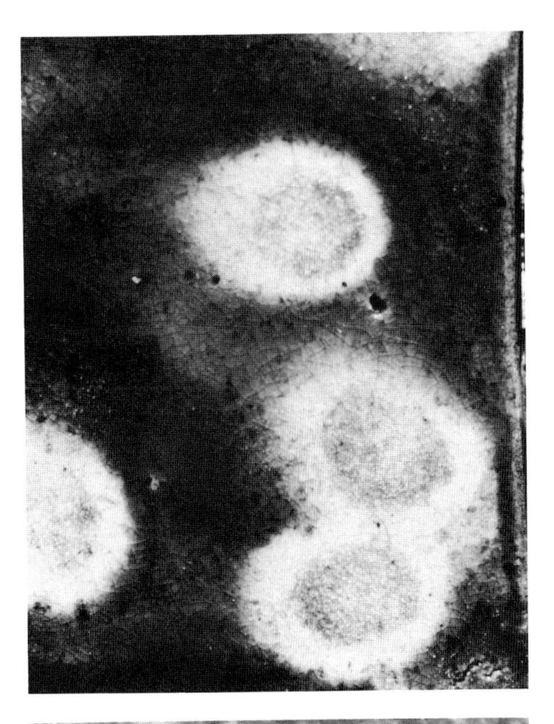

Figure 9.6
Ceramic pillow with dappled green and brown glaze. Tang dynasty or later. 12.5 × 9.9 × 6.0 cm. Photo courtesy of the Museum of Fine Arts, Boston.

Figure 9.7
Detail of clear spots in dappled ware. × 2.7.

Figure 9.8
Glaze on Korean celadon bowl (Ryuun-ri kilns, eleventh century A.D.). Magnified to show bubbles. × 50.

kiln by levitation—as indeed it was! The late Hans Wulff, who added immeasurably to our understanding of early technology, found the process still extant in Qum, Iran, where it is used to make donkey beads and other humble items. This is a remarkable survival of an early technique, and no hint whatever of its true nature had found its way into scholarly literature before the publication of his paper.[7] It is quite unlike the later and almost universal glazing process in which the ingredients are premixed and applied to the ceramic surface by dipping or brushing to be subsequently fused in the kiln with only minor changes of composition. The Qum beads are modeled from crushed quartzite mixed with an aqueous adhesive, then embedded in a mixture of lime and alkaline plant ash, with copper oxide for color, and heated to promote the formation of a reaction layer of sodium silicate glass containing about 73 percent silica. This composition is molten at 800°C and it wets the silica but does not wet the lime, so that the beads are virtually levitated by surface tension in the midst of the bed of powder without sticking to it (figure 9.9). The same principle was later used in making the first hollow objects of glass, the so-called sand-cored glass vessels, but the "sand," of course, was a nonwettable material, probably lime, certainly not silica.

Figure 9.9
Breaking open a pot of blue-glazed silica frit beads (Qum, Iran). The beads are about 1.5 cm in diameter. Photo by Hans Wulff.

Shortly after Wulff's paper appeared, Kieffer and Allibert described an independent duplication of the process in the laboratory and gave a dramatic photograph of a self-glazed "Egyptian" hippopotamus emerging from its calcareous blanket.[8]

Physically related to this is the texture of the Shigaraki ware of Japan.[9] This utilitarian, unpretentious, but in its early forms surpassingly beautiful ware frequently has areas in which wood ash from the fire has been deposited and fused onto the clay to give a wonderful rough irregular texture with white spots. Close inspection shows these spots actually to be small cavities surrounded by a raised ring of slightly cloudy glaze. Figures 9.10 and 9.11 are a general view and detail of a late example with a rather self-conscious spottiness. The earlier ware is far superior aesthetically, for in this the potter set up the conditions generally favorable to the effect but could not precisely control the environment within the kiln and patterns resulted from the interaction of gradients of many different physical and chemical factors effective at different scales. The clay used by the Shigaraki potter contained fragments of feldspar one to three millimeters in size. In the rare combination of circumstances where one of these particles lies at the surface of the clay and there is sufficient alkali present from the fortuitous deposit of plant ash, a fluid silicate is formed by reaction and the higher surface tension of the material with less feldspar in it pulls it away into the delicious rings.

The chemical basis of color in stained glass and ceramic glazes is well known. It has been well summarized by Hetherington,[10] but the physics behind the effects has been less appreciated. Thus, though the separation of crystals of iron oxide in the famed oil-spot *temmoku* ware (figure 9.12) of Sung dynasty China has been mentioned, not everyone who enjoys it realizes that the crystals remain visible at the surface because their surface energy against air is less than the interface energy between them and the underlying glaze. The sequence of crystallization of the glaze probably begins by the separation of silicate and rejection of iron. The iron, concentrated in the residual glassy phase, diffuses over long distances to wherever it can find a nucleus on which to

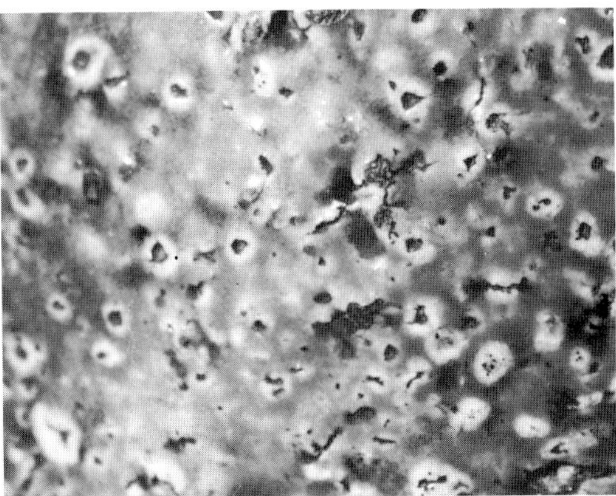

Figure 9.10
Salt jar, Shigaraki ware, nineteenth century. Height, 9.1 cm. Photo courtesy of the Smithsonian Institution, Freer Gallery of Art, Washington, D.C.

Figure 9.11
Detail of jar shown in figure 9.10. × 3.

Figure 9.12
Bowl with "oil-spot" *temmoku* glaze. Sung dynasty. Diameter, 12 cm. Photo courtesy of the Fogg Art Museum.

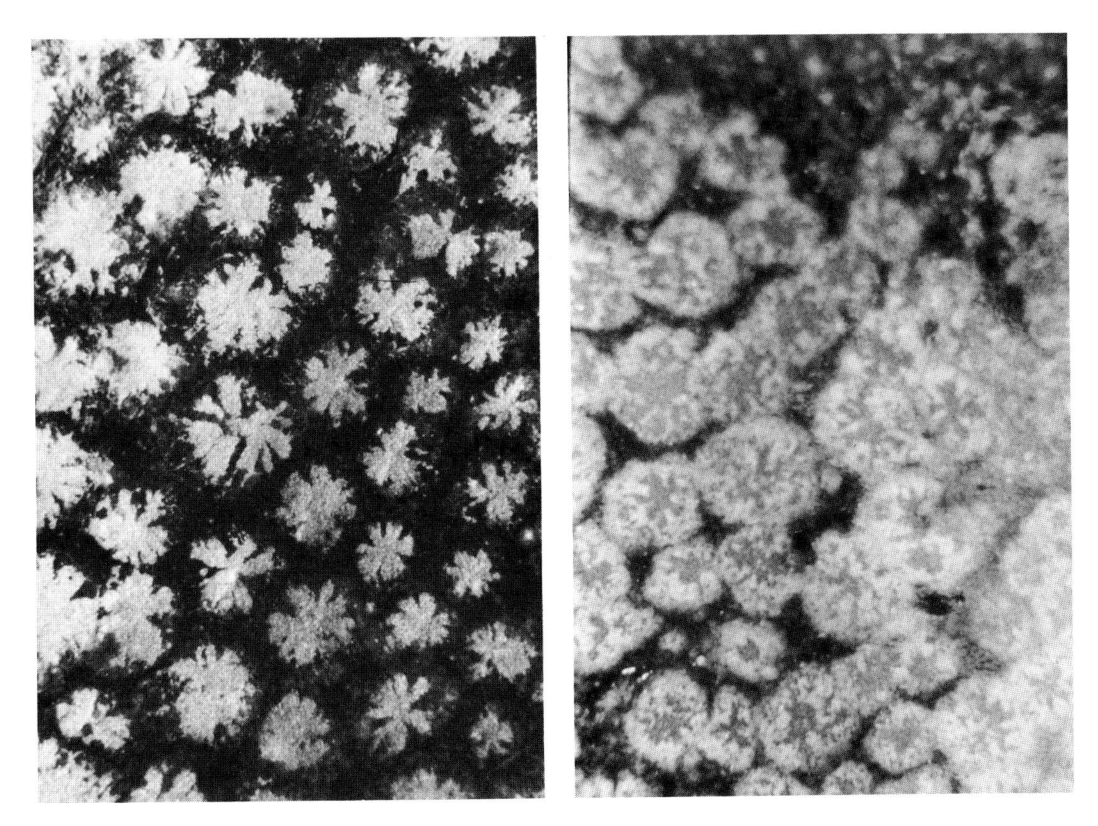

Figure 9.13
Detail of spots in *temmoku* glaze, with illumination reflected from surface. × 7.

Figure 9.14
Same as figure 9.13, under oblique illumination. × 7.5.

crystallize, and this is far more likely to be at the surface. Figure 9.13, taken with light reflecting from the surface, shows the rafts of iron oxide crystals, cracked peripherally as a result of shrinkage, and figure 9.14 is a similar area under oblique illumination to emphasize the light-colored clusters of silicate crystals beneath. The beauty arises from the physical processes, and the skill of the potter lies in setting up the conditions so that they can operate.

Hetherington has suggested that the bursting of bubbles is involved in the formation of these spots. Though bubbles often do produce interesting effects, I see no microscopic evidence for their being centrally involved in the usual oil-spot *temmoku* ware. On the other hand, the very rare *Yohen temmoku* bowls have a pattern of shaded rings that strongly hints at a thin layer of a liquid or a thin slurry containing a few large air bubbles laid on the smooth surface of a prefired glaze.[11] The bubbles were clumped by the surface tension of the liquid, which was subsequently dried to leave a deposit on the underlying glaze; on refiring, this caused devitrification and whitening of the iron-bearing glaze in direct proportion to the local thickness of liquid between and around the bubbles. The spots in Yohen ware are only on the inside of the bowls, indicating manual application by the potter in contrast to the thermal crystallization cycle that caused the random nucleation of the "oil spots" of the normal type.

Molten Metals and Mold Surfaces

All shaping of matter involves a competition between surfaces—surfaces sometimes deformed by their own interactive forces, sometimes by pressure transmitted from elsewhere and concentrated or dispersed by a tool. In shaping metal with a hammer, punch, or chisel hard steel is matched against metal which is soft either because it has been heated to a plastic condition or because it is by nature soft at room temperatures. The tool itself, however, had been shaped when it was hot and soft, or when it was cold by competition with something still harder, an abrasive. Both temperature change and chemical treatment can drastically alter the nature of materials by causing internal structural change, and so affect their ability to shape or be shaped.

Casting—the metalworking process that has been most closely associated with works of art—makes use of the easy shapability of moist clay and sand, combined with their refractoriness which makes them at some high temperature more rigid than the metal whose pressure they are to resist. Sufficient heat makes metal first soft, then liquid, but it hardens clay. Here, too, surface forces dominate. The same surface tension which joins glaze to the clay body of a ceramic pot prevents the adhesion of metal to a clay mold. Water is drawn into porous sand or clay by very high capillary forces; metal is pushed out of the same interstices with even greater force. The same principles are used by the lithographer who draws oil-attracting lines on a generally oil-repellent surface.

Casting is in its very essence a replicative process. Something shaped in one material is copied as a negative impression in another and replicated positively in a third. The final work reflects some of the qualities of the materials of the original model and of the mold as well as of the metal of which it is composed. Of all methods of shaping, casting is at the same time both the most and the least dependent on the individual properties of materials.

Casting by the Lost-Wax Process

One of the most intriguing casting techniques is also the earliest one used to produce any but the simplest of castings—the lost-wax method. This exploits the easy shapability of wax to produce a model of any desired shape, which is then embedded in water-softened clay. By heating, the wax is melted and burnt out to leave a cavity in the hardened clay into which molten metal at a higher temperature can be poured and allowed to solidify in a shape that duplicates that of the original wax in all its detail.

The example shown in figure 9.15 is the famous censer at Trier, made in the twelfth century. I have chosen this less for its beauty and its meaningful symbolism than for the fact that there is a detailed contemporary written description of making something very much like it. This is in the treatise *De Diversis Artibus* by the pseudonymous Theophilus, who was a Benedictine monk writing at the end of the first quarter of the twelfth century. He gives full detail of all the tools and operations in the crafts involved in the embellishment of God's house, painting, glass, and metalwork, the last being in most detail.[12]

Theophilus begins his chapter on the cast censer thus: "Take some clay, mixed with dung and well kneaded, and let it dry in the sun." He then tells how to shape cores from this clay for the upper and lower parts of the censer. These are mounted on a wooden axle and turned in a simple lathe to a stepped cylindrical shape, which is then cut into a cruciform section: "The three sides of each arm of the cross should be one and a half times the width, and on it you should also shape gables like roofs."

Theophilus goes on to tell how to roll a sheet of warm wax on a board, "first sprinkling water to prevent its sticking," and cut out pieces to fit the various areas on the core, applying wax all over and sealing the seams with a hot iron tool. He then cuts out the window and arches and "with a thin tool shaped like an arrow" carves and inscribes all the details, including figures of the apostles, the prophets and the Virtues. The work in wax is finished by the attachment of wax rods to form the gates and air vents, and the whole is then covered with a thin layer of fine clay. This is dried in the sun and two more layers applied and dried. In a charcoal fire the wax is then melted and poured out, and the mold heated until it is glowing hot throughout. Meanwhile, a crucible full of brass is melted in a forge fire urged with bellows and, when all is ready, the mold is quickly buried in a trench and the metal poured into it and allowed to cool. As a practical foundryman, Theophilus knows that castings do not always turn out well, and he adds instructions on how to replace a defective part by running on new metal in a local mold and, if need be, fitting patches and attaching them by soldering.

Even more vivid is Theophilus's account of the casting of large bronze bells, with a lathe to shape the core and cope, with procedures to lower the heavy mold into its place in the casting-pit and a detailed account of the furnaces for melting tons of metal, with strong men blowing bellows, and green flames playing above the furnace when the metal is hot enough. At the moment of casting, he says, "the work requires not lazy workmen, but agile and eager ones, lest someone's carelessness should lead to a mold being broken or one workman getting in the way of another or hurting him or provoking him to anger, which contingencies must be avoided at all cost." As the metal runs into the mold, he suggests that you

Figure 9.15
Cast bronze censer, German, twelfth century. Height, 22.0 cm. Cathedral Treasury, Trier. Bildarchiv Foto Marburg.

lie down close to the mouth of the mold and listen carefully to find out how things are progressing inside. If you hear a light thunderlike rumbling, tell them to stop for a moment and then pour again; and have it done like this, now stopping, now pouring, so that the bellmetal settles evenly, until that pot is empty. Then take it away and immediately bring another, put it in the same place and do the same with it as you did with the first, and likewise with a third one until the bellmetal can be seen in the gate. The [last] pot should not be taken away at once, but should be kept for a short time, so that if the bellmetal sinks, more can be poured on top.[13]

Such detailed descriptions are rare. Indeed, Theophilus is the first man in all history to record in words anything approaching circumstantial detail of a technique based on his own experience. Even in the Far East where craftsmen were more respected there is no comparable written source until much later. For most of the past, if one wishes to reconstruct the techniques, it is necessary to depend upon a critical examination in the laboratory of the objects themselves. The record preserved in their microstructure is actually enormously rich but it is only just beginning to be read in fine detail.[14]

The lost-wax process is not the only way of producing a casting from a model of a complicated shape with numerous cavities, overhangs, and reentrant parts. Foundrymen are proud of their ability to copy anything in molds of sand or clay by the use of deformable patterns, false cores, and molds with inserted pieces that can be withdrawn in various directions. Such methods are necessary if more than one casting is to be made from a complicated artist's model, and they have become common since the sixteenth century. Nevertheless, the simple but unavoidable geometry of mold division and withdrawal dominates everything. Any shape can be copied by casting, yet those shapes that have been designed with some concern for the mold-making processes have greater artistic integrity, at least to a metallurgical eye, than those that require a *tour de force* in their manufacture. They are also cheaper.

Though elaborate art castings in bronze in the Middle East and in Europe until the seventeenth century were usually made by the lost-wax method, simple or common objects were always designed so that a pattern could be used repeatedly and the clay for the mold simply pressed against it and withdrawn. Such an object must have some place where the dividing surface between the two parts of the mold, commonly a plane, can intersect the pattern with no overhanging parts. Biringuccio in his *Pirotechnia* (Venice, 1540) describes a brass foundry in Milan that was mass-producing belt buckles, harness fixtures, thimbles, window fastenings, and the like. As many as 1200 objects were cast simultaneously in a single mold composed of twenty superimposed layers of clay shaped over a composite metal pattern complete with gates and runners.[15]

Clay or sand molds, of course, can be used only once, though their material is recycled. Permanent molds with simple cavities shaped in stone or clay appear very early in metallurgical history, and even bronze molds for casting bronze, or iron for iron. They lead to easy mass production, especially of metals and alloys of low melting point. Innumerable pilgrim's tokens and toy soldiers have been so made, designed with considerable detail but usually with no overhanging parts. The pewterer's elaborate metal molds with replaceable parts carrying decorative detail were ready to be adapted in Gutenberg's hands to the casting of printer's type of standardized dimensions. Of all metallurgical operations, this one has had the greatest influence upon the minds of men.

One should note the basic aesthetic difference between a design that is made to withdraw from a mold and one that is shaped freely in wax. Compare the panels in the doors of Augsburg Cathedral (figures 9.16 and 9.17) with the slightly later ones, also cast in Germany, now at St. Sophia in Novgorod (figure 9.18). The Augsburg doors illustrate another thing, namely that the appearance of an object can be changed during its lifetime. It has been plausibly suggested that the differences between several rather similar panels on the Augsburg doors occurred in the foundry as a result of damage to duplicate wax patterns.[16] To my eye it seems far more likely that contact with the fingers of generations of pilgrims plus enthusiastic cleaning had obliterated the high points in initially identical castings made from a single pattern, and that at some later time sharp detail had been rechiseled into the design. Even in the foundry castings are commonly finished by extensive cutting and scraping, and lose some of their cast surface.

The requirement for low relief and free-drawing quality in the simplest castings is applied even more rigidly to the design of coins or medals regardless of whether these are to be struck or cast. Indeed much of the charm of this art form derives precisely from this technical limitation (figure 9.19).

Figure 9.16
Cast bronze panel (no. 66) in the great door of Augsburg Cathedral, eleventh century. The height of the figure is 22.5 cm. Photo courtesy of D. W. Laging.

Chinese Bronze Casting Techniques

Some incredibly intricate and finely finished castings have been made by the lost-wax process, especially in Japan. In other parts of the world the very lack of technical restraints on design for the lost-wax method sometimes has the curious result of making the

Figure 9.17
Another panel (no. 84) in the Augsburg bronze door. Photo courtesy of D. W. Laging.

Figure 9.18
Bronze panel showing the birth of Eve, in door at St. Sophia Novgorod. Cast in twelfth century by lost-wax method. Height, 40 cm.

Figure 9.19
Cast bronze medallion by Vittore Pisano, with portrait of Leonello d'Este. Dated 1444. Photo courtesy of the National Gallery of Art.

viewer more aware of the technical expertise of the foundryman than of the aesthetic insight of the designer. Technique is essential to produce a work of art, but without an artist's control it can produce ugliness. Chinese bronzes, especially the ceremonial vessels of the Shang and Chou dynasties, combine the requirements of design and technique more successfully than almost any other objects of metal. Through the labors mainly of Noel Barnard[17] and R. J. Gettens[18] in elaborating the earlier suggestions of Orvar Karlbeck and Li Chi, it is now possible to get a fairly good idea of how these bronzes were made. Until recently, Chinese and European connoisseurs uncritically accepted the account by Chao Hsi-ku,[19] writing in the first half of the thirteenth century A.D., who erroneously assumed that they were made from a positive lost-wax model—as, indeed, most things of such extreme intricacy would have been in the last millennium or so. When the skill of the Chinese foundryman was at its apex in the late Shang dynasty the bronzes had an almost brutal character with protuberances and flanges and with the design in several different levels of boldness and intricacy of texture (figure 9.20). Close inspection almost always shows marks slightly interrupting the details of the pattern which clearly originate in joints in the mold that would be unnecessary with wax (figure 9.21). However, still closer inspection reveals that the single mold sections extended around curved surfaces carry many details protruding at angles that could not conceivably draw without damage: this does suggest lost-wax, at least unless a flexible pattern material had been used such as the rubber or gelatine which have been used in art foundry work in the last century or so but are highly improbable in Shang China.

How then were the decorative details achieved? Some possibilities are suggested by some foundry practices in use in Japan today that are markedly different from traditional Western techniques. Neither the casting of bells for Buddhist temples, which the author has seen in the Takahashi foundry in Kyoto, nor the casting of *hibachi* and other art bronzes shown in a recent film of the Enjo foundry[20] involves the use of three-dimensional models, but a mold cavity in the form of a negative of the gross outlines of the casting to be produced is generated directly in space, and the decorative details are worked by hand in negative directly into the mold surface. In the case of bells, a moist plastic mixture of sand, grog, and clay is supported externally in iron rings, and the shape is swept out by a strickle board cut to the proper profile and rotated around a fixed vertical axis. The rounded rectangular shape of the *hibachi* is generated by the moving profile of a vertical strickle sliding against templates fixed to the top and bottom of the main mold segment. In both cases, all noncircumferential decorative details are cut, inserted, or, most commonly, impressed by stamps into the soft mold surface, and it is mainly to facilitate this that the molds are divided into horizontal sections supported by external interlocking rings of stone or iron (figure 9.22). The simple convex core of the bell is also shaped by a rotating strickle, but the core for the *hibachi* is made by pressing a damp clay mixture into the mold (after this has been baked to harden it), and, after drying it and removing the mold, scraping the core down to give space for the desired thickness of bronze.

The molds for large European bells are also made by sweeping out the form in loam (figure 9.23), but the mold is divided into only two massive parts, the core and the cope, and the latter, not easily accessible for

Figure 9.20
Bronze ceremonial vessel, type *kuei*. Late Shang or early Chou dynasty. Height, 14.8 cm. Photo courtesy of the Museum of Fine Arts, Boston.

Figure 9.21
Detail showing mold joint in flange in bronze vessel, type *chia*. Natural size. Photo courtesy of R. J. Gettens, Smithsonian Institution, Freer Gallery of Art, Washington, D.C.

Figure 9.22
Mold for casting a temple bell. The core, surrounded by bricks, is being dried out with a charcoal fire. Takahasi Foundry, Kyoto, 1969.

Figure 9.23
Eighteenth-century French bell foundry. Note the rotating strickle board, which is used to shape both the core and, when recut, the outside of the dummy bell on which the outer mold is formed. From Denis Diderot, *Encyclopédie*, plates vol. 5 (1767).

hand-finishing, is shaped over a dummy form of the bell which is built up with clay applied over the core. The inscription and all decorative details are shaped positively in wax or clay applied to the swept surface of the dummy. Most Japanese bells when hung still have on them one or more rough lines obviously arising in horizontal mold joints (figure 9.24). These lines are not removed in fettling the bell, and they seem to be regarded not as defects but rather as a reminder of the reality of the founder's interaction with his materials. One is reminded of the ceramics that are most treasured in Japan which usually have some unexpected tool marks or irregularity resulting from a kiln mishap.

Mold-joint "blemishes" are effectively displayed in many Japanese cast-iron tea kettles (figure 9.25). Molds for these are also made by rotating a strickle, and the mold is divided in order to provide for making the core, and to give access to the inner surface for decoration as described above.[20a]

Figure 9.24
Cast bronze bell, eleventh century. In the Byodoin, Kyoto. Height, 199 cm. Photo courtesy of Tsuboi Ryohei.

In both types of mold, the vertical panels, the chrysanthemums, dragons, inscriptions, textural patterns, etc.—that is, all the decorative details other than the circumferential features that are generated by the revolution of the strickle—are impressed with stamps into the soft surface of the mold, or if they are too deep for this, or nondrawable as in the case of knobs, they are premolded separately in loam, baked hard, and simply set into holes cut into the main mold surface and retouched. There is no problem of withdrawing the mold from beneath overhanging parts of the pattern, for there is no pattern, and since the stamps are curved or small they can always be withdrawn in a direction normal to the local surface.

The Shang and Chou bronzes become easier to understand with these techniques in

Figure 9.25
The Daikōdō kettle. Cast iron, ca. 1700 A.D. Maker, Miyazaki Kanchi. Photo courtesy of the Tokyo National Museum.

mind. It is not suggested that the molds for them were swept out negatively rather than being conventionally formed over smooth models (though they are mostly of shapes that could have been so made), but it does seem that the main vertical and horizontal divisions were needed as much, if not more, for the purpose of making the mold surface accessible to the hand of the artist-craftsman for final decoration as for removing the mold from a simple model. The ever-present feature, the *t'ao t'ieh* mask, and other major decorative features that are in medium relief were probably impressed in the fresh moist mold surface by the use of a stamp, and the fine lines of the background-filling *lei-wen* were supposedly carved by hand into the surface of the mold when it was just dry enough to be so treated. (In some cases it looks as if the smooth surface of the mold had been lightly baked and given a thin coat of moist fresh clay into which the fine carving was done.)

The inscriptions that appear in the interior of many vessels have a quality that is different from that of lines of similar weight in the decoration. They are often slightly undercut, and seem to have been molded from thin rods of soft clay applied to and shaped on the core surface. This was certainly done with the coarser intaglio lines in the relief areas in the main design as in the stylized *t'ao t'ieh* and round bosses of figure 9.20. A stamp for shaping a conventional *t'ao t'ieh*, found at Anyang, is shown in figure 9.26. Note that it has the linear decor only in outline, not in depth as in the bronze, and therefore linear details in relief must have been added to the stamped impression. The best evidence, however, comes from the mold fragments themselves (one from Anyang in the Sumitomo collection, shown in figure 6.6, and several in the Royal Ontario Museum)

Figure 9.26
Ceramic stamp for impressing the *t'ao t'ieh* mask in molds for bronze vessels, late Chou dynasty. Front and back views. Height, 10.5 cm. Photo courtesy of R. J. Gettens, Smithsonian Institution, Freer Gallery of Art, Washington, D.C.

which show unmistakably that the base clay had been prepared for the local overlay by appropriate scraping and scratched in outline by the final shaping. This clay broke away with the bronze when the filled mold was dismantled, leaving the mold quite different in appearance from the part that formed the finer *lei-wen* where only the higher parts of the cut-in clay have broken off and the remainder is clearly integral with the material of the mold.

Part of the attractiveness of the early Chinese bronzes undoubtedly arises from the fact that the bronze surface is a direct replica of the surface shaped by the artist, without the repeated inversions in material of differing textures and plasticities that are needed in most other methods of mold production. Moreover, it was interaction with an advanced indigenous ceramic tradition that enabled Chinese bronze metallurgy to develop almost explosively after its late start, and to achieve an individual character, quite distinct from that in the Middle East, reports of which supposedly inspired it.

Max Loehr has published photographs of some excellent bronzes of all the principal types that he selected for the Asia House Exhibition of 1968.[21] Looking at these (or still better at the bronzes themselves) it is obvious that a diversity of techniques had been used, and that the aesthetic effect of the final product is a direct result of the mold-makers' technology. Both the major divisions and flanges and the different qualities of detailed decor, though now in bronze, originated in clay, and the design and the shaping technique intimately reflect each other.

The later bronzes depend more often on some form of mechanical replication of detail (figure 9.27), but it is wrong to assume that all bronzes even at one time were made by the same techniques. A critical study of the record preserved in the detailed form of the vessels themselves suggests that the following techniques, in various combinations, were used in making the bronzes of the Shang or Chou dynasties:[21a]

1. The shaping of a mold cavity by applying clay against a preshaped model having the gross outlines of the bronze but no detailed decor.
2. The division of the mold into two or more parts to permit its removal from the model or the core or to provide access to the mold surface for decoration.
3. Decoration carved or incised directly into the mold surface after partial drying.
4. Decoration of the soft surface by impressions with a punch or stamp, either of a small area repeatedly impressed or a larger design feature replicated fewer times.
5. Bold intaglio linear features formed by thin bars or strips of clay applied locally to the mold surface and given accurate shape by tooling.
6. Details, especially undercut ones such as knobs and handles, provided by rigid submolds preshaped and set into cavities cut into the main mold surface.
7. Areas of molded detail applied as thin premolded slabs or plaquettes applied to the mold cavity in a condition soft enough to conform to its curvature.[22]
8. Bronze pieces such as legs or handles precast with protuberances and inserted into the main casting mold to become locked in as the bronze solidifies.
9. Bronze pieces separately cast and joined to the main casting with run-on molten bronze restrained by a local mold.
10. Precast pieces attached by pressing into a lump of bronze of very high tin content heated hot enough to be pasty but not mol-

Figure 9.27
Bronze vessel, type *hu*. Late Chou dynasty. Height, 44.8 cm. Many details of the pattern are exactly duplicated in different parts of the vessel. Photo courtesy of the Smithsonian Institution, Freer Gallery of Art, Washington, D.C.

ten. (See Gettens's book cited in note 18, pp. 84, 134–139. Metallurgically this is similar to the old plumber's wiped joint. Unlike their Western counterparts, joints of types 8, 9, and 10 in Shang or Chou bronzes do not always have true metallurgical junction of the parts by interfusion, but often depend only on mechanical interlock for their strength.)

11. Loosely mating parts such as hinged handles and chain links cast in molds with interpenetrating cavities that do not join.

12. Some vessels of the late Chou period with large areas of rather geometric inlay in silver and copper (such as nos. 100 and 101 in Gettens's book) have been shown on the basis of radiography by Pieter Meyers to be actually cast in, not mechanically inserted. The inlay was presumably cut to shape in a sheet ca. 1 mm thick and attached to the mold surface before the bronze was run in. The total wall thickness is only about 3–4 mm, and the technique must have required superb control of every detail of alloying, melting, molding, and casting.[22a]

13. The vessels were usually finished by polishing with a graded series of abrasive stones to remove small fins, and generally sharpen the design by eliminating local variations in height. A black pigment was often pressed into intaglio lines to contrast with the bronze.

14. In late bronzes the surface was chemically patinated (sometimes very deeply mineralized). Gold, silver, or copper inlay may be applied. In Han times decoration was carved or engraved into the surface of the cast bronze. (The presence of lead, commonly about 5 percent but ranging up to 25 percent in Chinese bronze, makes cutting with a chisel or graver very easy.)

The casting of mirrors is basically much simpler than casting vessels, for the designs are in low relief and, like medals, are easily drawable. Most of these were cast in clay molds, but many were cast in carved molds of soft stone that could be used repeatedly in mass production. In later mirrors the background texture obviously arose by pressing the clay for the mold against a patterned surface which had been built up from an assembly of units as in method 7 above. The texture of embroidered cloth is duplicated in some mirrors.[23] The main symbols or designs in bronze relief are carved into the textured surface of the mold, and there are commonly circular grooves and rims which suggest turning on a lathe of some kind but which were more probably produced by a tool rotating around a vertical axis as in the making of the mold for a bell or tea kettle.

While on this subject, I cannot resist mentioning the "magic" mirrors of China and Japan. Though the polished surface of these look normal, sunlight reflected from one of them onto a wall several feet away shows a bright image similar to the pattern that is cast in relief on the back (figure 9.28). The magic lies in a pattern of very shallow depressions a few millimeters wide and of a curvature to focus the sun at the proper distance, but how this concavity is so elegantly produced remains to some extent unexplained.[24] The effect could rather easily be produced on a uniform surface by local abrasion, but, at least in some Japanese mirrors of the Edo Period, the magic arises in the local mechanical distortion produced by a tracing tool used to outline the main features on the back, which introduces enough internal stress so that the opposite face locally springs back and becomes concave as it is ground and polished against flat surfaces.[25] This cannot, however, be the case with the earlier Chinese mirrors which are cast in a true speculum metal with about 25 percent

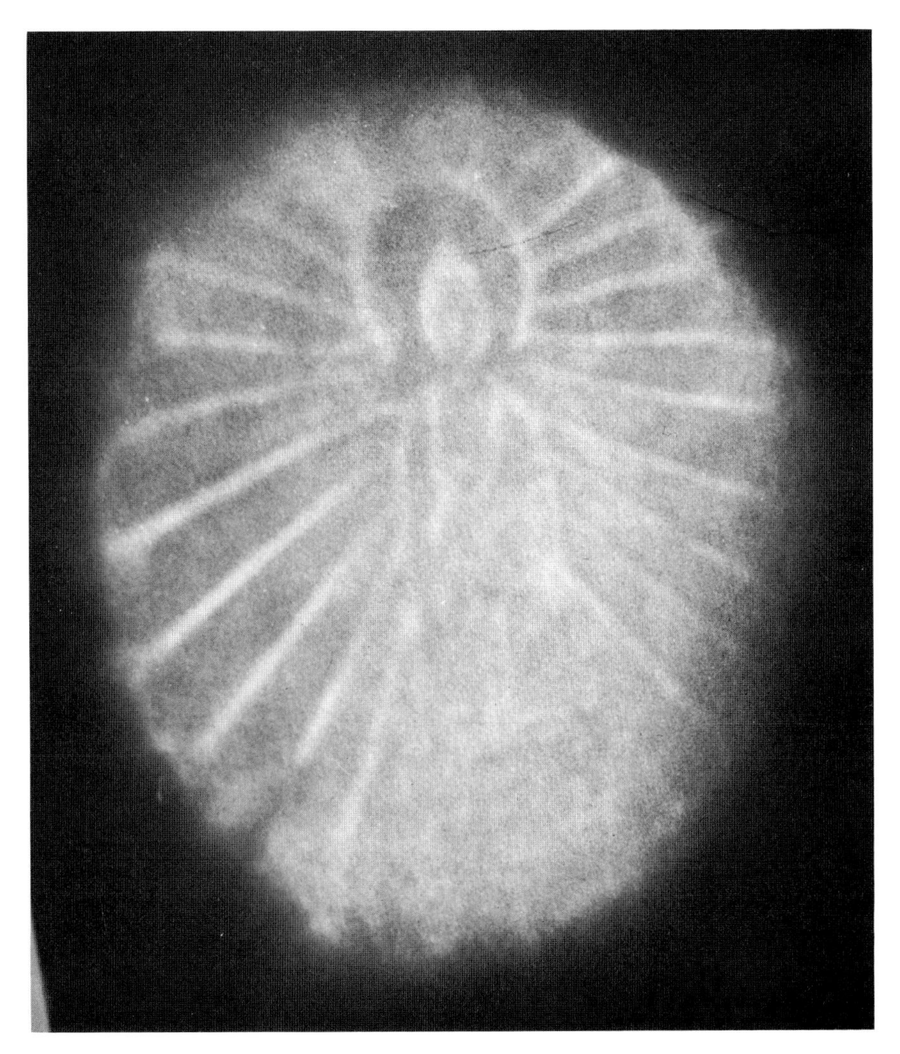

Figure 9.28
Image reflected by Japanese magic mirror, reproducing the pattern cast in relief on its back but invisible on the front surface. Height, 21 cm. Photo by H. Maryon in the Royal Ontario Museum.

tin, too brittle to withstand deformation. Perhaps the thick and thin parts of the metal react differently to polishing because of differing local springiness, or perhaps the thicker parts are more easily abraded because, having solidified more slowly, they are both softer and have more microporosity. The backs of Japanese mirrors sometimes show no evidence of tracing, but this is because they are in two parts, joined invisibly at the edge (see Turner's paper cited in note 24).

Casting in Sand Molds

The simplest of all methods of casting, and by far the commonest today, is that done in green (i.e., damp) sand molds. This is an astonishingly recent discovery despite the fact that good slightly argillaceous molding sand that is both strong enough to hold its shape against the metal and porous enough to allow the escape of steam and other gases evolved by heat exists in many localities. A foundryman using the traditional clay molds risked his life if he poured hot metal into one that had not been fully baked to dryness. He would be little inclined to experiment with undried materials, and they were not used until the end of the fifteenth century. Leonardo da Vinci mentions with surprise, as a novelty, the making of castings rapidly and cheaply in river sand dampened with vinegar. Biringuccio in 1540 remarks that "it has been discovered, contrary to the natural order of art, how to cast in moist earth in order to save labor and expense." Cellini, writing twenty-eight years later, remarks that there is a very rare kind of sand found in the bed of the river Seine which is

> very soft, and has the property, quite different from other clays used for moulding purposes, of not needing to be dried, but . . . you can pour into it while it is still moist, your gold, silver, brass, or any other metal. This is a very rare thing, and I have never heard of it occurring anywhere else in the world.

The making of such molds may perhaps have been suggested by the casting of sows and pigs of iron by running it directly from the newly developed blast furnace into open troughs simply scraped into the surface of the damp sand floor of the casting shed. (However, green sand molds were known much earlier in the Arab world than in Europe—they are mentioned in 1206 by Al-Jazari.)[26]

The great ingenuity of foundrymen in copying highly complex artists' models for statuary of various kinds cannot be treated in detail here. The plates in Diderot's *Encyclopédie* showing the Keller brothers' enormous construction for casting the heroic equestrian bronze statue of Louis XIV in a single piece give a vivid picture of such work in the seventeenth century. The publication of the Madrid codices of Leonardo da Vinci has added greatly to our knowledge, for they contain a full description of the making of the mold for the great Sforza horse that was completed but never cast. Most large castings are made in pieces and joined. Lechtman and Steinberg have shown the importance of joining methods in giving freedom of design, which the Romans sometimes overexploited.[27] A close inspection of any but the smallest bronze statues will usually reveal several lines or spots of welds made by running-in metal of the same composition. The process became widespread only after the composition of the bronze had been modified by the addition of lead to make it easier. In modern sculpture, the welds are usually visible and often form an integral part of the design conceived by the sculptor, and in all the composition of the metal needs to be such that it will facilitate the joining and not be damaged by the local heat.

Bronze vs. Iron Castings

Ever since its discovery, bronze has consistently been the preferred metal of the sculptor, for it is relatively permanent, has an attractive color both when polished and when patinated, and is easily shaped in molds. The last has some uneasy consequences, for sculptors too often deny the nature of metal and give to the foundryman shapes to be copied in bronze that are natural only to softly plastic substances which yield to the slightest touch and acquire nonmetallic textures and curvatures. Of course, the raison d'être of the whole process of casting is to exploit the fact that model and mold are much more easily shaped than is the final solid bronze, but good design arises only when the properties of all three materials are kept in mind. Lowly cast iron has in this respect been mistreated less often than has bronze. Not only does a typical iron casting have a surface with a soft texture that is partially inherited from the sand, but the warm grayness conferred by the tiny particles of graphite in its microstructure is unusually attractive, especially in functional objects.[28] Again, the best examples are from the Orient—such as the Japanese tea kettle (figure 9.25)—but in Europe the qualities of the metal have been excellently used in things like fire-backs, lamp posts, grills, columns, and other architectural details. It was eminently adapted to the nineteenth century, both technically and aesthetically (figure 9.29), and was even used to make jewelry—in the well-known "Berlin" jewelry of cast iron and its perhaps more attractive Bohemian counterpart (figure 9.30). This jewelry has added interest in that in the 1860s it inspired studies of metal quality that led to the microscopic studies of the German engineer Martens which served to bring the structure of metals at last to the attention of a scientific world that had ignored earlier hints.[29]

Jewelry from Sheet and Wire

This cast-iron jewelry was mainly an attempt to copy in a cheap medium the products of a more expensive one, that of filigree, and thus serves to introduce a discussion of the second basic property of metals that is central to their use in art and industry—their ability to be deformed without losing coherence. Malleability and ductility are intimately related to the nature of the metallic state itself, for the electronic state that gives rise to electrical conductivity and the bonding together of atoms in a metallic crystal is such that imperfections in the crystals can, under sufficient stress, move and multiply without destroying crystalline order as a whole. Metals are strong enough to resist the stresses of service but not too strong to yield smoothly to high local forces applied by tools used. Man's very first use of metal in the ninth millennium B.C. involved this property of malleability. Since most natural inorganic materials are brittle, the discovery of a malleable stone, native copper, must have been as exciting as it was portentous. Nuggets of native copper were beaten with stone hammers into sheets, then bent or cut to give beads and dress ornaments. Sheet copper was soon followed by wire, although wire was for a long time made by cutting narrow strips from sheet and rounding these by hammering and abrasion, not by drawing

Figure 9.29
Cast-iron columns on building at 46 Green Street, New York. Photo courtesy of the New York Landmarks Preservation Committee.

Figure 9.30
Cast-iron jewelry, Kamarov foundry, Slovakia, ca. 1830. The oval medallion is 5.4 cm high. Photo by Véra Kodetova, courtesy Dr. J. Kuba and the Narodni Techniké Museum, Prague.

through dies. (See below, figures 9.50 and 9.51 and the discussion thereon.) Though it is understandable that drawing, a very efficient way of making wire, should have been a late discovery, it is remarkable how extensively wire was used when it had to be so laboriously fashioned. One suspects the influence of a combination of material scarcity with aesthetics, perhaps inspired by an earlier use of biological fiber and sinew. In any case, the use of wire is a tribute to the strength and optical properties of metals.

It was natural that gold, the most ductile metal, should have been the first to be drawn. It seems to have begun by the smoothing and compaction of a helically twisted strip into a round section. Such wire is common in Greek jewelry, and its manufacture requires neither a great force in drawing nor an exceptionally hard material for the die.[30] It has been suggested, plausibly, that a bronze plate with punched holes would be adequate.[31] The evening-up of braided cable and interlinked chains for jewelry by drawing through a wooden or other die is also quite old. Such shaping would lead rather naturally to a true drawing operation in which greater force and a graded sequence of holes in very hard dies are used to produce a considerable reduction in cross-sectional area with a corresponding increase in length. When this was first done is uncertain; the first description of the operation is given in the early twelfth century by Theophilus. Wire continued to be made by slitting metal strips long after this, especially of iron which is far more difficult to draw than the nonferrous metals. Indeed, throughout much of the period during which mail armor (which is often presumed to be evidence for the drawing of iron wire) flourished it was made by slitting and rounding.[32]

There are many more ways of using the plasticity of metals than in the making of wire. Art objects utilizing malleability can be divided roughly into three classes:

1. Forged objects where lumps of smelted or cast metal are beaten between hammer and anvil or with special tools into massive solid elongated shapes possessing an almost organic relationship between their three-, two-, and one-dimensional aspects.
2. Objects composed of a single piece of metal in an essentially two-dimensional form, that is, a sheet, which may be cut to shape in a plane or given general or local curvature into a bowl or cup, and/or worked in detail by local hammering or punching (raising, repoussé, or chasing). Prior to the introduction of rolling in the sixteenth century, all sheet was made by hammering and was essentially a subclass of group 1.
3. Objects in which pieces of semifabricated sheet, rod, or wire of uniform section are cut, bent, and joined into a composite structure by mechanical interlock or by soldering or welding. The component pieces may themselves be of complicated section made, for example, by drawing or extruding through a profiled die.

The first is fundamentally the freest in form, but the last two have given rise to the majority of small works in metal and have a charm that arises in no small measure from the dimensionality of the process. Only in the present century has work on a large scale of the third kind been regarded as sculpture, but, inspired by the civil engineer's bridges and buildings, artists now weld rolled steel sheet and rod on a monumental scale. Yet neither the Chicago Picasso (figure 9.31) nor the gigantic stainless steel arch in St. Louis can have evoked either the skill or the national interest demanded by the past giant castings such as the Colossus of Rhodes or the *Daibutsu* at Nara and Kamakura.

Wrought Iron

Forged objects are best exemplified by wrought iron, which has provided most of the useful and many of the most beautiful objects of metal since the end of the second millennium B.C.[33] Iron has one quality unique among the common metals—the ability to be hammer-welded—and this profoundly affects what is done with it.[34] The difference between iron and other metals is subtle. All metal surfaces form oxides that prevent their joining at low temperatures, but with iron at a sufficiently high temperature the oxide will melt, and surface tension, by causing the liquid oxide to collect into relatively harmless isolated patches, facilitates the joining together of grains of iron whenever they come into contact. Hammering squeezes out excess oxide and leaves a strong joint of continuous metal. It is quite different with copper, for when this is heated to a temperature where it absorbs oxygen and begins to melt, the liquid penetrates between the microcrystals, completely destroying their coherence and causing the metal to crumble under the slightest attempt to deform it. This can best be seen in two photomicrographs. Figure 9.32 is a photomicrograph of a piece of wrought iron containing an oxide slag—actually a piece of 2800-year-old iron from Luristan—and figure 9.33 is a piece of copper containing about 0.03 percent oxygen heated just above the melting point of the eutectic and quenched. In the latter, the liquid has spread between the grains and completely destroyed metallic contact, whereas in the former the oxide which coated many of the particles of

Figure 9.31
Sculpture made from rolled-steel plate by cutting and welding. Design by Pablo Picasso, construction by U.S. Steel Corp. Photo courtesy of U.S. Steel Corp.

iron in the initial crude sponge has now contracted into a relatively harmless form. Not only the particles in a reduced iron sponge but also whole pieces can be joined together by hammer welding, provided only that the temperature is hot enough to melt the oxide that inevitably forms on the heated metal in air.

This intimate balance of interfaces is central to the existence of the entire iron industry up to the time of Bessemer. Since iron melts at a temperature far above that attainable by early metallurgists, it would have been utterly unusable were it not for the fact that iron sponge, so easily reduced from the ore, can be completely consolidated by hammering at a moderate temperature without ever becoming molten. And even Bessemer's first troubles came from the fact that sulphur in steel, like oxygen in copper, does form an embrittling liquid at hot-working temperatures. Mushet's addition of manganese simply changed the energy of the iron/sulphide interface and rendered sulphur relatively harmless.

If iron is sufficiently carburized it can, of course, be melted with only slightly more difficulty than copper. Such iron, cast iron with about 3 percent carbon, is easily made from oxide ore or ductile iron by prolonged heating when deeply buried in a charcoal fire, and fragments are found in many ancient smelting sites. It is, however, brittle. A mystery of cultural diversity is why brittle cast iron came to be widely appreciated in the Far East from the time of its early discovery, while in the West it was regarded as useless or at most as a transitory intermediate stage in the manufacture of more malleable steel. Westerners seem to have thought that a metal should be ductile and so failed to recognize the utility of cast iron.

Iron is easily reduced from its ores to form

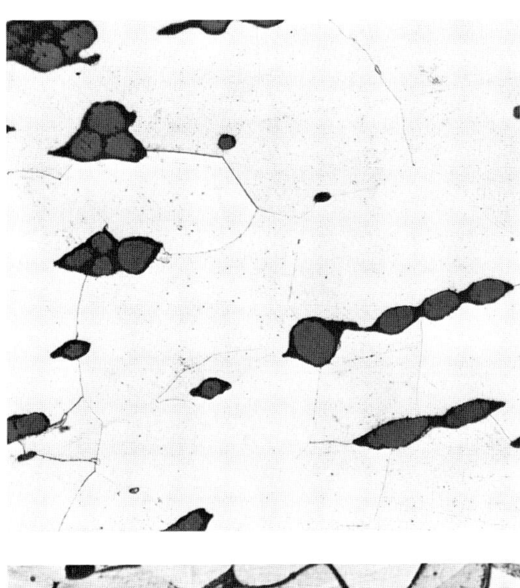

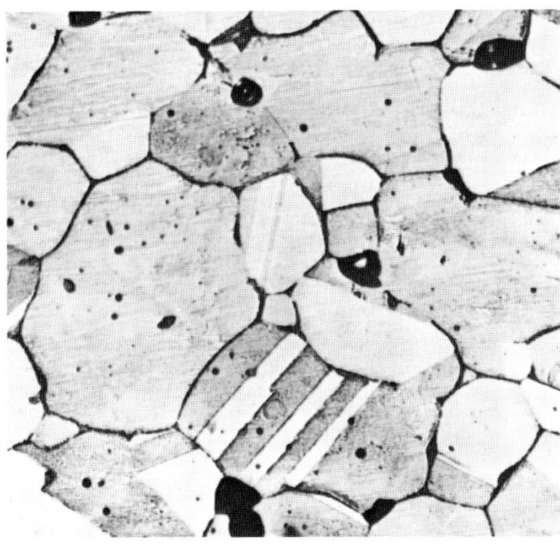

Figure 9.32
Microstructure of forged-iron sword from Luristan, ca. 800 B.C. Shows the indifference of the grain boundaries to the inclusions of once-molten oxide phases. Etched. × 300.

Figure 9.33
Microstructure of wrought copper containing about 0.03 percent oxygen, heated to the melting point of the oxide phase, which has completely destroyed cohesion between the metallic grains. Etched. × 200.

a low-carbon metallic sponge which, without melting but after preliminary consolidation, is extremely tractable when hot and reasonably strong when cold. It was first made near the beginning of the third millennium B.C., but it did not become economically significant until about 1200 B.C. Even as late as 800 B.C. superbly shaped objects were being made by forging, but with details that show that the makers—the smiths of Luristan—were ignorant of the welding of iron but laboriously assembled preshaped pieces with intricate mechanical joints.[35] Later smiths, however, used the weldability and the forgeability of iron to good effect, and there are few works in which aesthetics and the properties of metal are more naturally combined than in the best wrought iron of almost any period.[33] The flow of metal under the hammer and the necessity of working fast while the metal is hot and in contact with the anvil, swage, or fuller confer grace to every change of section. The sparse details that the smith impresses with a blunt chisel or punch have a simple boldness quite different from the precise markings on cold metal. The strength of iron when cold enables parts to have a slenderness that could not be tolerated with other metals. Moreover, iron leaves the forge with a tight black coat of oxide which is thoroughly appropriate both to the metal itself and to the manner of its working and also provides a good shield against rust. Figure 9.34 shows a wrought-iron grill at Puy. Note how the smith has achieved an organically suggestive design by combining shaping with a hammer against the anvil and hammer-welding with some finishing marks applied with that simplest of decorative tools, the punch. Later smiths often used preshaped dies for forging knobs and other details.

Though iron is most naturally a metal to be worked hot on the smith's anvil, its strength also permits other quite different types of decorative effect to be used based on thin sections and sharply defined edges. Iron chiseled or otherwise cut into shape has yielded much fine work, especially that of the locksmiths of seventeenth- to eighteenth-century Spain and France (figure 9.35). In this, all of the final surfaces have been formed by filing, cutting, or punching, and the whole presents slim sections, nearly geometric surfaces, and sharp outlines quite inappropriate to softer gold or silver.

Iron is also, when reasonably pure, beautifully responsive to cold working, to raising and repoussé, to punching, tracing, and chasing, and to similar techniques that are seen perhaps at their best in gold and silver. Some fine repoussé plaques in iron found in the eighth century B.C. levels at Hasanlu in Iran are among the earliest.[36]

More elaborate are the heavily embossed shields of the sixteenth-century Italian armorers (figure 9.36). Such decoration is inconsistent with the main purpose of armor, for it would engage and not deflect the point of an attacking weapon: more appropriate uses of the properties of iron lie in the smoothly curved surfaces of the simpler armors, or in the purely decorative repoussé work of the Miochin family in Japan (figure 9.37).

Figure 9.34 (*left*)
Wrought-iron grille, Puy Cathedral, twelfth century.

Figure 9.35
Panel in a lock made of wrought and chiseled iron. French, fifteenth century. Height, 21 cm. Photo courtesy of J. M. Hoffeld, Metropolitan Museum of Art.

Etching

Armor introduces etching, that is, the use of localized chemical attack either to change the surface texture or to remove metal in a predetermined pattern. This not only gave some beautiful pieces of decorative arms and armor (figure 9.38) and led directly to the graphic artist's printing process, but it also

Figure 9.36
Embossed iron shield, signed George Sigman. Dated 1552. Diameter, 61 cm. Photo courtesy of the Victoria & Albert Museum.

Figure 9.37
Pair of wrestling bears, formed by hammering from a single sheet of iron. Japanese, nineteenth century. Height, 19 cm. Photo courtesy of the Victoria & Albert Museum.

Figure 9.38
Etched hunting sword made by Ambrosius Gemlich, Munich, 1540. Detail. Natural size. Photo courtesy of the Trustees, The Wallace Collection, London.

inspired the two most important discoveries in the history of metallurgical science—first in 1774 (via a study of Damascus steel) the discovery of carbon in steel, and then, in 1863 after a period of work on meteorites, the disclosure of the crystalline microstructure of terrestrial steel. These two discoveries, chemical and structural respectively, underlie most of the modern science of materials, which is intrinsically the understanding of the relationship between structure and properties.[37]

The earliest decorative etching is that done on carnelian beads in Harappa (ca. 2000 B.C.), seemingly done by coating with alkali, with reserved areas, followed by gentle heating.[38] The stone is locally whitened and opacified, without any change of level (figure 9.39). Etching on metals first appears in China at the end of the Chou dynasty or the beginning of the Warring States period, a time of considerable experimentation in techniques.[39] Figure 9.40 shows a short sword with a design obviously produced by deep chemical attack, though the surface of the corroded material had been finished coplanar with the adjacent bronze. It is mineralization in situ rather than removal of the metal by the etchant. The process is clearly related to the "black bronze" of contemporary swords and mirrors. The black mirrors are cast in a bronze with about 25 percent tin. They have a surface layer of dark but transparent material about 0.05–0.1 mm thick lying over a region two or three times as thick in which only the alpha phase in the microstructure has been replaced. (In normal corrosion it is the delta phase, here unaffected, that is preferentially removed.) Analysis shows an accompanying depletion of copper and an increase in tin content, while both iron and silicon have been introduced from outside by some chemical treatment whose nature is at

Figure 9.39
The first etching: Carnelian beads with designs produced by local chemical attack. From Mohenjo-Daro, ca. 2000 B.C. The bead with herringbone pattern is about 15 mm long. Museum of Fine Arts, Boston. Photo by author.

present unknown beyond the fact that it did not involve high temperatures. Chase and Franklin, who have published the most extensive study of the mirrors, believe the surface layer to be "a glassy amorphous silicate containing Sn, Fe and some Cu."[40] However, since the silicon content is only on the order of 1 percent and the tin about 45 percent, it could perhaps better be described as an impure amorphous form of tin oxide. Whatever it is, it provides a beautiful finish that is highly resistant to corrosion.

Another process involving deep superficial chemical modification of metal objects was that used in pre-Columbian Ecuador and Peru to produce a layer of pure gold on the surface of objects made of a dilute alloy of gold with copper or with copper and silver.

Not long after this, etched shells appear—in Arizona of all places. In Europe there is no hint of etching in any classical literature and, although some La Tène iron swords were etched,[41] the process thereafter disappeared for many centuries except for etching to develop the overall texture of pattern-welded and Islamic swords. Clear descriptions of reagents for etching iron first appear in the *Mappae Clavicula* (late eighth century), but these are intended for covering iron surfaces with redeposited copper so that they will hold gilding applied by the amalgam process.[42] Supposedly it was experiments in localized gilding that led to the production of directly etched patterns on armor, which are first documented early in the fifteenth century.[43] Experiments with the making of prints from copper plates etched in positive relief in the manner of the armorer were made by Leonardo da Vinci in 1504.[44] Shortly thereafter etching of intaglio lines to hold ink for printing as in the earlier engravings was put to good use by many artists. The greatest, Albrecht Dürer, began

Figure 9.40
Chinese bronze sword with etched decoration. Warring States period, ca. 475 B.C. Detail. Width of blade, 5.0 cm. From the Brundage collection, Center of Asian Art and Culture, San Francisco.

with iron plates around 1515, and copper did not return for some decades. Cellini was the first (in 1568) to print a mordant recipe specifically for copper.

The decorative use of etching on nonferrous metal is not common. The only Western example I know of is the early nineteenth-century *moiré métallique*, etched crystallization patterns of tin on iron sheet, which by local heating were sometimes controlled to produce semblances of flowers or landscapes. Late in the seventeenth century Japanese makers of *tsuba* (swordguards) who had previously used etching to bring out the wood-grain-like texture in forged iron (figure 9.41) began to use etching on nonferrous metals, both deep attack after the application of a stop-off to give intaglio designs (figure 9.42), and (more often) as a light overall etching to achieve fine surface textures.[44a] The most widely used effect of this kind is on the alloy *shibuichi*, a two-phase alloy of copper and silver, which, when etched, has a nearly visible pattern of silver segregation which confers a quality and permanence to the patina unmatched by any other finish (figure 9.43). Other *tsuba* textures arise from the etching of cored dendritic crystals in cast bronze or the different orientations in fine-grained cast brass, sometimes used in contrast with a minutely punched surface texture (figure 9.44). One of the greatest tsuba-smiths of all time, Tsuchiya Yasuchika (1670–1744), achieved effects of great subtlety by these means.

Figure 9.41
Tsuba, Japanese, iron *mokumé*, nineteenth century. Private collection. Photo by Jock Gill.

Hammer and Punch on Cold Metal

The malleability of metals when cold is a property more widely used on the softer nonferrous metals than on iron. In fact, although copper, gold, silver, and many of

Figure 9.42
Detail of Japanese swordguard, signed Yasuchika (1670–1774). Made of a brasslike alloy with a deeply etched intaglio design. The texture in the etched area is the dendritic structure of cast metal. × 5. Museum of Fine Arts, Boston.

Figure 9.43 (*below, left*)
Detail of swordguard made of the copper-silver alloy called *shibuichi,* signed Nagayuki. Nineteenth century. The surface texture arises in the duplex microstructure of the cast alloy. × 15. Museum of Fine Arts, Boston.

Figure 9.44 (*below, right*)
Detail of swordguard, brass, signed Takeaki. Early nineteenth century. Note the pattern of etched grains in the ground, contrasting with the punched texture at the right. × 5. Museum of Fine Arts, Boston.

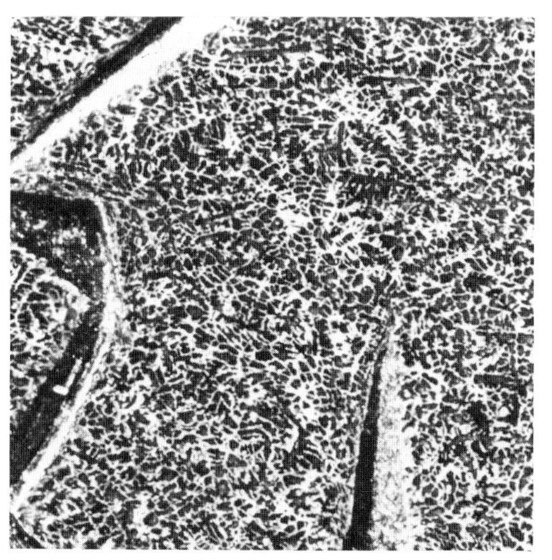

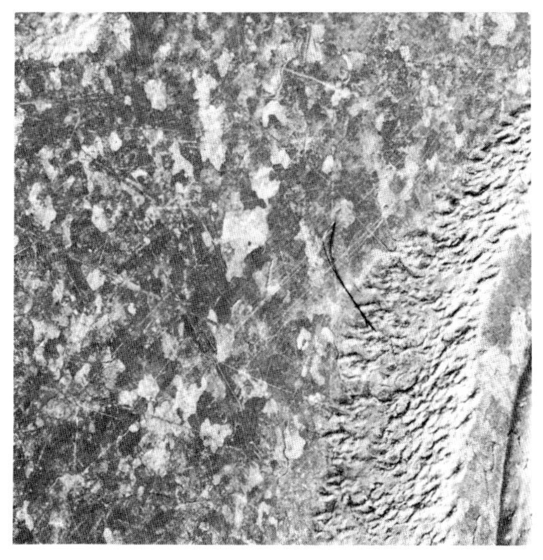

their alloys can readily be forged hot, it is rare to do so in making works of art. The qualities of the best dinanderie like that of goldsmith's or silversmith's work arise from the cumulative interaction of a large number of small local deformations caused by the hammer against a smoothly curved anvil (stake) in forming the main outlines and by the shaped punch in finishing details. In raising, the metal as a whole is not forced into the concave shape of the vessel, but the shape arises by a continuing balance between high local plastic strains and longer-range elastic ones. Innumerable local hammer blows are integrated in a way which permits sensitive shaping with little residual long-range stress in the metal. Figure 9.45, a sixteenth-century engraving by an anonymous master, shows many of these operations being carried out and the necessary furnace for annealing at the side.

The aesthetic quality of this kind of work comes partly from the balance of the short- and long-range aspects (both spatial and temporal) of the technique, but it is far more dependent on the skill of the craftsman than on the nature of the metal itself, and it will not be discussed in detail here. I illustrate it only with the superb gold cup from the Royal graves at Ur (figure 9.46), 4500 years old and about as beautiful an object as ever left the hands of the sons of Hephaestus. This probably began as a thick cake of gold, which was hammered to a round flat sheet, from which the main cup shape was made by raising, that is, by a sequence of local hammer blows from slightly convex hammer against the sheet metal held at a slight inclination against the rounded edge of an anvil, so producing innumerable little bends in concentric circles or close spirals which eventually integrate into the formation of a progressively deepening cup. Because metal is hardened by cold work, it would have been necessary to soften it repeatedly by annealing to allow the work to proceed. The fluting originated by local punching, that is, by a combination of repoussé and chasing done, not against a hard anvil, but against an internal filling of pitch or some other material of strain-rate-dependent properties which provides moderately strong distributed support for the metal under a local blow, followed by slow relaxation to conform to the new shape in preparation for the next stage of deformation. During each stroke the internal and surface atomistic properties of the punch, the gold, and the pitch are intimately matched against each other, while at the same time the eye of the smith relates the local changes in shape to the lines of the piece as a whole, both as it exists at each moment and as he envisages it when finished.

Mechanical and thermal treatments have a profound effect on the microstructure of a piece of metal. Compare, for instance, the photomicrograph of a cast bronze (figure 9.47) with figure 9.48, which shows the microstructure of a fragment of hammered sheet bronze armor from Crete similar to the superb mitra shown in figure 9.49, alongside of which it was found.[45] Not only has the segregation of tin been eliminated by the intermediate heat treatment, but the microcrystalline grains have been entirely reformed and are of a different shape from those in the casting. The narrow straight lines crossing the grains are strain markings (slip bands) produced by terminal cold work. Figure 9.50 is a cross section of the rim of a similar piece of bronze armor, which shows how it had been given stiffness by curling the edge of the bronze sheet (about 0.7 mm thick) around a piece of thick wire. The wire, however, had not been drawn as wire

Figure 9.45
St. Eligius in his workshop. This engraving, made about 1500 A.D. by the anonymous Master of Bileam, shows the raising of a silver chalice, the striking of a design from a punch, the drawing of wire, and the setting of gems. The tools are much like those in a workshop today. Courtesy of the Rijksmuseum, Amsterdam.

Figure 9.46
Fluted gold cup from the Royal Graves at Ur, ca. 2600 B.C. Height, 15.5 cm. Photo courtesy of the University Museum, Philadelphia.

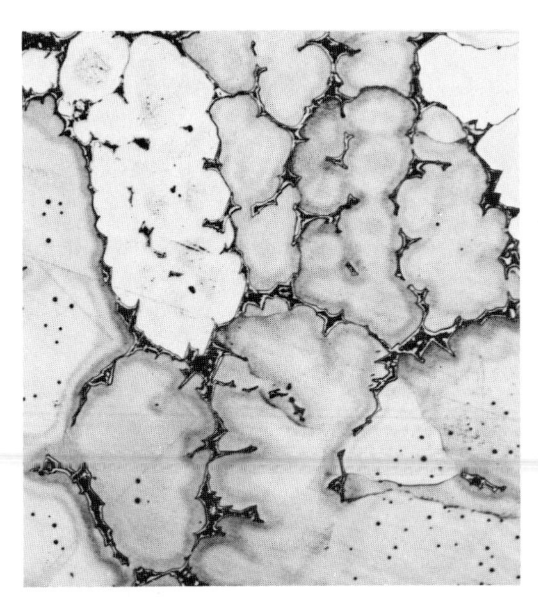

Figure 9.47
Photomicrograph of cast bronze (90 percent copper, 10 percent tin). Etched. × 100.

Figure 9.48
Photomicrograph showing the structure of sheet-bronze armor from Crete, ca. 600 B.C. About 10 percent tin. Etched. × 200.

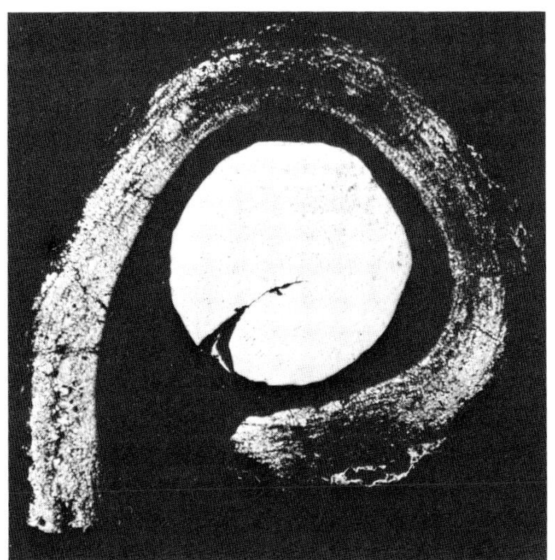

Figure 9.49
Mitra made of sheet bronze with repoussé decoration. From a cache of armor found in Afrati, Crete, ca. 600 B.C. Width, 24.2 cm. Photo courtesy of the Schimmel Collection.

Figure 9.50
Cross section showing the construction at the edge of a piece of Cretan armor. The structure of the central stiffening wire shows that it was bent from strip and hammered round, not drawn. The surrounding sheet metal is heavily corroded, especially where most severely bent. Etched. × 18.

would be today, but by longitudinally bending and rounding up a strip of flat metal of rectangular section, clearly cut from sheet that had already been considerably thinned by hammering. That the cross-sectional area had not been much changed in the final rounding is shown by the fact that the inclusions are no more elongated in a longitudinal microsection of the wire than in a transverse one: in a drawn wire there is, of course, an enormous difference in distortion in these two directions. After removal of the surrounding sheet metal the surface of this wire (shown in figure 9.51) can be seen to have preserved many coarse abrasion marks obviously made by some kind of a file used in its shaping. These were, of course, invisible in the finished armor. Wire was also made simply by forging a rod to successively smaller diameter, but for the finer wires the hammering of a sheet, followed by slitting, twisting, or longitudinally folding and rounding, was far less laborious.[46]

It is often difficult by external examination to determine precisely what means were used in shaping a metal object. The finishing of the surface may obliterate all earlier tool marks, and ancient objects were not infrequently made in a way that does not appear natural to us. Shapes which would easily be made from soldered wire were sometimes cast, and designs most appropriate for working in repoussé in thin sheet metal were made by carving from a thicker slab. Examples are the incredible gold filigree pieces from the Sinu culture of Colombia (figure 9.52) and the famous Sasanian silver dishes (figure 9.53). In both cases, however, there are details that reveal the technique. In the first, the filigree is actually that of threads of wax or other plastic and volatile substance that were joined together to make the pat-

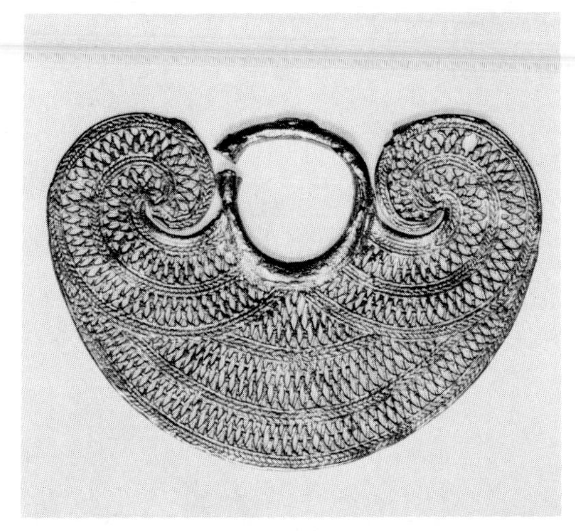

Figure 9.51
Bronze wire removed from the turned-over edge of piece of armor from Crete. Note the longitudinal seam and the file marks. × 10.

Figure 9.52
Nose ornament of gold filigree, made by lost-wax casting. Sinu culture (Colombia). Pre-Columbian, after 1000 A.D. Width, 6.0 cm. Photo courtesy of the Field Museum, Chicago.

Figure 9.53
Silver dish with decoration produced by tracing, carving, and inlaying of both cast and hammered detail. Sasanian, fourth century A.D. Diameter, 23.9 cm. Photo courtesy of the Smithsonian Institution, Freer Gallery of Art, Washington, D.C.

tern, subsequently cast by the lost-wax method. This leaves a surface quality that has a faint granularity incompatible with drawn wire and a curvature of the locally melted material to make the joints that is not characteristic of metal.[47] The Sasanian work has a smoothness of surface and a sharp merging of convex design parts into the background that is unusual in repoussé, and the fact that the back of the plate (which is solid) shows no concavity matching the convexity of the front is, of course, proof that it was not so made. The surest way of revealing such methods is to examine the objects under the microscope. If a cross section of the metal can be cut out for examination (which it rarely can be on objects of high quality for obvious reasons), the microstructure usually instantly reveals what has happened.

A magnified cross section cut from a damaged area of a similar Sasanian silver plate reveals, after etching (figure 9.54), many parallel lamellae that had unmistakably arisen in the structure of the cast material (containing enough copper to form a second phase), but these had been considerably extended by the flattening of a thick ingot into the thinner plate. Note, however, that these lamellae do not run parallel to the curves of the surface but frequently intersect it in an angle: The present surface has clearly been formed by the actual removal of metal from a thicker piece. This part of the design had therefore been mainly shaped by chiseling, scraping, or abrasion (or a combination of all three) and not by the simple bending and stretching of repoussé which leaves the grain parallel to the surface. In common with much Sasanian silverwork, this plate displays a masterly use of different techniques for different parts of the design. Its richness comes, in fact, from this diversity. The king's head in full relief is a casting that is attached mechanically into a grooved space cut to receive it; the horse's thigh and other parts of medium convexity are previously shaped by bunting and by carving and then inlaid into spaces the edges of which are cut and turned up by means of a chisel to receive them (the missing piece at 7 o'clock allows this to be seen). Portions of the design are continued in the form of lines simply punched with a tracing tool into the smooth surface of the ground. Finally, some areas are amalgam gilded, others left in white silver to contrast.[48]

Let us return to the bronze armor. Figure 9.55 shows a section through a fragment with shallow repoussé decoration like that around the border of the mitra in figure 9.49. Compare this with figure 9.54. Here the grain of the metal obviously follows the contortions of the surface, for the shaping has been done by the use of tools that have locally bent the thin sheet to conform to the design. (It was just this final work, done on sheet that has been annealed, that gave rise to the slip bands discussed in connection with figure 9.48 above.) One side of the piece has considerably less relief than the other, and the partial sectioning of the microstructural lamellae by the present surface suggests that the high spots had been worn down. If the edges were turned over as in figure 9.49, this worn surface was the outer one, and one can confidently surmise that the metal had been abraded by frequent polishing of the armor to make its wearer more conspicuous on parade or battle ground.

Up to the present, the microstructure of castings, though frequently used to confirm that an object was cast, has not been studied for further detailed information that it might

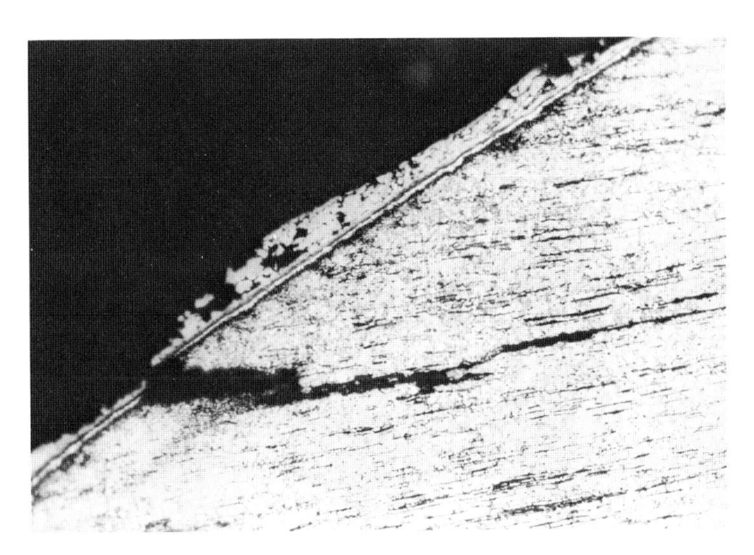

Figure 9.54
Section of Sasanian silver dish, through a part of low relief similar to the boar's foreleg in figure 9.53. Etched. × 250. Photo by Katherine Ruhl, Case-Western Reserve University.

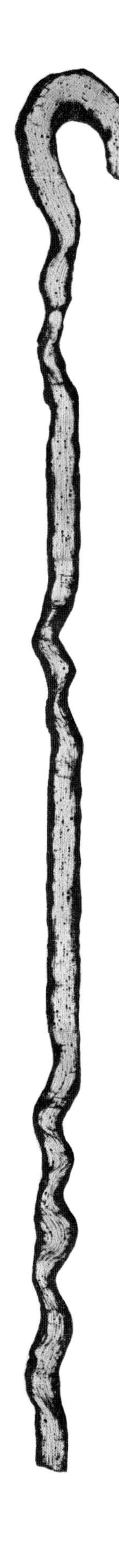

Figure 9.55
Section of bronze armor with repoussé bands similar to those at the border of the mitra (figure 9.49). Etched. × 8.5.

yield on historical techniques. The most important tool of all remains radiography, which has been so finely exploited by R. J. Gettens in his studies of Chinese bronzes.[18]

The Japanese Sword

The best of all examples of a satisfactory art form based upon the inner nature of a metal is provided by Japanese swords.[49] *Nippon-to* have been called the supreme metallurgical art, but they are not widely appreciated in the West because the richness of their beauty is only apparent on rather close and informed inspection. They are also impossible to photograph satisfactorily. Our example (figure 9.56) was obtained with illumination adjusted to enhance the contrast, and has a quality that does not fully represent the subtle texture of the blade. Both the beauty and the serviceability of Japanese sword blades arise in their micrometallurgy.[50] Their efficacy as weapons depends upon the intense hardness of the cutting edge, but this is backed by massive softer metal in a way that prevents the whole sword from being brittle. Their aesthetic quality resides in many things, including the shape of the interface between hard and soft metal, the texture of the metal itself, and the final polishing operation, which gives precision of shape and surface while revealing the local differences in the character of the metal as subtle changes of reflectivity.

Our perception of beauty seems to involve the interaction of several patterns having origin and significance at many different levels of space, time, matter, and spirit. In the Japanese sword blade there is heterogeneity in both the macrostructure and the microstructure. The manner of forging, the

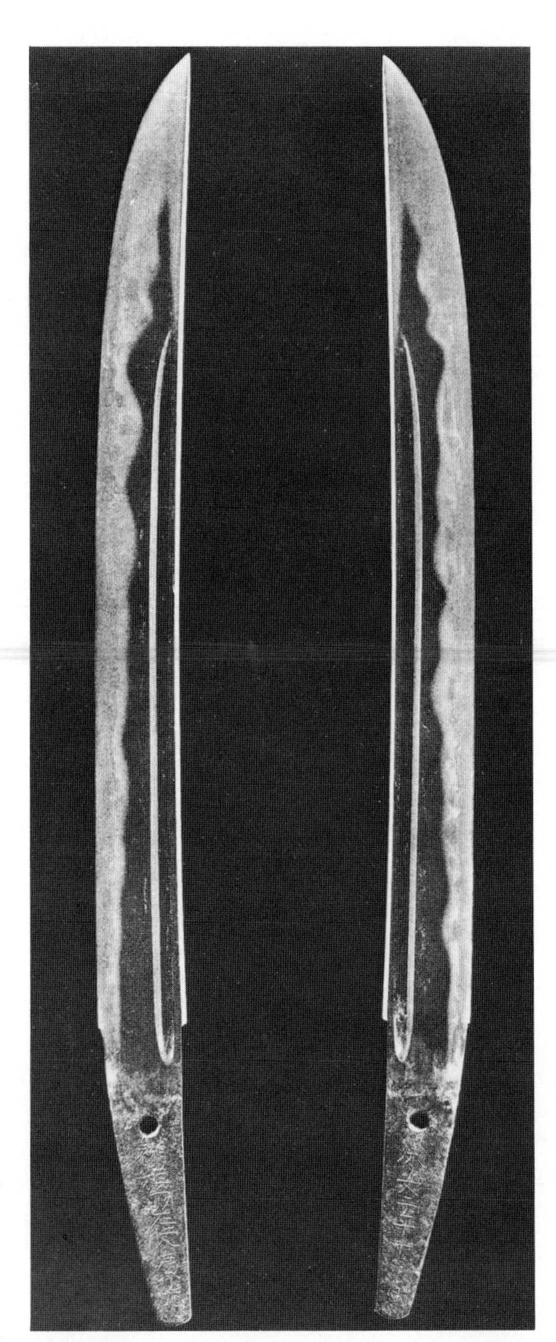

Figure 9.56
Sword blade forged by Hiromitsu Sagami, mid-sixteenth century. Note, beneath the conspicuous smoothly waving outline of the polisher's decoration, the intricate pattern of hard and soft areas with slightly differing reflectivity. Photo courtesy of Walter A. Compton.

heat treatment, and the final polishing operation are all uniquely Japanese techniques, and all make necessary contributions to the final quality of the blades. The shape alone would be simplistic form; the forged texture of the steel without heat treatment would at best faintly echo the beauty of grained wood; the outline of the quench-hardened zone at the edge would be sharp and uninteresting if it depended only on the control of cooling rate during quenching; and the polish would be uniform glitter if the metal were homogeneous. With true artistry all these are made to interact.

Steel, of course, acquires its full hardness only when heated above a critical temperature and very rapidly cooled. In the West today local hardening in steel is usually produced by locally heating and quenching. The Japanese swordsmith did the opposite: He heated the entire blade to the appropriate temperature and quenched it, but the part that he wished to remain soft was coated with a thin adherent layer of a thermally insulating compound to delay cooling locally. The final outline (*hamon*) of the hardened area on the surface of the blade marks the narrow zone in which the steel cooled at a critical rate, just slowly enough to allow the microcrystals of austenite (the high-temperature phase of steel) to begin to decompose, but not slowly enough for them to have fully changed into the soft form of steel, called pearlite by metallographers. It is a subtle combination of circumstances, depending partly on the hardenability of the steel itself (which, since no alloying elements were used, in turn depends mainly upon the carbon content and the grain size, which in turn is influenced by prior heat treatment and the distribution of minute inclusions as affected by the forging operation), and partly on the thermal gradient, which is controlled by the thickness of the insulating layer and the sharpness with which its thickness and outline change. The gross outline and general form were fixed by the hand of the smith as he applied this thermal barrier, but many fine details and nuances of pattern come from the local modification of heat flow resulting from variations of thickness and density of this layer as well as its behavior under the shock of immersion in the quenching bath. In addition there is the partly controlled heterogeneity of the steel itself. The diffusion of both heat and carbon in the steel produces pleasing contours, for, as in all flow phenomena, locally abrupt gradients become smoothed out by interaction with regional conditions.

Figure 9.57 is an entire cross section of a sword clearly showing the rich internal pattern of hard and soft layers that are revealed by intersection with the surface. This structure illustrates most of the features that appear on any good blade, although the mass of soft low-carbon steel in the center of the blade is more homogeneous than usual. The cutting edge and the entire visible surface is of a carefully forged steel containing about 0.6 percent carbon, with only traces of nitrogen and oxygen as alloying elements. This steel is laminated on an invisible scale arising from the forging process mentioned below, but usually it also has some layering on a potentially visible scale coming from the welding together of slabs of two or more kinds of steel of different hardening characteristics at a late stage of forging—not more than about ten stages from the end or they become too thin to be visible to the naked eye. Hammer blows locally displace the lamellae from a true plane, and the final

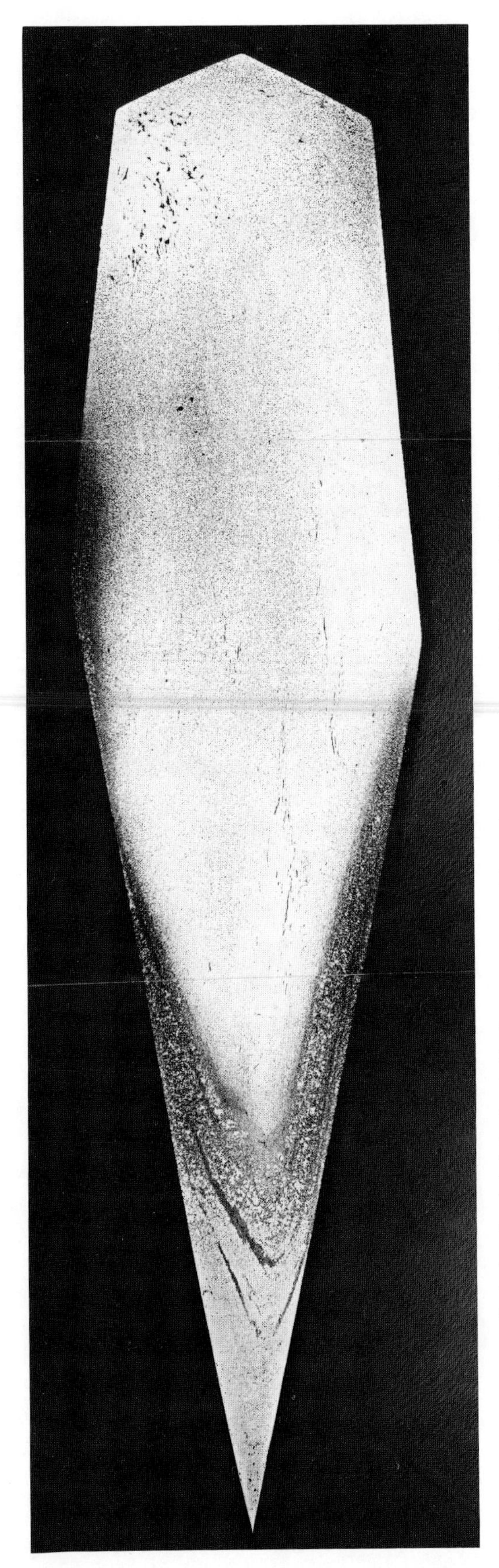

scraping of the blade to shape cuts through some of these lamellae at the surface, giving outlines like a contour map of a complicated terrain of hills and valleys. Unless the inclusion contents of these layers differ, they are not distinguishable from each other in those parts that cool either fast enough to be fully hard or slow enough to be fully soft, but in the zone of intermediate cooling rate there will be large local differences in the hardening of the steel in a pattern that depends upon the local contortions of the metal in the last few stages of forging. When steel with such heterogeneity on a micro and semimicro scale is heat-treated under conditions that give a complicated pattern of thermal gradients, the result is a texture of unmatched richness. Like a potter with his glazes, the smith sets up the environment for hardening his sword but leaves the actual outcome locally to the balance of physical factors—composition and thermal gradients, nucleation and time.

The rich patterns of the *yakiba* do not arise either in the complicated regimen of the forging or in the intricacies of the heat treatment. They are interactive in their very essence. In some measure, their generation is analogous to the generation of a physical landscape, which is the result of the interaction between the geologically determined stratification of the rock, the interactive gradient-determined flow of erosive water, and the nucleation, growth, and interference of individual patches of vegetation—all with

Figure 9.57
Cross section of blade attributed to Nagamitsu (1222–1297). Note the white core of low-carbon steel, surrounded by somewhat lamellar darker metal at the face and cutting edge, and the hardened zone at bottom. Etched. Height of section, 2.7 cm. (Later swords usually have a core with a more complex structure than this.)

some man-determined features superimposed and everything softened by the effects of time.

The best properties of a modern steel usually come when the grain size is kept far below the range of visibility to the naked eye, but the Japanese swordsmith intentionally produced grains just large enough to be seen, undoubtedly because he could then use the visible pattern to determine the success of his work—though perhaps in recent centuries he has turned his eye more on the aesthetics than on technique. In any case, these large grains of hard steel embedded in a mass of softer transformed material constitute a well-known feature, the white spots called *niye* (figures 9.58, 9.60). These cluster in narrow bands along the edge of the *yakiba*. They also often occur in isolated patches in the soft metal beyond this and sometimes as streaks extending into the generally soft area and marking the presence of an isolated lamella of steel of high hardenability (figure 9.59). If these hard grains are too small to be resolved by the naked eye, the cloudy line that they form at the edge of the *yakiba* is called *nioi*. It is interesting to compare the microstructure of the transition region in a Japanese sword, shown in figure 9.60, with the microstructure of a steel of similar composition heat-treated in a temperature gradient in a modern research laboratory in a standard test to determine hardenability (figure 9.61).

The smiths contrived and controlled the details of the heat treatment to make the zone of structural transition relatively wide, and various masters introduced innumerable individual complexities with a double motive—mechanical and aesthetic—that has made these objects unique both as weapons and as works of art. The patterns of different swordsmiths are as diverse and as distin-

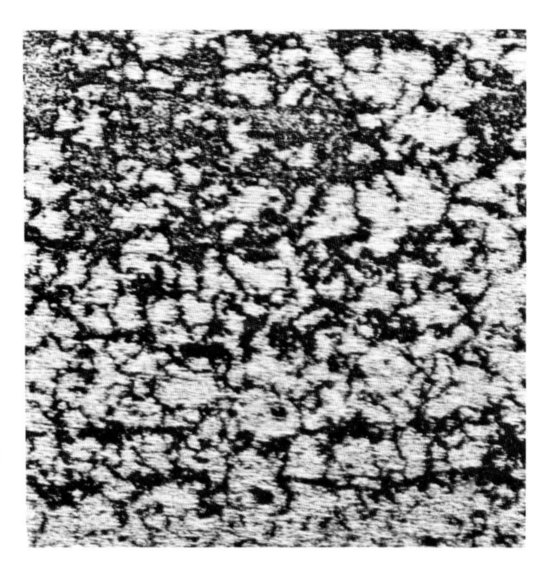

Figure 9.58
Microscopic view of surface of sword blade signed Sukehiro Tsuda and dated 1677 but probably an eighteenth-century copy. Area showing *niye* spots in transition zone at edge of the hard area, revealed by the Japanese abrasive polishing process. × 50.

Figure 9.59
A sixteenth-century Japanese sword blade. Surface etched in nitric acid to intensify contrast. (Cutting edge at bottom.) × 1.8.

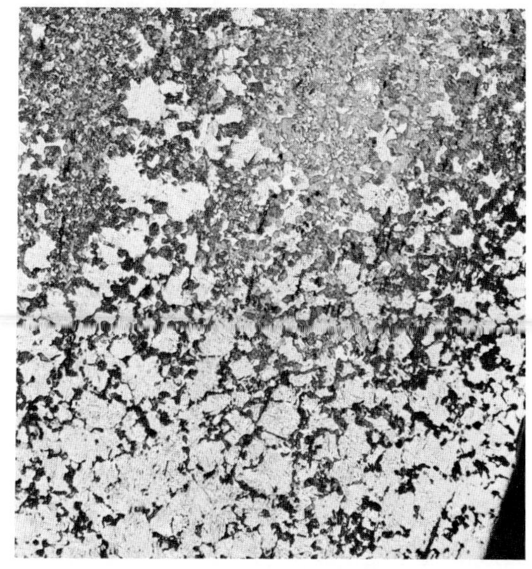

Figure 9.60
Microstructure of the transition zone in an unsigned sixteenth-century sword. Cross section, with surface of blade at right. Etched. × 65.

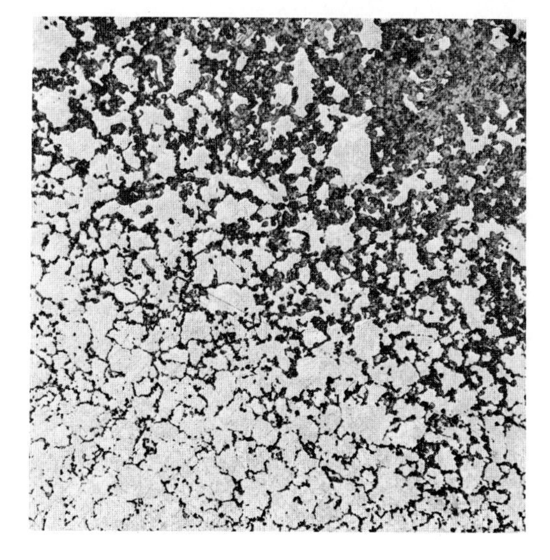

Figure 9.61
Transition zone in a steel rod (0.6 percent carbon), end-quenched in a modern test for hardenability. × 140. Photo courtesy of U.S. Steel Corp.

guishable as are the brush styles of different painters—a recent book[51] illustrates over 90 different *hamon*—and the patterns of heat flow can be combined with differences in steel texture and sword shape to give almost infinite variety.

The specimens in figures 9.59 to 9.61 were etched, which leaves the hardest areas (the metallurgist's martensite) light, and the relatively soft areas (pearlite) dark. In figure 9.58 the structure was revealed by the normal Japanese polishing process—actually an abrasive finish rather than a true polish—which has left the hard parts with a finely matte finish to scatter the light. The polishing is done with great precision by the use of a series of graded abrasive stones. When finished, the different surfaces of the blade are covered with microscratches of different degrees of coarseness, all nearly invisible to the naked eye but producing subtle differences in reflectivity. In proper light the *yakiba* seems to have an inner glow, which arises in the roughness produced by the penultimate stone persisting while the areas of softer metal acquire a better polish. The small *niye* spots, however, shine like tiny mirrors set in the darker, softer ground. The back of the sword is given a high polish by burnishing.

Figure 9.62 shows other areas of the sword shown in figure 9.58, *c* and *b* being the fully hard and fully soft zones (the *yakiba* and *jigane* respectively), with their different scratch textures, and *a* the fully burnished area (the *shinogi-ji*) between the ridge and the back of the blade. In recent years collectors (including even some great museums) have come to prefer a rather gaudy false *yakiba* produced by local manual application of abrasive. Though this does yield a more striking *yakiba,* it does so by overwhelming the rich metallurgical texture produced by

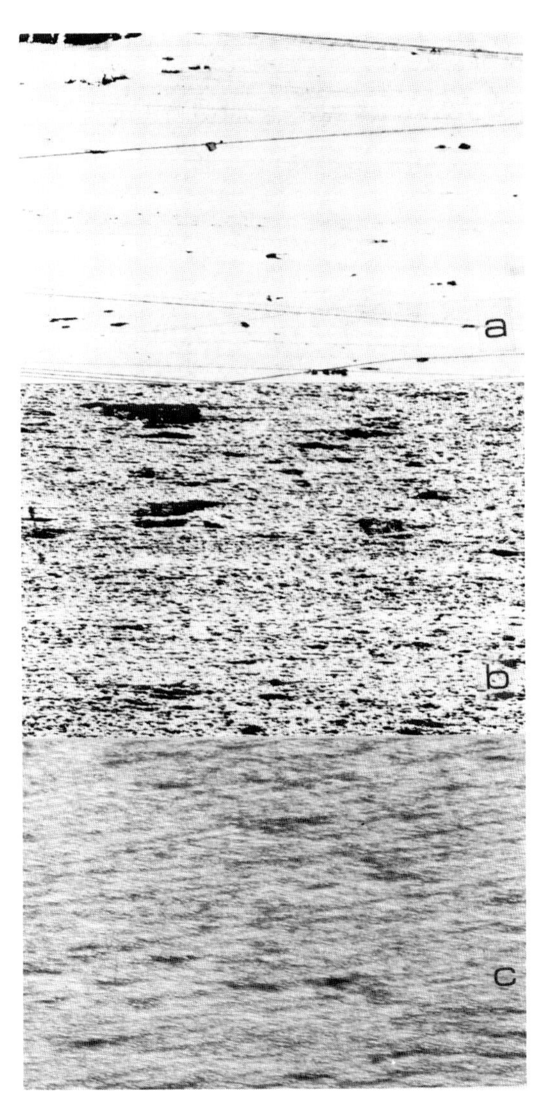

Figure 9.62
Surface textures of a blade by Sukehiro dated 1677. Japanese polish, probably done in early twentieth century. × 50. (*a*) Burnished area in *shinogi-ji*. (*b*) Soft blade surface *jigane*. (*c*) Hard metal near cutting edge (*yakiba*). Parts *a* and *b* differ mainly because of the polishing process; *b* and *c* because of the reaction of metal of different hardness to the same abrasive treatment.

the smith by the simpler finger-marks of the polisher, and it should be condemned.

The quality of the steel in the blade, as well as its visible texture in the areas away from the edge of the *yakiba,* comes from the special process of forging. In common with most steel made before 1860, the steel for Japanese swords was not melted. The partly consolidated metal sponge that came from the ore-reduction furnace was highly heterogeneous. Not only did the carbon content locally vary between none and about 2 percent, but there were large voids and inclusions of rocky matter from the ore, partly converted to slag. By the repeated extension, folding, welding, and drawing-out, the steel was rendered homogeneous in carbon content, much of the slag was worked out, and whatever slag remained was reduced to minute well-dispersed particles. Though ancient smiths everywhere forged iron to make it more homogeneous and to break up the slag, the Japanese swordsmith repeated the folding, welding, and reforging operation many times, sometimes as many as twenty, which would give 2^{20} or slightly over a million layers of steel with an equivalent increase in area and decrease in thickness of the original laminations. If the deformation were microscopically homogeneous, this would result in steel with no visible texture, but visible texture does arise partly because the slag resists deformation differently from the surrounding steel and partly because it gathers into small drops under surface tension, but mainly because some additional slag is inevitably picked up at each stage in the forging operation and additional texture was achieved by the intentional lamination of different kinds of steel during the last stages of forging the metal for the critical parts of the blade.

The best well-forged steel was used only for the outer part of the blade and the cutting edge. The inside, never seen on a finished blade, was composed of soft steel, piled in various ways depending upon the smith's preference. This core was always of low carbon content and frequently contained rather large inclusions of slag, which are not mechanically harmful in this location and may actually contribute to the absorption of shock when the blade is used.

When one looks at the whole sword, these microscopic details take their place. The sword is finished by polishing on a series of graded abrasive stones, done by hand but controlled with a precision that results in almost perfect geometric surfaces intersecting to give the clean-cut outlines. The final texture and reflectivity of the surface reveal the very essence of the steel, for the changes of roughness on a microscopic scale directly follow the metallurgical features and make them visible to the naked eye. The association of the rich gradients of metallurgical texture with the austere geometry of the shape gives an aesthetic effect of the highest order. In addition, the Japanese see in their swords a rich symbolism with overtones of Shinto traditions, Zen aesthetics, and national history, the full extent of which a Westerner can hardly appreciate. The whole thing, however, depends upon technical factors—slag, grain size, crystallographic change, the diffusion of carbon and heat in temperature gradients, and the reaction of a steel surface to an abrasive.

The best blades of all time are those forged in the Kamakura period (1190–1337). The technique of making a sword capable of passing severe tests was then undoubtedly more important than the aesthetics, but the evident relationship between the visual quality and serviceability enabled the swordsmith to improve and to control his mechanical and thermal procedures. A good sword would have a pattern that would not appear without proper control of texture, carbon content, and cooling rate. There are more spectacular examples but no better ones to demonstrate the aesthetic relationship between parts and wholes, of cultural overtones interacting with the properties of matter, which is the theme of this paper.

Conclusion

In all of the examples that have been cited, the physical properties of materials underlie some aesthetic quality. They may be effective on a scale that cannot be resolved, or may give a grossly visible feature with characteristics comparable to an abstract painting. As with Croesus's gold, Portia's leaden casket, or the Prussian iron cross, the material itself is sometimes symbolic. Usually much of the appeal of a material is a natural one, not dependent upon iconography and, frequently, not even on the form in the artist's mind or its manipulation by his fingers. However, the material always relates in subtle ways to the artist's concept. Though at his greatest moments an artist may be concerned only with transmitting the highest levels of human experience and insight, the details of his finished work will always depend intimately upon its technique. And in a period such as the present when there is a definite trend to nonobjective art, the formative qualities of the materials can account for a larger fraction of the total effect than in the past.

In the Orient, far more than in the West, the appreciation of a work of art involves consideration of its maker's sensitive use of the properties of material. Indeed, the physical qualities and the sometimes not seen but always sensed microstructural characteristics of materials such as paper, ink, glaze, lacquer, wood, and jade are essential components of all the great art of the Far East. As art in the West turns away from the representational, it has led increasing numbers of collectors and museum visitors to a deeper enjoyment of those objects of the past that are casually cataloged as "minor" or "decorative" arts.

In the twentieth century, creative artists seem no longer to be leading the discovery of new materials or physical phenomena, though they study the new products of technology with considerable interest. Is this an inversion of the age-old pattern in which the artist discovered, the engineer used, and the scientist explained? Both the scale and the philosophy behind the works of many artists seem to be moving away from the structural diversity of materials that has been emphasized in the body of this paper. Many artists are moving entirely outside of museum classification and are using their skills to make people look at the new environment provided by science and technology. Lecture demonstrations originally used by nineteenth-century physicists to capture the interest of students are being presented as an aesthetic experience to a wider audience, while intriguing scientific toys abound. If more people experience moving patterns based on polarized light, diffraction, surface tension, diffusion, flow and stress, and stroboscopy, or otherwise react to environment based on technological or psychological gadgetry, it will certainly lead to wider understanding of the nature of visual perception and the operation of the physical world, but it seems unlikely that these devices can ever do for the spirit of an individual what the contemplation of a fine pot or painting does.

Artists, always searching for things to manipulate for visual enrichment, are led by the present world of artifice to examine relationships on a larger scale than those that could be captured in objects fashioned by their own hands for an elite market. Their innovative role seems to be turning toward the design of environment, large structures that can only be cooperatively (or corporately) produced and democratically experienced. Nevertheless, there will always remain a need for objects to give private delight, and the resonance between the properties of matter and the perceived qualities of an object will continue to provide a human experience worth cultivating. The properties of matter are real and in atomistic detail exact and repetitious; yet they are diverse enough in combination and separation to suggest (as, of course, they actually produce) the immense richness of the universe. The entire record of history is preserved ultimately in the structure of materials at various levels. Man himself is perhaps at the scale of maximum significant complexity based on the interaction of atoms, for much larger aggregates have no specific communication between all their individual components. Much of man's exploitation of the richness of materials has come about because he can see them at a scale where a new quality is emerging from the structural hierarchy of quantity. Materials, when manipulated into an aesthetic form in resonance with their finer structure, can interact with the hierarchical properties of man's senses and his mind to convey both deep emotional significance and logical meaning.

Acknowledgments

This paper would have been impossible without the willing help provided by the staff of many museums and libraries, especially the Boston Museum of Fine Arts, the Fogg Museum, and the Freer Gallery. The author's research in the history of technology has been supported in part by the National Endowment for the Humanities, by MIT's Sloan Basic Research Fund, and by the Wenner-Gren Foundation. James Howard helped with photography, and the author's colleagues Heather Lechtman and Arthur Steinberg have provided valuable criticism of an early draft of the paper.

Notes and References

1
See chapter 8 in this volume.

2
A good summary from the craftsman's point of view of most modern and traditional ways for working metal on a small scale will be found in Oppi Untracht, *Metal Techniques for Craftsmen* (New York, 1968). For ancient techniques see Herbert Maryon, "Metal working in the ancient world," *Amer. Jour. Archaeology* 53:93–125 (1959); and H. Maryon and H. J. Plenderleith, "Fine metal work," in Charles Singer et al., eds., *A History of Technology* (Oxford, 1954–1958) vol. 1, pp. 623–662. Among the copious literature on ceramics the best welding of technical detail and art is in H. H. Sanders, *The World of Japanese Ceramics* (Tokyo and Palo Alto, 1967).

3
The effects of surface tension are commonly referred to as capillarity, for they were first mea-

sured in glass tubing drawn to the size of a coarse hair. They can be envisaged as an equilibrium between stretched elastic membranes in the surfaces, expanding or contracting until their forces normal to the lines where they join achieve equilibrium, while elsewhere balancing pressure differences by their curvature. When, as in this example, two of the surfaces are rigid and plane, the adjustment can only be in the liquid, which spreads or contracts (its volume being fixed) until its surface meets the solid at just the angle at which its resultant force, projected in the plane of the solid, exactly matches the difference in surface tension between the wetted and nonwetted surfaces. A liquid will spread over a solid only if the contact angle is zero, i.e., only if its surface energy plus that of the interface is less than that of the solid itself. In systems where all three phases are fluid, all three angles adjust until the ratios of their sines to the tension of the opposite interfaces become equal. In a single surface such as ceramic glaze with local differences in surface tension, lateral flow will occur locally at such a rate that the shear force, acting against the pertinent thickness and viscosity of the material, becomes equal to the local gradient in surface tension.

A surface free to conform to surface tension between fixed boundaries adjusts its shape to achieve the minimum area possible under the restrictions placed upon it. Even if, because of viscosity, the surfaces do not have time to adjust completely to the configuration of minimum area, local approach to equilibrium will occur more rapidly than long range adjustment and quickly eliminate jagged junctions or outlines. Consider the outlines of the splashed paint in a work by Jackson Pollock, or (with different physics and a much longer time scale) the natural rounding of a landscape to adjust to a volcanic event or road builder. All these patterns are basically time-related. The local rate of change varies with the local curvature, and surfaces of gentle curvature merge into and balance the requirements of more quickly established local geometries. They may have complicated curvature, but in their balance between local and regional requirements they achieve a natural appearance that is almost invariably pleasing to the human eye. This is perhaps a result of locking-in with a similar hierarchy of size- or complexity-dependent reaction times in the mechanism of human perception.

4
Mappae Clavicula, Latin manuscript originating ca. 800 A.D. Text of expanded twelfth-century version edited by Thomas Phillipps in *Archaeologia* 32:183–244 (1847); English translation by J. H. Hawthorne and C. S. Smith (Philadelphia, 1974).

5
Theophilus, *De Diversis Artibus*, manuscript ca. 1125 A.D. Text and translation by C. R. Dodwell (London, 1961); translation with technical notes by J. H. Hawthorne and C. S. Smith (Chicago, 1963).

6
Benvenuto Cellini, *Due trattati, uno intorno alle otto principali arti del l'oreficeria. L'altro in material dell'arte della scultura* (Florence, 1568). English translation by C. R. Ashbee, from the Italian edition edited by Milanesi, 1857 (London, 1898; reprinted, New York, 1979).

7
Hans Wulff et al., "Egyptian faience: A possible survival in Iran," *Archaeology* 21:98–107 (1968).

8
C. Kieffer and A. Allibert, "Pharaonic blue ceramics. The process of self glazing," *Archaeology* 24:107–117 (1971). See also J. V. Noble, "The technique of Egyptian faience," *Amer. Jour. Archaeology* 73:435–439 (1969).

9
Fujio Koyama, *Shigaraki otsubo* (Tokyo, 1965). [See also Louise Cort, *Shigaraki, Potters' Valley* (Kyoto and New York, 1980). Both books have superb plates.]

10
A. L. Hetherington, *Chinese Ceramic Glazes* (Cambridge, 1937), p. 47.

11
F. Koyama and K. Yamasaki, "The *Yohen temoku* bowls," *Oriental Art* 13:89–93 (1967).

12
Theophilus, *De Diversis Artibus,* 1963 edition, pp. 132–139.

13
Ibid., pp. 167–179.

14
See, for example, chapter 4 in this book.

15
Vannoccio Biringuccio, *De la Pirotechnia* (Venice, 1540). English translation by C. S. Smith and M. T. Gnudi (New York, 1942), pp. 72–74.

16
Adolf Goldschmidt, *Die Deutschen Bronzeturen des Frühen Mittelalters* (Marburg, 1926); D. W. Laging, "The methods used in making the bronze doors of Augsburg Cathedral," *Art Bull.* 49:129–135 (1967).

17
Noel Barnard, *Bronze Casting and Bronze Alloys in Ancient China* (Monumenta Serica Monograph XIV, Canberra and Nagoya, 1961); "Chou China: A review . . . ," *Monumenta Serica* 24:207–458 (1965).

18
R. J. Gettens, *The Freer Chinese Bronzes,* vol. 2, *Technical Studies* (Washington, D.C., 1970).

19
Chao Hsi-ku, cited in Barnard, *Bronze Casting*, pp. 95–96.

20
The Enjo Foundry [Matsue, Japan]. 16 mm film, in color, available from the Desert Research Institute, Tucson, Arizona. Photographs and drawings showing the construction of bell molds and diagrams of the principal Japanese temple bells with dimensions of several hundred more are given in Ryohei Tsuboi, *Nippon no Bonsho* (Tokyo, 1970).

20a
Tetsuji Nagago, *Asiya-kei No Kama* [Tea kettles of the Asiya District] (Kyoto, 1957); with 58 superb plates.

21
Max Loehr, *Ritual Vessels of Bronze Age China* (New York, 1968). Excellent photographs of Chinese bronzes are to be found in John Alexander Pope et al., *The Freer Chinese Bronzes,* vol. 1, *Catalogue* (Washington, D.C., 1967). A wider range of types is shown in William Watson, *Ancient Chinese Bronzes* (London, 1962).

21a
(*Note added in proof, 1980*) *The Great Bronze Age of China*, edited by Wen Fong (New York, 1980), the catalog of an exhibition sent by the People's Republic to the Metropolitan Museum of Art and other museums in the United States, provides an excellent and well-illustrated discussion of all aspects of Chinese bronze technology. This opportunity for further study of the objects has made the present writer uneasy about his earlier rejection of the possibility that something like the lost-wax process was used. Shang bronzes certainly have more restrained design than would be associated with the plasticity of wax, yet there are many details that overhang too much to be compatible with simple withdrawal of the curved mold sections of which they are integrally part. Could there perhaps have been some relatively hard but thermally evanescent material used—something that would lose coherence on heating but not melt and evaporate as does wax? A plastic composition of some fine refractory powder such as levigated loess mixed with an organic adhesive could be applied over the prepared core, shaped by molding and carving, dried, and covered with clay. Firing in a kiln would oxidize the organic matter and leave a loose pile of ash to be removed after the mold had been opened. This would necessitate divisions in the mold, which are not required with wax. The inscription would have been previously formed in relief on the surface of the core, and this would be responsible for its manifestly different quality from comparable intaglio features on the outer surface of the vessels. Experiments with both fine silica and with charcoal dust mixed with flour paste have produced extremely sharp detail in clay investments after firing in air at a red heat.

22
A late Chou dynasty bronze *chien* in the Freer Gallery (No. 39.5) has two bands of decoration, one concave and one convex, each reproducing exactly the same detail. Supposedly this was premolded in thin flat slabs, cut and bent to conform to the curved local surface when applied to the main mold. [This *chien* has been studied in detail in a fine paper by Barbara Keyser, *Ars Orientalis* 11:127–162 (1979). She believes that the premolded slablets were applied to the positive pattern rather than to the negative mold. Perhaps both techniques were used: A Chou dynasty bell at the Fogg Museum, a twin to that at the Freer

shown in our figure 8.5, has joints in the decor on its sides where bronze has unmistakably run into negative crevices.]

A similar technique is sometimes used to give positive decoration in ceramics. There is a large eighth- to ninth-century Persian water jar in the Boston Museum of Fine Arts (No. 58.92) which has about 40 thin small slabs with premolded design applied to give a continuous band of decoration on the convex shoulder of an otherwise smooth pot. In many late Chou bronzes, a small unit of pattern has been repeatedly replicated and these pieces assembled into larger complex blocks which were used for impressing the clay slab applied to the mold to give the final surface in bronze. This results in sectionalism in the design that is not identical with any physical division of the mold.

22a
This paragraph is new (1980).

23
Herbert Maryon, "The making of a Chinese bronze mirror, 2," *Archives, Chinese Art Soc. of America* 17:23–25 (1963).

24
Recent papers, which include summaries of the earlier literature on the subject, are G. L'E. Turner, "A magic mirror of Buddhist significance," *Oriental Art* 12:1–5 (1966); M. Watanabe, "The Japanese magic mirror," *Archives, Chinese Art Soc. of America* 19:45–51 (1965).

25
Herbert Maryon, "A note on magic mirrors," *Archives, Chinese Art Soc. of America* 17:26–28 (1963); discussion by C. S. Smith, ibid., pp. 29–31.

26
Al-Jazari (Ismail ibn al Razzaz), *The Book of Knowledge of Ingenious Devices,* translated from the Arabic by Donald R. Hill (Boston, 1974).

27
Heather Lechtman and Arthur Steinberg, "Bronze joining: A study in ancient technology," in S. Doeringer et al., eds., *Art and Technology: A Symposium on Classical Bronzes* (Cambridge, MA, 1970), pp. 5–35.

28
John Gloag and Derek Bridgewater, *A History of Cast Iron in Architecture* (London, 1948); R. Lister, *Decorative Cast Iron Work in Great Britain* (London, 1960).

29
Cyril S. Smith, *A History of Metallography* (Chicago, 1960), chapter 15.

30
H. Hoffmann and V. von Claer, *Antiker Gold und Silber Schmuck* (Mainz, 1968); H. Hoffmann, "Greek gold reconsidered," *Amer. Jour. Archaeology* 73:447–451 (1969).

31
D. L. Carroll, "Drawn wire . . . in ancient jewelry," *Amer. Jour. Archaeology* 74:401 (1970). [See also W. A. Oddy, "The production of gold wire in antiquity," *Gold Bulletin* 10:79–87 (1977).]

32
C. S. Smith, "Methods of making chain mail (14th to 18th centuries): A metallographic note," *Technology and Culture* 1:60–67 (1959); discussion pp. 151–155, 289–291.

33
Dominique de Menil, *Made of Iron* (Houston, TX, 1966); F. Contet, *Documents de Ferronnerie Ancienne* (7 vols. of plates, Paris, 1923–1929); Otto Hoever, *A Handbook of Wrought Iron from the Middle Ages to the End of the Eighteenth Century* (London, 1962).

34
All of human history rests upon very subtle properties of matter. Life itself is built on the rigid fourfold coordination of the carbon atom. Rock-building is the slow sorting of atoms into aggregates of crystals best suited to the idiosyncracies of whatever atoms come close enough to interact within the time available. The dependence of lakes and life on the unique properties of H_2O near the freezing point has often been noted. Technology similarly exploits such differences in the behavior of matter, and its history is an interaction between the basic physicochemical possibilities and the more complex probabilities of man discovering them and seeing value in their exploitation. If the heat of formation of aluminum oxide had not been so high, aluminum metal could have been reduced with carbon as the first useful metal millennia before its actual production by potassium reduction in 1825. This would have changed history, for aluminum is a major component of the earth's crust, and once it has been

reduced to metal it has many properties that make it almost ideal for working under primitive conditions. With all easily reduced copper ores already depleted, perhaps the new civilization that will rise from the radioactive ruins of the present one will be aluminum-based! Even now, scrap aluminum from the West is being used in a simple technology in the bazaars of the Middle East.

35
Albert France-Lanord, "Le fer en Iran . . . ," *Revue d'Histoire des Mines et de la Metallurgie* 1:75–127 (1967); C. S. Smith, "The techniques of the Luristan smith," in Robert H. Brill, ed., *Science and Archaeology* (Cambridge, MA, 1971), pp. 32–52; R. Pleiner, "The beginnings of the Iron Age in ancient Persia," *Annals of the Naprstek Museum* (Prague, 1967), pp. 1–63.

36
Robert H. Dyson, "The Hansalu Project," *Science* 135:637–647 (1962).

37
Smith, *History of Metallography,* chapters 12 and 13.

38
H. C. Beck, "Etched carnelian beads," *Antiquaries Journal* 13:384–398 (1933). [See also P. Francis, "Etched beads in Iran," *Ornament* 4(2):24–38 (1980).]

39
Max Loehr, *Chinese Bronze Age Weapons . . .* (Ann Arbor, MI, 1956).

40
W. T. Chase and U. M. Franklin, "Early Chinese black mirrors and pattern-etched weapons," *Ars Orientalis* 11:215–258 (1979).

41
R. Wyss, "Belege zur keltischen Schwertschmiedekunst," in Elisabeth Schmid, L. Berger, P. Bürgin eds., *Provincialia: Festschrift für Rudolf Laur-Belart* (Stuttgart, 1969), pp. 664–681. The author is indebted to Radomir Pleiner for information on this work.

42
Mappae Clavicula, chapters 146, 219, 291.

43
James G. Mann, "The etched decoration of armor," *Proc. British Academy* 28:17–44 (1942).

44
Ladislao Reti, "Leonardo da Vinci and the graphic arts: The early invention of relief-etching," *Burlington Magazine* 113:189–195 (1971).

44a
For details see item 193 in the bibliography.

45
Herbert Hoffmann and A. E. Raubitschek, *Early Cretan Armorers* (Cambridge, MA, Fogg Art Museum, 1972). An illustrated appendix by the present writer (pp. 54–56 and pls. 53–56) gives the results of a detailed metallographic examination. The armor, which is in the Schimel collection, was shown in the exhibition *Master Bronzes from the Classical World* at the Fogg Museum in December 1967, and briefly described in the catalog thereto edited by David Mitten and Suzannah F. Doeringer (Cambridge, MA, 1968).

46
See the references in note 31. The slitting of sheet into rectangular strips remained a useful technique long after true wire drawing became common. The slitting mill using rotary cutters for making iron nail rod from forged flat bar was one of the most ingenious power-driven machines of the late sixteenth century and was an important antecedent to the mill for hot-rolling iron shapes. [The earliest illustration of, indeed the earliest definite technical reference to, the slitting mill is the engraving (figure 8.22) in the little-known book by Jean Errard, *Le Premier Livre des Instruments Mathématiques [et] Physiques* (Nancy, 1584).] Some fine illustrations appeared in Diderot's *Encyclopédie* (Paris, 1765), which also show brass wire being made by drawing a hand-slit strip.

47
I have examined a pre-Columbian fluted bead made of a surface-enriched gold-silver-copper alloy which was so light and thin—0.3 mm—that it was confidently assumed that it had been made from thin sheet metal hammered into bead form. The microscope revealed, however, the outlines of huge dendritic crystals, unmistakably the structure of undistorted cast metal that had been solidified slowly without disturbance in a hot mold. Even thinner flat sheet (0.15 mm) of the alloy tumbaga was made by casting, not hammering. These incredibly delicate castings were supposedly produced by the lost-wax method in molds of powdered charcoal with a little binder.

The mold, however, was probably heated above the melting point of the metal, which simply ran in from a "crucible" space forming an integral part of the mold. Dudley T. Easby, Jr., "Prehispanic metallurgy and metalworking in the New World," *Proc. Amer. Philos. Soc.* 109:90–98 (1965); André Emmerich, *Sweat of the Sun and Tears of the Moon. Gold and Silver in pre-Columbian Art* (Seattle, 1965).

48
W. T. Chase, "Technical examination of Sasanian silver plates," *Ars Orientalis* 7:75–93 (1968). This paragraph also summarizes, in part, the unpublished consensus of three conferences on Sasanian silver held in Washington, Cleveland, and Cambridge, 1968–1971, under the chairmanships of Oleg Grabar and D. F. Gibbons. [See D. F. Gibbons, K. C. Ruhl, and D. G. Shepherd, "Techniques of silversmithing in the Hormizd II plate," *Ars Orientalis* 11:163–176 (1979).] Other types of Sasanian silver, including raised vases and some plates in two parts joined around the edges, are finished by repoussé.

49
The copious literature in Japanese is poorly reflected in Western languages. The best works are B. W. Robinson, *The Arts of the Japanese Sword* (2nd ed., New York, 1970), and the fine catalog of the Japan House exhibition: W. A. Compton et al., *Nippon-to: The Art Sword of Japan* (New York, 1976).

50
The most complete study of the metallography of Japanese swords is Kuni-ichi Tawara, *Nikon-to no Kagakuteki Kenkyu* [Scientific study of Japanese swords] (Tokyo, 1953). In English there is only C. S. Smith, "A metallographic examination of some Japanese sword blades," *Documenti e Contributi,* Centro per la storia della Metallurgia, Milan, Quaderno II (1957), pp. 41–68, and Smith, *A History of Metallography,* chapter 3. A summary article by Tawara in *Kikai Gakkai Shi* no. 54:1–39 (1918), was translated in the *Bulletin of the Japanese Sword Society of the U.S.* no. 6:1–20 (1966). [See also Hiromu Tanimura, "Development of the Japanese sword," *J. Metals* 32(2):63–73 (1980).]

51
W. M. Hawley, *Japanese Swordsmiths* (Hollywood, CA, 1967), vol. 2, pp. 669–683.

10
Reflections on Technology and the Decorative Arts in the Nineteenth Century

Introduction

Art and technology have always been intimately linked; indeed in their origins they were almost indistinguishable. In an earlier paper, I discussed at some length the role of aesthetic curiosity in leading to the discovery of various classes of materials and processes for productive use in engineering or industry.[1] Examples were cited dating from prehistory through the late nineteenth century. In the twentieth century art no longer precedes technology, for invention has become more consciously purposeful, and art enters at a late stage as package design, a marketing aid, not as a conscious inspiration at the beginning. Yet, discovery is an aesthetic activity even today, although its patronage has changed and although for over a century art has led to the discovery of fewer new phenomena than has science. Nevertheless, it is worth pointing out that the ingenuity involved in devising mass production itself began as an urge to multiply pleasant objects, not just utilitarian ones. The craftsman working for an elite trade is, of course, mainly concerned with individual beauty, but many of the eighteenth-century technologists who aimed at more production for less cost were consciously concerned with what might be called art for the masses.

There are probably earlier expressions of this viewpoint in the literature on printing and ceramics, but here let us note the motive behind the metallurgical work of the versatile French scientist R. A. F. de Réaumur. The second part of his famous book, *L'Art de Convertir le Fer Forgé en Acier,* published in 1722, is devoted to the art of making cast iron malleable for the specifically stated purpose of giving cheap cast-iron work a finish comparable to the wrought iron used for elaborate door knockers and artistic *serrurerie* (fine ironwork) in general. In his preface Réaumur remarked, "It is the foremost duty of all of us to work for the general good of society." His sixth memoir begins, "The production of more beautiful work, without sacrifice of quality and at lower cost, is the route to progress along which we must endeavor to guide the arts." Later, "the strikers, or knockers, on carriage gates and other entrances are nowadays almost devoid of ornamentation but they cost as much as quite ornate cast-iron door knockers will cost in the future. . . . From now on, escutcheons, large and small bolts, hinges—in other words, any ironwork pieces that are not subject to strain—can be made most artistically and yet will cost hardly more than the plain ones do today."

Réaumur, however, was worried about the effects of his invention in cheapening art. Though it is important in general to produce more refined, more decorative work, he said,

> One might ask, however, what would be gained by the human race if the number of objects we call "beautiful," and which are simply beautiful, were increased beyond a certain limit. If we knew the secret of how to build palaces as cheaply and as quickly as cottages, if small houses were suddenly changed into magnificent buildings, we should be struck by the novelty of the spectacle. But soon it would be just as well if our common houses had remained unchanged. We should look with less pleasure and interest at the paintings of the great masters if daubers discovered how to paint similar ones. We can judge what we call "beautiful" only by comparison, but we can judge at all times the things directly connected with our occupation and decide whether they are good. There is always something with which to compare them.[2]

Against Réaumur's caution one can advance the view of an Englishman a little over a century later who, inspired by the superb "Berlin" castings, advocated the use of cast iron for sculpture "to supply the place of many less interesting but indispensable architectural and other supports: As the material is so cheap, its application in the way proposed might tend to increase the taste for, and thus foster the patronage of the arts of sculptural design with reference to our mortuary and other monuments."[3]

Réaumur also worked to improve the manufacture of tinplate and porcelain. In a paper read in 1740 to the Académie des Sciences, he stated:

> It is not by their masterpieces, by their rarest products, that [potters] are at their most useful to us, it is by their less perfect works that they provide for our ordinary need. The potter who gives us only glazed pots, made from the commonest and grossest clay, but who gives them to us for practically nothing, is more useful to us than a workman who would have us buy vases at a great price even though these rival in beauty the precious porcelain of China itself.

By using his new process of devitrifying glass, Réaumur said proudly, "There is no one who cannot transform all the bottles in his cellar into bottles of porcelain."[4] Réaumur's poor-man's porcelain never became an important product. Though often discussed by eighteenth-century scientists interested in crystallization, it was used mainly in laboratory vessels, and even this ceased before long.

Reflections on Replication

Another aspect of the relationship between art and technology is the way in which subtle properties of matter basically affect the quality of a work of art. Any artist knows that sensitive selection and treatment of materials is critical to the realization of his intent. In retrospect, I notice that most of the examples I used in an earlier historical paper, which discussed this point in some detail, were from the decorative arts, not the "fine" arts.[5] This may reflect the fact that the latter are less bound than the former to the universal sensual qualities of shape, color, and texture and are more directly concerned with suggesting human overtones within a particular culture. Yet the decorative or useful object in one culture may be regarded a very fine art in another, as ceramics in particular show.

The fine arts are conscious and essentially individual in tradition. Though the artist may make a number of attempts before he is satisfied or satiated with his concept, he usually aims at making a unique object, one in which the material is subjugated to the idea and on which no amount of labor is misplaced. His purpose is discovery and singular statement. He will use existing materials, sometimes in radically new ways. Though he may discover new materials, such discoveries are incidental to his aim.

The quantitative and economic aspects of the decorative arts, on the other hand, make them intrinsically repetitive. Because of this, their aesthetic qualities have a very intimate relationship to the technology of materials, and their design is thereby basically affected. Does not the discovery of symmetry lie in the decorative arts? Something akin to knowledge of the nature of the crystal lattice inevitably arises from experiments in which decorative details are repeated as the simplest way of filling space. This shows in simple form in Greek geometric pottery and, much earlier, in the famous mosaic walls at Uruk. The same property appears in block- or roller-printed fabrics, and it is no accident that most introductory discussions of crystal symmetry and space filling from Kelvin to the Braggs and Weyl reproduce wallpaper designs.

The use of stamps in decorating clay tiles—most delightfully in the early Han tombs of China (figure 10.1)—prefigured both the printing press and the aesthetic value of repetition with differences. From a design standpoint, the complex repetitive patterns in Islamic tilework show how small local variations can integrate into symmetry on ever-larger scales. Even the superb Byzantine mosaics at Ravenna and elsewhere [illegible] ments, involving tedious work of a kind that few self-conscious artists today would be willing to do.

Repetition affects the selection of matter as well as form. The shaping operations must not fight the material but must conform naturally to it. Whether stone, metal, wood, clay, or glaze, the physical properties and structure of the material have their say in determining both surface finish and overall shape. Nowhere does this show to better effect than in folk art, especially in ceramics and above all in the common wares of Korea and Japan (figure 10.2), where the "no-mindedness" of the potter has led to the production of forms with surface decoration of the simplest kind, and yet the overall effect of superb beauty is based on a balance between purpose, process, and the properties of the material.[6] There is a natural relationship in the yielding of the aggregate of clay particles to the micro and macro forces applied in shaping, in the flow and adhesion of the glaze as it is applied, and in the effects of fire in rendering the glaze partly fluid while it shrinks and hardens the clay. The Japanese tea masters of the seventeenth century knew this intuitively, though one may doubt that they gave any thought to the properties of thixotropy, surface tension, viscosity, vitrification, nucleation, crystallization, and diffusion upon which all these effects depend. The whole environment of the tea ceremony is provided by simple natural surfaces of wood, bamboo, lacquer, paper, cast iron, copper, carbon, stone, plaster, clay, and vitreous glaze, almost as if to present examples of classes of quantum-mechanical bonding between atoms to illustrate a modern course in the science of materials. When Oriental materials reached Europe in the seventeenth and eighteenth centuries, they inspired many attempts at duplication by manufacturers—not the least Meissen and Wedgwood—as well as research on mineralogy, chemical analysis, and high-temperature reactions. The effect of contact with the Orient was at least as great on science as it was on the decorative arts.

In addition to the qualitative need for repetitive detail in design, the decorative arts have a quantitative requirement, namely the imperative of covering large areas or making large numbers of individual objects. Historically, both of these were strong incentives to mechanization. Covering the expanse of plaster walls led to molded bricks and tiles. The mechanical concept employed in the rotary drill for beadmaking moved through the potter's wheel to the lathe; stamps for body decoration in clay moved to punches for metalwork, dies for coinage, Arentine relief-molded pottery, drop stamps for pinheads and buttons, and eventually to power-driven presses for nearly everything.

Consider also the development of mass-production methods involving the casting of molten metals. Though the finest castings were made individually by the lost-wax process, the majority of castings from the earliest days have been designed expressly to facilitate molding. As with punches and dies,

Figure 10.1
Clay tiles from the wall of a tomb. Lo-yang, Honan Province, China, Han dynasty, 206 B.C.–220 A.D. Note the symmetry arising from the repeated use of stamps. Photo courtesy of the Royal Ontario Museum, Toronto.

Figure 10.2
Ceramic tea bowl, named *Kizaemon Ido* by Japanese tea masters. Korea, sixteenth century. Height, 3.5 inches. From Sōetsu Yanagi, *The Unknown Craftsman: A Japanese Insight into Beauty*, ed. and trans. Bernard Leach (Tokyo and Palo Alto, 1972), p. 11, pl. 1, courtesy of Kodansha Publishing Company.

most foundry processes have the characteristic that the careful work of the master designer is involved only once, whereafter replication takes over. A lost-wax casting can be almost any shape the designer wants, for the mold is destroyed each time. Repetitive casting, on the other hand, requires either that a cheap plastic mold material be quickly shaped by pressing against a permanent model or that the mold itself remain intact for repeated use. In either case the design of the object must be such as to enable the model or the casting to be withdrawn cleanly from the mold; that is, there must be no undercuts or overhanging features in at least two directions. Though more freedom can be attained by making the mold in several pieces fitted together, the two-piece mold is the simplest, and the requirement of a parting line running around the object at its maximum width to mark the surface of separation of the mold confers a special quality to the design of anything made in this way. Sometimes the parting line marks the equator of a convex object, but it can also meander up and down irregularly. It can be seen in the cast axes of the Early Bronze Age, in terra-cotta figurines, toy soldiers, belt buckles, pressed glass, plastic kitchenware, and a host of other objects. It forms the undefined edge of early coins and becomes defined and double in later coins and medals struck in retaining rings. The rich low relief of coins and medallions provides some of the finest examples of the relationship between design and technique, for their dies are subject to the geometrical limitations needed to allow the withdrawal of the work even more stringently than in the case of molds for castings, and the metal cannot flow laterally to a very large extent.

The virtue of thin sheet metal in giving the greatest glitter for a grain of gold was exploited in the earliest days of metallurgy. However, before the days of rolled sheet and drawn wire, most metal objects were made by hammering and were basically three-dimensional in form. Theophilus, for example, described the making of a chalice in the twelfth century A.D. by hammering a silver ingot into a thin disk with a projection left in the middle, and the fully rounded form then raised with offset hammer blows on a small anvil called a stake.[7] Flutes were nearly as easy to produce as rotational forms. The introduction of rolled sheet as a starting material did not change the second stage of the operation (as many beautifully hand-wrought silver objects made today testify), but it did naturally invite a boxy form—again fine if done straightforwardly.

Thin sheet metal, particularly gold, had been repetitively shaped since very early times by simple pressing or hand hammering against three-dimensional forms (dies) of wood, stone, or metal (figure 10.3). Theophilus described the multiple die-stamping of saintly images with decorative borders in sheet silver and gilded copper for use on caskets, book covers, and the like.

Although water-driven hammers for rough forging were used earlier, the mechanization of metal stamping begins with Leonardo's wedge and screw press for striking coins; by the seventeenth century the latter had become the rigid frame press with heavy swinging arms bearing inert masses on their ends, still to be seen in the mint at Paris. A smaller version, the fly press, with a single swinging arm was then used for cutting coin blanks; later this was adapted to precise blanking, cupping, and forming operations. Flashier decorative work in thin metal was done in the simpler drop press. All these devices can be seen in André Félibien's engraving of 1676 (figure 8.28).

The advantage of the screw press for precise work was not immediately obvious, and coins were also struck in engraved rolls for some time, for example, in Sweden.

The drop hammer falling between vertical guides was used for pile driving at least as early as the fifteenth century. Its use with two dies to shape metal seems to be a seventeenth-century innovation for shaping the solid heads of pins and nails. A 1769 British patent describes a guided falling hammer with a lead face—later called the force—used to strike thin sheet metal against a single die, "more especially to be used in the production of coffin furniture." The process soon spread to stamping buttons and all kinds of brummagem ware from thin sheet metal, and it partially displaced casting for the formation of cheap three-dimensional contoured surfaces.[8] Look at the simple drop press (figure 10.4)—its unmodulated blow striking in a single direction symbolizes much of nineteenth-century mechanized production. To make multiple stampings, stacks of very thin metal sheets were superimposed under the hammer, and the final profile with moderately high relief was gradually achieved as finished sheets were removed from the bottom and new ones added at the top (figure 10.5). A corner-cutting operator could remove two or three at a time, sacrificing detail for quantity.

W. C. Aitken described the process in 1865 with rather perceptive remarks on its effect upon design.[9] It was an ideally cheap means for the multiplication of surfaces in low or medium relief. Innumerable decorated objects stamped of sheet metal in poorly cut or worn dies formed a characteristic feature of Victorian decoration, and these inferior examples brought stamping to a disrepute that the technique per se does not deserve. (One might remark that most handwork also was and is extremely sloppy; only the best survives or is noticed, and the worst is forgotten.)

The coining press working on blanks of thick metal fitted well into traditional design. But when the drop press was used to shape large areas of thin sheet metal, the aesthetic qualities of the surface became divorced from the underlying substance, and decoration became independent of the body needed to support it. In any object there is natural relationship between the surface and the bulk, that is, between its one-, two-, and three-dimensional aspects. The fakery involved in applying gold or silver plating on a solid copper object is quite different from the deception of an ornately stamped piece of thin sheet brass. Compare a magnificent ormolu furniture fitting or even a gilded plaster picture frame with a cheap lamp base embossed in thin sheet brass. In the former the surface is simply and honestly applied for its optical effect alone; in the latter the fakery is fundamental for it is dimensionally misleading.

Objects in relief designed to be shaped by replicative processes of any kind have *yin-yang* overtones arising from the spatial inversions involved. The matching convexity and concavity of the dual die/coin relationship or the triple pattern/mold/casting sequence profoundly influence the appearance of small details of the object. A narrow channel cut into a broad surface differs from the equivalent projection not only in its strength but also in the quality of imprecision resulting from tool movement and the rounding of edges caused both by tool wear and by limitations of material flow. A line in intaglio can be a single cut; one in relief is what remains as the terminus of many cuts. The aesthetic overtones of both the manner in which surfaces join and the texture of surface details shaped by the craftsman's tool are profoundly different when inverted. The scratch yields a projection; the punch mark becomes a positive replica of the sharp edges of the punch, not a sharp-bottomed concavity with rounded metal flow leading into it. An external projection subject to rounding cannot be the same as a cut intaglio groove. Of course a good diesinker will avoid these effects in finishing his work, but if tool marks remain, the inversion looks unreal, for the eye remembers that a tool can only approach a surface from the outside and senses discrepancies that are hard to define (figure 10.6). The effect is related to the perceptual figure-ground inversion in two dimensions that has been used by artists, most notably by Maurits Escher, and has led to many interesting experiments on the psychology of perception. In decoration consider a stenciled resist-dyed fabric (figure 10.7). One's first reaction is to expect the color to be applied in the cutout areas of the stencil, and there is a faint tension when color covers the areas of the stencil itself, still joined rather than separated by the little supporting strips.

The graphic arts abound with examples of inversion. In the West most xylographers' blocks are made to follow closely the positive lines of an artist's sketch, which calls for extremely precise and intricate cutting. However, some artists (notably Paul Gauguin) work simply and directly on the block, the tool forming blank spaces on the print between larger variable areas of ink coming from the residual untouched block. From the beginning, Japanese woodblock prints made much use of subtly varying line width, and by the late seventeenth century they were using large areas of block surface. In many modern Japanese color prints (*hanga*) the grained texture of the block itself is proudly replicated.

In intaglio printing processes such as etching and engraving, the artist working on the plate makes a negative line directly to carry the ink, and the only inversion on printing is chiral, not that of figure and ground. The quality of the print nevertheless depends intimately on the depth, width, and edge sharpness of the lines as well as on their spacings, crossings, and junctions, permitting an infinity of effects.

The Dimensionality of Hand- and Machine-Shaping Processes

There is an essential difference between handcrafted and machine-made objects that is based on dimensionality. The manual worker can apply his tool to the material in any direction, with any force, on any part, in any sequence. Though the basic movement of a hand tool is unidimensional, the three-dimensional shapes resulting from the integral effect of many motions are unlimited. Machine operations, conversely, are essentially one-dimensional or, at most, combine a very few sequential motions. Computer-controlled machines may well void this limitation, for they allow complete three-dimensional freedom if the working tool itself is small. Intermediate examples of the products of machine-controlled combined motion are the guilloché snuff boxes of eighteenth-century France and the Lissajous figures engraved on bank notes and bond certificates.

The very essence of nineteenth-century ornamental work in metal and other materials lies in shapes that are generated by one-dimensional translation of a tool with an extended working surface, geometrically either simple or complex. The drawbench, the rolling mill, and the extrusion press all give shapes that are swept out by the simple linear translation of the profile of the working surface. Pressed and stamped work marks the end point of the one-dimensional advance of the three-dimensional contoured surface of the die, or it may be (as in lithographic printing) a transfer of locally different surface qualities that are three-dimensional only on a microscopic scale. The requirements are similar to those of easily made castings discussed above, and indeed, all mass-production methods suffer similar dimensional restraints. There is a philosophical similarity between the reduction to essentials needed for mechanical production and the exclusion of extraneous factors in a good scientific theory; by invoking a sufficient number of variables and motions and by controlling their sequence anything can be duplicated, but in practice, economy of both thought and motion leads to dimensional simplification. Only if the abstraction is sensitively done will the product be intellectually satisfying and beautiful to behold.

This distinction between "the workmanship of chance" and "the workmanship of precision" is the basic theme of the excellent little book by David Pye.[10] And materials are central to the difference. The craftsman can compensate for differences in the qualities of his material, for he can adjust the precise strength and pattern of application of his tools to the material's local vagaries. Conversely, the constant motion of a machine requires constant materials. The gradual standardization of shapes and sizes of prefabricated materials as well as the control of their composition, internal structure, and properties was second only to quantity and cost of production among the principal themes of nineteenth-century metallurgy.

The machine, of course, also froze design, very often with unsatisfactory results because the designer improperly copied features appropriate to earlier techniques. Fifteenth-century typographic printing is one of the very few examples of a satisfactory close imitation of a handmade product by mechanical means, and it was possible only because the surface being shaped was purely two-dimensional; simple replication was possible without deformation or inversion of concavity and convexity. As one surveys the history of printing and the graphic arts, it is impossible to claim that replication is necessarily the enemy of aesthetics.

The quality of design imposed by machine inversion of relief decoration permeated the Victorian environment, or at least the popular vision of it today. Consider the molded terra-cotta facings for municipal buildings; the cast-iron fluted columns on warehouses or the more elaborate molded iron details affixed to the outside of department stores and domestic stoves; the innumerable stamped or cast parts of brass for the newly flourishing gas or electric lighting industries; or the molded panels in terneplate ceilings or pressed-wood furniture. Such designs are usually too exuberant to be beautiful by today's aesthetic standards, but it can be noticed that their unpleasant aspects lie less in the overall form than in the details. These were not usually designed from the beginning with a replicative process in mind but were modified from the rich hand-carved profusion of eighteenth-century metalwork just enough to allow withdrawal from mold or die. The same process of die stamping that produced Victorian excesses also produced superb coins and medals. The nineteenth-century stampings were usually made from thin sheet metal locally stretched

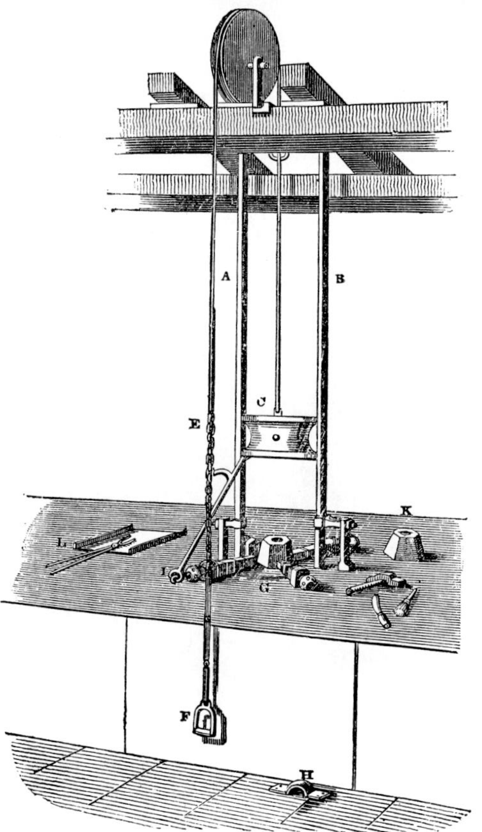

Figure 10.3
Appliqué dress ornaments. Iran, fourth century B.C. Thin sheet gold shaped by pressing against a die. Width, 1.5 inches. The photograph shows both obverse and reverse. Photo courtesy of the Oriental Institute, University of Chicago.

Figure 10.4
Simple drop press for use with lead "force." This type of press was used mainly for cheap stamped metalwork in the nineteenth century. From [John Holland], *A Treatise on the Progressive Improvement and Present State of the Manufactures in Metal* (London, 1834), vol. 3, p. 220.

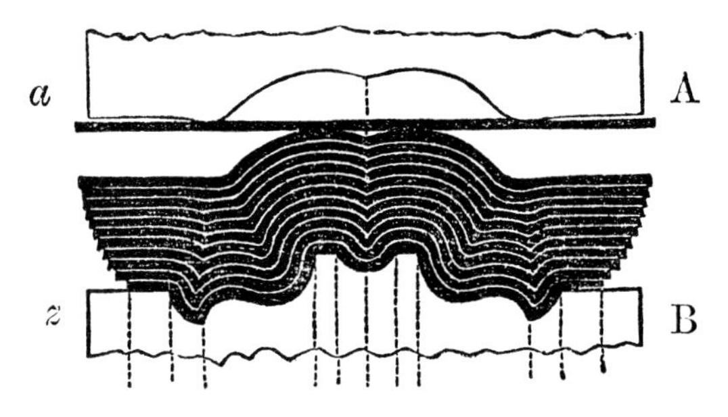

Figure 10.5
Diagram of stack of metal sheets being stamped in a press like that shown in figure 10.4. From Charles Holtzapffel, *Turning and Mechanical Manipulation* (London, 1846–1884), vol. 1, p. 409.

Figure 10.6
Impression of a cylinder seal. Sumerian, ca. 3200 B.C. Note the relief details of the design, which are inversions of the intaglio shapes that arise naturally from the use of gravers and rotary drills. Height, 1.25 inches. Photo courtesy of the Museum of Fine Arts, Boston, H. O. Cruft Fund (34.192).

Figure 10.7
Child's pinafore with design formed by stencil-printing a resist and subsequently dyeing in indigo. Modern. From Chai Fei et al., *Indigo Prints of China* (Peking, 1956), courtesy of the Foreign Languages Press.

Figure 10.8
Covered pail. American, 1800–1850. Sheet tin plate joined by soldering. Height, 7 inches. Photo courtesy of the Winterthur Museum (58.3029).

and bent to conform to the die, while the coins were of heavy metal and underwent volumetric flow. The thin sheet could neither accommodate sharp changes in surface contour, except uncomplicated bending in a single direction, nor draw smoothly to yield high local relief. There was a return to the rather unpleasant quality of some of the earliest sheet-metal work shaped over forms, such as the so-called mask of Agamemnon in the Athens museum in which the malleability of gold is exploited to get a large area of glitter at the cost of unpleasant folding, quite different from more careful repoussé work in thicker metal such as the mask of Tutankhamen. The differences in design required for repoussé work in sheet as compared with stamping are not great; nevertheless, they influence the whole quality of the work.

When it had to be made by hammering, thin sheet metal in any material was too expensive to be widely employed; the introduction of rolling mills of sufficient power and precision made sheet metal one of the cheapest ways of covering exterior surfaces of commercial buildings as well as making common utensils for the home. Thin sheet iron, whether tinned, galvanized, oxidized ("Russia iron"), japanned, or just painted, lends itself remarkably well to making objects of simple design. The cylindrical, conical, or planar surfaces that naturally arise from the cutting, bending in a single direction, and joining of sheet metal can be used to form objects such as boxes, buckets, lanterns, funnels, and trays that are most appealing in their geometric restraint (figure 10.8).[11]

Totally different in spirit are works raised from sheet metal by hand hammering. In these, carefully offset local hammer blows integrate to give three-dimensional changes of shape, slowly forming the necessary curvatures while maintaining nearly uniform thickness. This process, which gives almost infinite freedom of shape, is an old and beautiful one and still a favorite method of the silversmith (figure 10.9) but highly demanding of time and skill. In quantity production it has been largely displaced either by stamping or deep-drawing sheet metal between dies or by spinning, in which a disk of sheet metal is pushed by a burnishing tool until it conforms to the surface of a wooden form mounted on a lathe head and revolving at high speed. Spinning, which is subject to less geometric restraint than drawing, first became important early in the nineteenth century, when it was widely used to shape objects from the tin-antimony alloy known as Britannia metal. It resulted in a revival in popularity of objects made of solid tin, following a century of neglect of pewter, which had been displaced by the availability of cheaper but attractive ceramics and glassware.

Good design in the decorative arts must take account of the properties and the prefabricated shape of the material as well as the limitations imposed by the replication process on tool shape and tool motion. Though in principle it might be possible to make by mechanical means an exact copy of any individual work of art with all its nuances, the process would need to have the same sensitivity as the original handwork to the locally variable responses of the material, and to achieve this requires a method of reproduction with resolving power, response, and control on a microscopic scale. The success of a mechanic's, or a machine's, reproduction of a thing depends on his, or its, sensitivity to whatever qualities are important, just as the skill of the designer lies in the proper appreciation of surface qualities in terms of structure and shape variation that come from the intended means of production.

Decorative Textures

The process by which an object is finished affects its appearance mainly by changing the microcontours of the surface and hence the visual clues that arise in the variations of light reflected at different angles from different parts of the surface. However, mechanical preparation of surfaces may change the structure of the immediately underlying material, and if there is subsequent coloring by chemical treatment, it can cause even greater changes in surface quality. The unmatched beauty of some Japanese metal finishes, notably on the alloy *shibuichi*, depends on the preservation of a discrete though almost invisible microstructure that burnishing would entirely destroy—as indeed it did on European metal surfaces, none of which revealed their underlying rich texture to the scientist's microscope.[12] The coarser crystals in tinplate were decoratively used in the ware called *moiré metallique* that was briefly the rage early in nineteenth-century Europe and later appeared in America (figure 10.10). The preferred finish for a metal seems usually to have been a highly burnished reflective surface, but finely tooled or abrasively scratched lines to give unexpected reflections have been and are popular. Chemical treatment may give color by producing superficial layers of carbonates, oxides, sulphides, and the like, with a texture, if visible at all, not much related to the underlying structure of the metal. From the sixteenth century on, innumerable western publications of the practical handbook type include recipes for pickling solutions of various types to give such effects.

Figure 10.9
Frederick A. Miller, silver bottle. Cleveland, Ohio, 1972. Made by raising from sheet silver. Height, 6.5 inches. Photo by John Paul Miller.

Figure 10.10
Painted tin tray. Possibly Stevens Plains, Maine, 1874. The center area (L, 8″; H, 4″) is of *moiré metallique,* a pattern formed by the controlled crystallization in tin, subsequently developed by etching in acid. OL, 12⅜ inches. Photo courtesy of the Winterthur Museum (65.1717).

On Materials in the Nineteenth Century

Material innovation in the nineteenth century lay mainly in the province of the organic chemist. Bleaching textiles with newly discovered chlorine and dyeing with synthetic dyes permitted all classes of people to have some of the show of color that, at least in Europe, had previously been limited to the rich. For centuries organic materials such as bone and various pasty mixtures had been softened by moisture and heat and molded by pressure into buttons, knife handles, and the like, but the nineteenth century saw a transition toward a quite new type of material. Rubber and gutta-percha were first used in more or less their natural states; then came rubber changed by sulphur cross-linking into both elastic and hard forms, the latter called vulcanite, and pressure-molded to shape. The first fully artificial polymer was celluloid or xylonite, cellulose nitrate stabilized with camphor. This was introduced in 1862 by Alexander Parkes, an English metallurgist, and provided a kind of synthetic ivory that could be shaped either by pressing or by cutting (figure 10.11). Parkes's firm failed, but an improved version of the material offered in the early 1870s by John Hyatt, of Albany, New York, quickly established a stable market. Billiard balls and piano keys were its first large applications, but celluloid was widely used in men's collars and in innumerable small die-molded objects. As the first artificial thermoplastic material, it clearly foreshadowed the synthetic polymer industry of the twentieth century.[13]

Among nonmetallic inorganic materials in the nineteenth century, that vastly important material cement was beginning to modify architecture, and moldable synthetic stone had some vogue in the arts. The production of glass was mechanized to some extent, but its nature remained basically unchanged except in glass for optical and laboratory use. The ceramics industry also was mainly concerned with exploiting the changes of the eighteenth century rather than seeking new compositions.

In metals, the nineteenth century saw great changes in shaping methods (mainly the introduction of machinery to reduce labor costs, often fiercely resisted by the older workers). Slight changes in old alloys were accompanied by new names, but on the whole the only sign of the new science was in the better control of composition and the introduction of electroplating. There was, however, a change in the pattern of use of the old metals. The alloy paktong, previously imported from China, became cheaper and more available when manufactured in Europe under the name German silver or nickel silver. Metallic zinc moved from a laboratory curiosity or a component in brass to a place of importance in its own right.

Tin took on renewed importance in the form of Britannia metal, an alloy of tin with about 4 to 8 percent antimony and 1 to 2 percent copper. This alloy is not only brighter and more tarnish-resistant than pewter but is nontoxic, harder, and has the resonance expected of a metal. Moreover, it has superlative working properties, and though easily cast to shape, objects that weigh and cost less than cast ones can also be formed out of rolled sheet. Though tin-antimony alloys had been used sporadically on the continent long before, the industry really began to develop in Sheffield around 1770. Innumerable objects imitating both the glitter and the form of the best silverware appeared in the second half of the nineteenth century, but even before this the alloy itself had found

Figure 10.11
Collection of objects molded from Parkesine (celluloid), the first artificial polymeric material. Made between 1860 and 1866 by the inventor, Alexander Parkes, they mark the very beginning of the great plastics industry of today. Photo courtesy of the Science Museum, London. Crown copyright.

widespread use. Writing in 1834, John Holland remarked, "Although . . . almost every article manufactured in silver now has its counterpart in Britannia metal, the greater part of the material used is in the production of candlesticks, tea pots, coffee-biggins and all kinds of measures for liquids."[14] The sheet metal was soft enough to be shaped easily on cast-iron dies, cheaply made from plaster of paris molds simply formed on the fine silver objects to be copied. Britannia metal ware is particularly associated with spinning, a process using simple dies of wood or other soft material that do not need elaborate die-sinking operations in steel as do the tools for most competitive processes. Britannia metal as a poor-man's silver came into competition with nickel silver, and in one of the many unpredictable interactions that have so strongly influenced the development of technology, both of these white alloys were later to provide an excellent base for electroplated silverware.

Zinc was a metal that took on new life in the nineteenth century. It was, of course, a component of brass, which had been made by cementation of copper with zinc oxide and charcoal for centuries before the metal itself was discovered. Zinc had been imported into Europe from the Orient to some extent from the seventeenth century on, but its extensive use dates from the early nineteenth century when industrial methods of rolling it into sheet were introduced. (The coarsely crystalline cast metal is brittle and needs to be heated to between about 100 to 150°C before it can be extensively deformed.) The sheet metal came to be widely used in both Europe and America for roofing, gutters, chimneys (figure 10.12), cisterns, dairy utensils, and the like. Embossed zinc sheets for ceilings and external sheathing for buildings were common. Alfred S. Bolles in 1879 extolled sheet zinc: "It is so cheap, too, that it has brought handsome cornices within the means of all; and the invention has really been the means of improving the architectural appearance of our formerly exceedingly plain business-streets, as well as their security."[15]

Castings of zinc alloyed with 3 to 4 percent of copper, known as bidri metal, were widely used in India for decorative objects. Pickling gave bidri an attractive black matte finish to contrast with the usual silver inlay (see figure 12.9). The Topkapi museum contains two roomfuls of cast zinc objects elaborately inlaid and encrusted with gold, diamonds, rubies, and emeralds. Nevertheless the use of zinc alloys for casting was slow to develop in the West until the 1930s, when the making of die castings of alloys stabilized against corrosion became a large industry. Before this, commercially pure zinc was used in making cheap ornamental hardware, statuettes, and the like by the process of slush casting, but the principal use of the metal had been as a minor constituent in alloys or as rolled sheet. Zinc played an invisible but essential role in electroplating and a visible and important one in galvanized sheet iron.

The Beginnings of Electrometallurgy

Most of the developments in nineteenth-century materials production involved analytical chemistry, which revealed new sources of raw materials and identified impurities responsible for undesirable properties in the products. It gave the better control of smelting and refining operations that was necessary for the increased scale of operation, but chemistry did not directly produce major changes in the working of materials—these came from the mechanical approach, mainly the application of power to the processing of raw materials and their final shaping. The new methods of chemical analysis did disclose previously unknown elements that had lain hidden in unrecognized minerals or invisibly present in minor amounts in common ones, and they led inevitably to the discovery of the quantitative laws of combination between atoms.

The discoveries of nickel and aluminum (new metals that were to become most important in the decorative arts) were thus a natural outcome of the old chemistry. But a new route to discovery—indeed a newly found aspect of nature that would change society more than all the social revolutions of history—was coming out of the physics laboratory. Just before 1800 voltaic electricity was discovered. It opened to study innumerable properties of matter of which there had been almost no hint in earlier science, still less in the decorative or useful arts. Its discovery followed observations of the twitching of a severed frog's leg, hardly an aesthetic object. The next four decades produced a superb series of discoveries relating the now measurable currents to chemical reactions, to magnetism, and to magneto-electric forces and movement. This was one historical case in which science led directly to applications in technology rather than vice versa. There was to be a lapse of almost four decades, however, before any significant useful applications of electricity emerged; then, between 1837 and 1839, both the telegraph and electroplating became thoroughly practical.

Much has been written about the development of the telegraph, whose utility and

Figure 10.12
Zinc chimney pots. Drawing attributed to *Punch*, reproduced from J. H. Pepper, *Playbook of Metals* (London, 1866), p. 498.

profitability were obvious enough from the beginning; electroplating was less spectacular, but it played a greater role in spreading knowledge of electricity. It was both pleasant and democratic. While the telegraph enlisted the interests of governments and mammoth corporations, electroplating was at first only a popular hobby, a means of copying art objects and printing pretty pictures. The replication of printing surfaces by cheap and exacting electrotyping had somewhat the same effect on popular culture that Gutenberg's typography had had almost exactly four centuries earlier, but its greatest significance lay in the fact that it led directly into the development of commercial generators and so to the great power systems of today.[16]

Notes and References

1
See chapter 8 in this volume.

2
R. A. F. de Réaumur, *L'Art de Convertir le Fer Forgé en Acier, et l'Art d'Adoucir le Fer Fondu ou de Faire des Ouvrages de Fer Fondu Aussi Finis que de Fer Forgé* (Paris, 1722); English trans. Anneliese Grünhaldt Sisco (Chicago, 1956), pp. 340, 342, 344, 352.

3
[John Holland], *A Treatise on the Progressive Improvement and Present State of the Manufactures in Metal,* ed. D. Lardner, 3 vols. (London, 1831–1834), vol. 3, p. 210.

4
R. A. F. de Réaumur, "L'art de faire une nouvelle sorte de porcelaine, . . . ou de transformer le verre en porcelaine," *Mémoires de l'Académie Royale des Sciences,* 1739 (published, 1741), pp. 370–388.

5
See chapter 9 in this volume.

6
Sōetsu Yanagi, *The Unknown Craftsman: A Japanese Insight into Beauty,* ed. and trans. Bernard Leach (Tokyo and Palo Alto: Kodansha, 1972).

7
Theophilus, *De Diversis Artibus,* manuscript treatise, ca. A.D. 1125; Latin text and trans. C. R. Dodwell (New York: Thomas Nelson, 1961); trans. with technical notes by J. G. Hawthorne and C. S. Smith (Chicago, 1963).

8
Holland, *Treatise,* vol. 3, pp. 218–224.

9
W. C. Aitken, "Brass and brass manufactures," in Samuel Timmins, ed. *The Resources, Products, and Industrial History of Birmingham and the Midland Hardware District* (London, 1866), pp. 225–380. Also issued as a separate publication with its own pagination.

10
David Pye, *The Nature and Art of Workmanship* (Cambridge, England, 1968).

11
Good discussions of sheet-metal work can be found in Oliver Byrne, *Practical Metal-Workers Assistant* (Philadelphia, 1851); Charles Holtzapffel, *Materials; . . . Various Modes of Working Them without Cutting Tools,* vol. 1 of *Turning and Mechanical Manipulation,* 5 vols. (London, 1846–1884); and in Joseph K. Little, *The Tinsmith's Pattern Manual* (Chicago, 1894).

12
On Oriental textured metal see chapter 9 and item 193 in the bibliography. On the scientific importance of texture (microstructure) see C. S. Smith, *A History of Metallography* (Chicago, 1960).

13
M. Kaufmann, *The First Century of Plastics* (London, 1963).

14
Holland, *Treatise*, vol. 3, p. 104. See also "Pewter and Britannia metal trade," in Timmins, *Resources, Products, and Industrial History of Birmingham*, pp. 617–623.

15
Alfred S. Bolles, *The Industrial History of the United States* (Norwich, CT, 1879), p. 368.

16
(*Note added 1980*) As originally published, this paper continued with a detailed account, here omitted, of the beginnings of industrial electrochemistry. An expanded version of this account appears as item 190 in the bibliography.

11
On Art, Invention, and Technology

Nearly everyone believes, falsely, that technology is applied science. It is becoming so, and rapidly, but through most of history science has arisen from problems posed for intellectual solution by the technician's more intimate experience of the behavior of matter and mechanisms. Technology is more closely related to art than to science—not only materially, because art must somehow involve the selection and manipulation of matter, but conceptually as well, because the technologist, like the artist, must work with many unanalyzable complexities. Another popular misunderstanding today is the belief that technology is inherently ugly and unpleasant, whereas a moment's reflection will show that technology underlies innumerable delightful experiences as well as the greatest art, whether expressed in object, word, sound, or environment.

Even less widely known, but important for what it tells of man and novelty, is the fact that historically the first discovery of useful materials, machines, or processes has almost always been in the decorative arts, and was not done for a perceived practical purpose. Necessity is *not* the mother of invention—only of improvement. A man desperately in search of a weapon or food is in no mood for discovery; he can only exploit what is already known to exist. Discovery requires aesthetically motivated curiosity, not logic, for new things can acquire validity only by interaction in an environment that has yet to be. Their origin is unpredictable. A new thing of any kind whatsoever begins as a local anomaly, a region of misfit within the preexisting structure. This first nucleus is indistinguishable from the few fluctuations whose time has not yet come and the innumerable fluctuations which the future will merely erase. Once growth from an effective nucleus is well under way, however, it is then driven by the very type of interlock that at first opposed it: it has become the new orthodoxy. In crystals undergoing transformation, a region having an interaction pattern suggesting the new structure, once it is big enough, grows by demanding and rewarding conformity. With ideas or with technical or social inventions, people eventually come to accept the new as unthinkingly as they had at first opposed it, and they modify their lives, interactions, and investments accordingly. But growth too has its limits. Eventually the new structure will have grown to its proper size in relation to the things with which it interacts, and a new balance must be established. The end of growth, like its beginning, is within a structure that is unpredictable in advance.

The Shape of Universal History

The "S" curve in figure 11.1 (adapted from a paper on the transformations of microstructure responsible for the hardening of steel) can be used to apply to the nucleation and growth of anything, *really* any "thing" that has recognizable identity and properties depending on the coherence of its parts. It reflects the underlying structural conflicts and balance between local and larger order, and the movement of interfaces in response to new conditions of components, communication, cooperation, and conflict.

Applied to the growth of either individual technologies or to the development of whole civilizations based upon interactive technologies, the "S" curve reflects origin in art, growth in social acceptance, and the eventual limitation of growth by interactions within a larger structure which is itself nucleated in

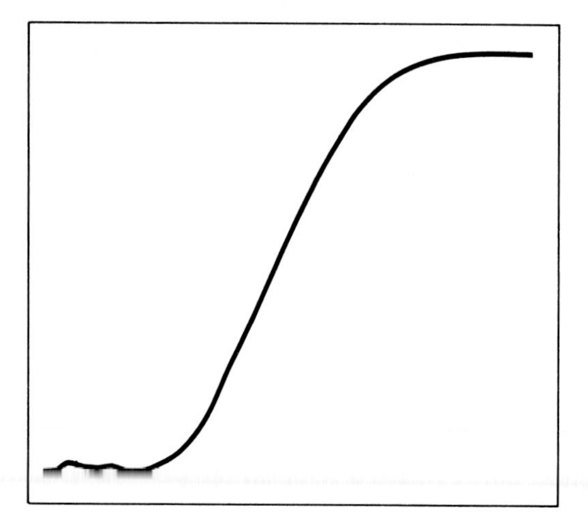

Figure 11.1
Curve depicting the beginning, growth, and maturity of anything whatever. Adapted from a paper on the hardening of steel, it is here used to show the beginnings of most branches of technology in the decorative arts, their industrial growth in response to a social demand, and their maturity in conflict and balance with other things. Both the beginning and the end depend on highly localized conditions and are unpredictable in detail.

the process. The conditions of beginning, development, and maturity are very different from each other.

Though a computer program can duplicate such curves, it is only by looking at the whole hierarchical substructure and superstructure that intuitive understanding can be gained. All stages involve a balance between local structure and overarching regional restraints. All change involves a catastrophic change of connections at some level while topological continuity is maintained, though perhaps with strain, at levels both above and below. Human history follows the same general principles of structural rearrangement as a phase change in a chemical system, though most teachers of history ignore the nucleating role of technology and concentrate on the social changes that are engendered by it.

The transition from individual discovery and rare use of techniques to the point where they affect the environment of Everyman and the content and means of communication between people and peoples underlies virtually every great social or political change and every fundamental change in man's view of the world. Few general histories reflect this. An understanding of the proper place of technology within the whole human experience is desperately needed in order that society can wisely decide what to develop and what to discourage. Technology needs to be seen in the perspective that humanists have traditionally applied to man's other activities. Personally, I believe that the life of a craftsman, indeed of any man making something to be enjoyed or used, is a fine example of what it is to be human: mind, eye, muscle, and hand interacting with the properties of matter to produce shapes reflecting the purposes and cultural values of his society, and sometimes extending them.

The verbal records conventionally used by the historian reflect this very poorly indeed. Conversely, works of art, when seen at every level from the atom to the whole, provide excellent records of almost everything about man. Usually they are enjoyed for their outer form and symbolism alone, and appreciated as a statement of the artist's ideas on some aspect of the world, an expression of the forms and feelings that he selectively absorbed from the culture of which he was part. However, his work is also an object and as such a product of technology. Thus the famous bronze statue of Poseidon (figure 8.6) involves technology both submerged in the emotional and cultural meanings carried by its glorious form, and also more tangibly in the actual operations of smelting, alloying, casting, and welding that produced it. The statue was cast in several parts by the lost-wax process, and the parts were joined together by running in superheated molten metal of the same composition. The techniques had themselves developed through earlier (non-Greek) history and they were to have an influence on the subsequent development of the Western world comparable, in my view, with that of Greek ideas in aesthetics and philosophy.

Historians of science, while properly emphasizing the development of unifying concepts, commonly overlook the fact that thoughtful intimate awareness of the properties of matter first occurred in the minds of people seeking effects to be used decoratively. Both Democritus's atoms and Aristotle's elemental qualities are expressions of what the philosophers could have observed on a stroll down Hephaestos Street, noting the changes in strength, plasticity, texture, and color produced by the treatment of materials as the artisans shaped things for their customers. Similarly the multivalent game tokens found in very early excavated sites in the Middle East as well as the space-filling interlock of features in decorative pattern must have some place in the earliest history of mathematics.

Archaeologists and art historians, of course, long ago learned to interpret human experience from the evidence of artifacts. But even they have concentrated upon iconographics and styles, on ideas external to the object, and only occasionally have they sought to understand the technical experience in its production. Yet in making a work of art, a man must select material having a "nature" that will conform to the larger shapes that he wishes to impose upon it. There is a continuous hierarchy of interactions: the object stands at the very point where the structures and properties of matter resulting from forces between atoms are in visible interaction with man's ideas and purposes. An artist's work preserves a record of both—one in the outer form and decoration, the other in the texture and color and fine contours that result from the interplay of atomic, molecular, and crystalline forces. The texture continues downward into a rich microstructure: hierarchical patterns of atomic order and disorder that change in recognizable ways as matter is subjected to thermal and mechanical treatment in its compounding and shaping.

Everything complicated must have had a history, and its internal structural features arise from its history and provide a specific record of it. One might call these structural details of memory "funeous," after the unfortunate character in Borges's story "Funes the Memorious" who remembered everything. The aim of respectable science through most of history has been to study afuneous details; it has been analytic, seeking the parts or ideally simplified wholes. Analysis is, of course, absolutely essential for understanding, but no synthesis based upon it can reproduce the funeous structures that provoked interest in the first place unless the essentials of their individual histories are repeated. History selects and biases statistics. The particular structures that do exist, however improbable they may be, must be given priority in man's studies. The messier sciences such as old-fashioned biology or my own metallurgy have always been concerned with complex structures, and they have emphasized the relation between real structure and properties, while the pure sciences have perforce avoided that which was structure-sensitive, and hence funeous. At least until very recently. Now, however, the material sciences have merged with solid-state physics and are showing how to study real structure with a wholly new emphasis upon imperfections in atomic order. Dissymmetry rather than symmetry is seen to be the key to many marvellous new materials—though dissymmetry is invisible except in a matrix of symmetrical order. New methods capable of revealing the whole structure at all levels have been developed—thereby incidentally opening a new level of funeous record for study by historians.

The Origins of Techniques

Many of the illustrations in this book show antique art objects and their structures as seen under the microscope. The microstructures differ more from each other than do the external forms—and they instantly reveal to a knowing eye the technical history of making the object. Such records are in a universal language, and they are free from the distortion that inevitably accompanies passage through a human mind. Through such records, I have communicated with dozens of craftsmen, including a Luristan smith of 800 B.C., a bronze founder of Shang China, an ancient Greek goldsmith, and a thirteenth-century Japanese swordsmith; and I have understood them better than I understand some of my English-speaking colleagues today! This newly found Rosetta Stone is making accessible records of a new world or, more correctly, an unnoticed aspect of culture in the old. As a metallurgist trying to understand the history of his profession, I had exhausted the literary sources without finding evidence of the beginnings of most of the techniques that interested me: only when I moved from libraries to art museums did I find the real origins of metallurgical (and other) techniques, and in doing so my whole view of man, matter, and discovery changed.

Practical metallurgy is seen to have begun with the making of necklace beads and ornaments in hammered native copper long before "useful" knives and weapons were made. The improvement of metals by alloying and heat treatment and most methods of shaping them started in jewelry and sculpture. Casting in complicated molds began in making statuettes. Welding was first used to join parts of bronze sculpture together; none but the smallest bronze statues of Greece or the ceremonial vessels of Shang China would exist without it, and neither would most of today's structures or machines. Ceramics began with the fire-hardening of fertility figurines molded of clay; glass came from attempts to prettily glaze beads of quartz and steatite. Most minerals and many organic and inorganic compounds were discovered for use as pigments; indeed, the first record that man knew of iron and manganese ores is in cave paintings where they make the glorious reds, browns, and blacks, while the medieval painter controllably used pH-sensitive color changes long before the chemist saw their significance. In other fields, archaeologists have shown that the transplanting and cultivation of flowers for enjoyment long preceded useful agriculture, while playing with pets probably gave the knowledge that was needed for purposeful animal husbandry. To go back even earlier, it is hardly possible that human beings could have decided logically that they needed to develop language in order to communicate with each other before they had experienced pleasurable interactive communal activities like singing and dancing. Aesthetic curiosity has been central to both genetic and cultural evolution.

Mechanical devices were less developed in the ancient world than were materials of comparable complexity. Perhaps this was because the aesthetic rewards of play with simple, and hence invariant, mechanisms are small. However, note that rotary motion was first used in the drilling and shaping of necklace beads and that the playful rolling of beads strung upon a wire is a likely antecedent to the idea of a pair of wheels turning purposefully on the axle of a cart or carriage. The automata based on hydraulic and mechanical tricks that were used in Greek temples and theatres were the prelude to the waterwheel and the clock. The lathe reached an apex of ingenuity in turning guilloché snuff boxes more than a century before heavy industry used it. The printing of pictures preceded purposeful type, and the use of rockets for fun came before their military use or space travel. The techniques of casting bells, like the material of which they were made, were ready to be directed toward a different kind of sound and purpose when princes wanted cannon.

Enjoyment of color has inspired the development of many alloys—for example, the famous Mycenean inlaid dagger in the National Museum in Athens, and the exquisite colored metal inlay of Japanese sword furniture. It is also related to the refining and purification of metals in early times because of the use of corrodants to change the color of native electrum. The color changes in metals, oxides, and sulphides discovered by far earlier artisans permeate medieval alchemy—a dead end of delightful but unproductive theory. The marvellous golds and blues of medieval illuminated manuscripts came from pigments made by processes that foreshadow modern powder metallurgy and the flotation process of ore separation. The desire for pigments, dyes, and cosmetics inspired much mineralogical and botanical exploration, while precious stones, dyes, spices, and jewelry formed the first base of commerce, for long-range trade did not start with necessities. Even bankers were once goldsmiths. The chemical industry later grew from the need for quantities of mordants, bleaches, and alkalies for use in the finer textiles and glass. Geology, chemical analysis, and high-temperature research all took a leap forward in eighteenth-century Europe under the impact of the potter's efforts to duplicate the marvellous wares coming from the Orient, which had started the craze for chinoiserie. The great French scientist Réaumur made a cheap, crystalline "porcelain" by devitrifying glass, and he also developed malleable cast iron in his search for a cheaper substitute for the handsome chiselled wrought iron work on the gates of the chateaux of the aristocracy.

More Like Love than Purpose

In all of these cases, and many more that could be cited, it was aesthetic curiosity that led to initial discovery of some useful property of matter or some manner of shaping it for use. Although the maker of weapons was quick to follow, it was nearly always the desire for beauty or the urge to make art available to the masses (or, if you will, the desire to exploit mass desire for pretty things in order to make profit) that led to advances in production techniques. The desire to beautify the utilitarian has always stretched the ingenuity of the mechanic, who made drawbenches, stamps, and screwpresses to shape trinkets before automobile parts or weapons. It is the same in building construction: temples and churches, greenhouses and Crystal Palaces, not necessary shelter, led to imaginative new structural methods. Even railroad rails and the steel girders for today's skyscrapers needed a precursor in the form of the little mill that rolled lead cames to be used in medieval stained-glass windows, and it was a French gardener who invented reinforced concrete because he wanted a larger flower pot for more magnificent display.

In the nineteenth century the milieu of discovery began to expand. Science created a new environment in which imaginative curiosity could operate. Though the discovery of voltaic electricity could have come from metal-replacement reactions used in the arts or from the delightful philosopher's toy, the *Arbor Dianae* (an electrolytic tree of crystalline silver), it actually came from an unaesthetic experiment on a frog's leg. It remained unused until 1837, when the electric telegraph and electrotyping were both seen to be useful. The utility of the latter, however, at first lay only in the arts: it was used as a process for electrolytically duplicating coins, plaques, statuary, and engraved or etched plates for the graphic artist. All the great illustrated newspapers stem from this—the *Illustrated London News,* the *Scientific American, L'Illustration,* and *Harper's Weekly.* Soon electrolytic baths were giving rise to monumental sculpture, some weighing over 7500 pounds. Many of the "bronzes" in the Paris Opera House are of electrodeposited copper, while a nice English example is the ten and a half foot high statue of Prince Albert behind Albert Hall in London. It was made by the firm of Elkington in 1862.

Almost immediately an even larger use for "galvanism" developed—the production for middle-class tables of metalware having all the glitter of the rich man's silver and gold. Within a decade, Sheffield plate was supplanted by electrodeposited silver plate, with a not always felicitous relaxation of restraint on design.

At first the electric current for these applications came from banks of small batteries (Daniell cells) in which nearly three pounds of zinc and acid were consumed for every pound of copper deposited. The larger uses of electricity could develop only after a steam-driven generator had grown out of an 1832 lecture-demonstration device made to intrigue physics students with the realities of Oersted's electromagnetic interaction. The first commercial electrical generator was constructed in 1844 to the 1842 design of J. S. Woolrich (whose patent includes also a plating bath); it was used in the shops of the Elkington Company for several decades before it was donated to the Birmingham City Museum, where it now stands. The giant electric power industry of today thus did not begin with a preconceived desire for its utility; the first suggestion came from the arts. Once power generation had been demonstrated, however, it was ripe for development and use by men of a different cast of mind; soon came lighting (beginning with arc lights for lighthouses), and then motive power. All big things grow from little things, but new little things will be destroyed by their environment unless they are cherished for reasons more like love than purpose.

Banquo expressed a deep human wish when he demanded of the witches, "If you can look into the seeds of time, and say which grain will grow and which will not, Speak then to me." But how do the seeds of human achievement form in the first place? Not just by taking logical thought, but rather by giving curiosity full rein and using all of a human being's capability—his holistic powers of understanding and aesthetic imagination as well as his analytical skills. I do not mean to imply that all technologists are sensitive aesthetes, but I do claim that the *beginning* of much useful technology (as indeed of most human achievements in the past) has arisen in aesthetic experience. The subsequent and more obvious stages of profitable development can occur only as a sequel to a quite different dynamics.

The simple picture of origins outlined above, which applies so well to the early stages of many early technologies, seems hardly applicable to the twentieth century. The experience of discovery in the laboratory is still an essentially aesthetic one (a fact rather thoroughly disguised by the accepted style of reporting the results), but the motivation is rarely a desire to create beauty. Why is this? Is it just that the patronage for creation has changed, or is it that most of what we notice today is not creation but merely a natural or unnatural refinement of the old, while the really new is around unnoticed, awaiting an environment that does not yet exist? In any case, neither art nor history can be understood without paying attention to the role of technology; and technology cannot be understood without history and art.

12
Some Constructive Corrodings

Corrosion is generally regarded as evil, destructive, or at least undesirable. But, like the electrolytic couple that underlies it, corrosion has two sides. A glance at history shows that corrosion has stimulated much useful science and has been central to many useful processes and the making of many useful objects. One of its principal applications is in the prevention of corrosion, for the products of initial reaction, when they have the right structure, block further attack. Indeed, since the underlying electrochemistry is already fairly well understood, and is immutable, future research in the field will increasingly find opportunity in problems involving the microstructural, interfacial, and mechanical aspects of passivity. The unresolved problems are more akin to diffusion-controlled mineralogenesis in the earth's crust than to basic electrode behavior.

Corrosion can be broadly considered as the movement of interphase interfaces—its chemistry is that of heterogeneous systems in general. The first corrosion was the weathering of rocks after the primeval formation of the earth's crust, with the accompanying redistribution of the available atomic species into new materials or new arrangements of the old, including the formation of beautiful landscapes and gemstones as well as the ores of useful metals.

Some of the constructive uses of corrosion have historically included the following, listed in no particular order:

- To obtain a solution for use in later chemical processing, as in making pigments, inks, and mordants.
- To obtain directly a useful mineral corrosion product, such as the pigments verdigris and ceruse and fine abrasives such as crocus or rouge.
- To remove base metals from precious metal alloys for surface coloration, for purification, and for assay.
- To put metals into solution for chemical analysis.
- To remove superficial layers of the products of prior corrosion; for example, in pickling tarnished or heat-blackened metals.
- To obtain an adherent superficial layer of corrosion product, either for decorative purposes or to confer resistance to corrosion, as in the bluing of steel, the chemical coloring of bronze, the anodizing of aluminum, and the formation of alloy coatings by diffusion.
- To roughen or to smooth a surface for decorative or technical purposes, or, when locally restrained, to produce controlled designs as in etched armor, plates for the graphic artist or photoengraver, the manufacture of integrated electronic circuits, and "chemical milling" to predetermined shapes.
- To expose the internal textures in materials for decorative purposes or to reveal their macrostructures and microstructures for scientific study and process control.
- To obtain electric current for uses involving electrochemical, electrothermal, or electromagnetic effects.

The earliest evidence of the use of many of these processes lies in archaeological objects, but the earliest literature on the chemical arts, for example, the Leyden manuscript of the third century A.D. and the *Mappae Clavicula* which began to take form early in the ninth century, are rich in recipes for producing color changes by corrosion.[1] The *Mappa* tells how to make the pigments ceruse and verdigris by corroding lead and copper, respectively, while iron is corroded to yield crocus powder, or put into solution for making ink, and it is etched with a copper-bearing solution to obtain a rough copper-plated surface as a basis for amalgam gilding. Fluxes are used to clean metal surfaces for both soft and hard soldering.

The selective removal of copper from its alloys with the noble metals is a most interesting process. Today it is used, in the Western World, only in refining and to improve the color of solid gold very superficially; in earlier times it was employed to obtain thick layers of pure gold on the surfaces of objects made of cheaper alloys. Chemically, this is identical with the inquartation and parting operations of the assayer—in mechanism, it is related to the dezincification of brass. For corrosion to proceed to a significant depth the presence of 50 atom percent or more of the baser metal is necessary. The Old World beginnings are uncertain. Wasteful of gold, the process could not compete with gilding processes in which thin layers of gold were externally applied in the form of foil or amalgam, but it remained in use as the basis of the alchemist's method of "multiplying" gold. A good summary of this and similar "alchemical" operations is given by Joseph Needham in his general discussion of colored alloys.[2] In the numismatic collection of the Kunsthistorische Museum in Vienna there is a fine "gold" plaque made by an alchemist named Seiler in 1677.[2a] About 25 cm high, it is actually an alloy of gold about 44 percent silver, almost identical in composition with the Japanese Oban coins of the seventeenth century and later.

It was in South America that this type of selective corrosion particularly flourished (figure 12.1). Most pre-Columbian "gold" objects are actually made of alloys that contain less than a third of the precious metal, but they were given a pure gold surface after shaping by treatment with corrosive natural minerals, probably one of the varieties of basic ferric sulphate such as copiapite, $Fe_3(SO_4)_4OH \cdot 13H_2O$, mixed with salt. Known to early metalworkers in the West as *misy*, ferric sulphate is almost as potent a corrodant as sulphuric acid itself. When used on Cu-Au alloys by the metalworkers of Peru and Ecuador, it left on the surface a submicroscopically porous layer some 50 to 200 micrometers thick of pure gold which was consolidated by burnishing or by annealing to close the pores. The process, which the principal student of it, Heather Lechtman, calls "depletion gilding," can be identified by the composition gradients beneath the surface of the object.[3] The gold content increases in a stepwise manner, quite distinct from the sinusoidal gradient left in the metal by cementation processes which are done hot and involve bulk solid-state diffusion.

The enrichment of silver on the surface of silver-copper alloys is a process with somewhat similar results, but this involves oxidation of the copper to form an external layer of scale which is subsequently removed by pickling. This process, known as blanching, was often used by mints to make debased coins appear to be of higher value.[4]

The reaction of copper with arsenic vapor to produce silvery coatings of Cu_3As was used in the third millennium B.C.[5] (See figure 12.2.) The earliest of such reactive coatings were the alkaline-glazed steatite and quartz beads produced in the fifth millennium B.C. In the third millennium B.C., patterns in contrasting colors were produced on carnelian beads by local reaction with alkali.[6] (See figure 9.39.)

The production of colored oxide films on steel by heating it in air is well known and

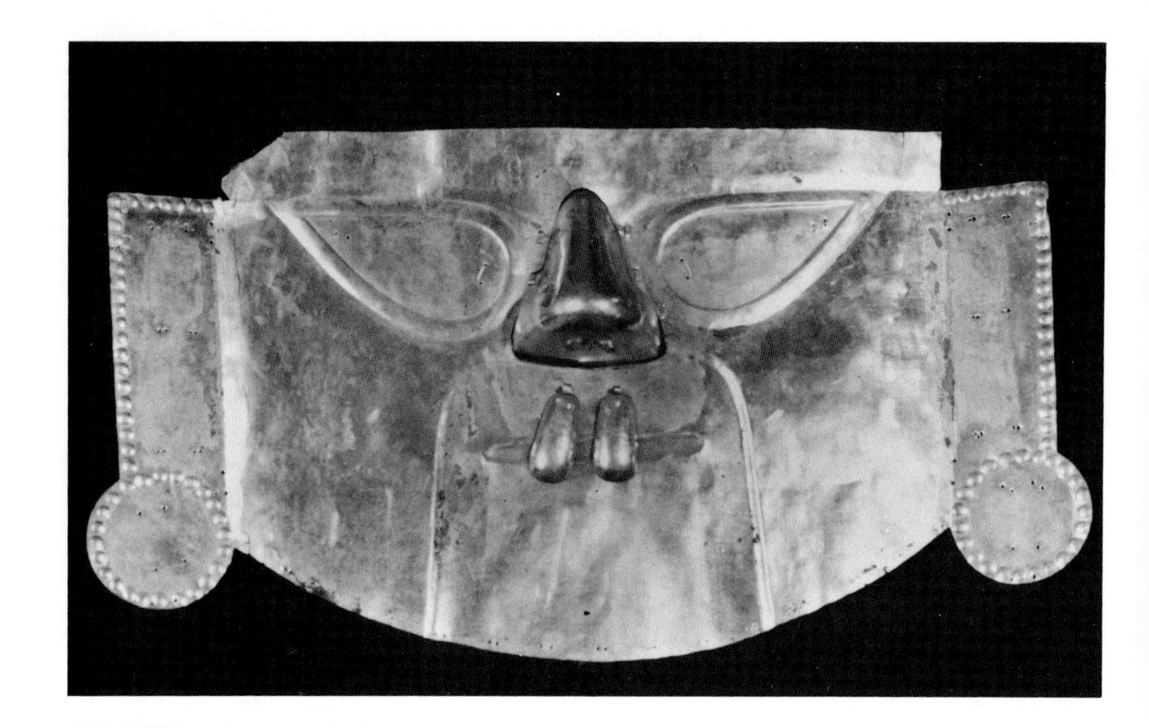

Figure 12.1
Mask of sheet "gold," Lambayeque, Peru, ca. 1100 A.D. The mask is actually about 35 percent copper, 36 percent gold, and 27 percent silver. It was treated chemically to yield a pure gold surface. Photo courtesy of the Museum of Primitive Arts.

Figure 12.2
Cast bronze bull from Horoztepe, Anatolia, ca. 2100 B.C. Length, 12.2 cm. The uncorroded area is a silvery copper-arsenic compound produced by a diffusion reaction at about 400°C. Photo courtesy of W. J. Young and the Museum of Fine Arts, Boston.

ancient. The fact that the kinetics of the reaction closely parallel the softening of quench-hardened steel enables the color to be used to control the tempering operation with considerable precision. Sixteenth-century recipes describe the necessary knowledge of the two colors, the appropriate red for quenching and the yellow through blue colors that characterize steel tools of the different hardnesses appropriate to different kinds of service.[7]

Less well known than the colors of tempered steel are the oxidation colors produced on the copper alloy foils that were used to back gems in the days before they were cut to obtain total internal reflection. There are excellent instructions for obtaining purple, green, blue, and other colors on foils of various copper-silver-gold alloys that were carefully burnished in a dust-free room and heated over a smoke-free charcoal fire. "It is very admirable how on a suddain these copper rays will change into several colours: Wherefore, when they have attained the colour you desire, take them off the fornace presently, for otherwise they will alter into another."[8]

The electrochemical replacement of one metal by another was noticed in antiquity. Pliny mentions that iron when smeared with "vinegar or alum" becomes like copper in appearance.[9] Replacement seems to have been used in Roman times for tin-plating bronze and it definitely underlies the recipes for the preparation of iron surfaces to receive amalgam gilding that are given in the ninth-century *Mappae Clavicula*. One of these (Chapter 146H) reads:

Rounded alum, the salt that is called rock salt, blue vitriol, and some very sharp vinegar are ground in a bronze mortar; the cleaned iron is rubbed with these [materials] using some other kind of soft little point. And, when it has taken on the color of copper, it is wiped off and gilded, and then, after the quicksilver has evaporated, it should be cooled in water and rubbed with a tool that is very smooth and bright until it becomes brilliant.

The conversion of iron objects into copper when they were immersed in certain mine waters was sometimes thought to be a proof of the possibility of alchemical transmutation of one metal into another. The Chinese writer Shen Kua in 1086 A.D. refers to a mountain spring whose waters contain a bitter "alum" which becomes copper when it is heated and an iron pan containing it is slowly changed into copper.[10]

In the seventeenth and eighteenth centuries, cement copper was commercially produced in Erzgebirge. At Herrengrund the copper was unusually pure and it was used to make charming dishes and cups, often incorporating models of mining and smelting buildings and bearing rhymed reference to their former life as a baser metal.[11] (See figure 8.14.)

The recovery of silver from waste parting-assay solutions by reaction with copper bowls was a standard procedure described in Ercker's great treatise on assaying of 1574.[12] By the early eighteenth century, assayers knew of the successive replacement of silver from solution by copper, of copper by iron, and of iron by zinc. In 1718 Geoffroy published his famed table of chemical affinities (figure 12.3), which was an important step in the development of the theory of chemical reactions.[13] The columns listing the experimentally observed order of the replacement of metals from aqueous solutions of their salts form what would today be called the electrochemical series, but it was,

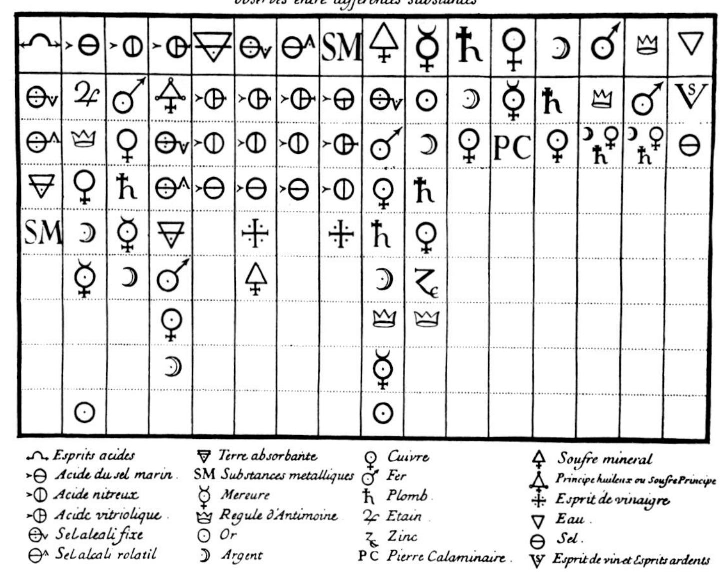

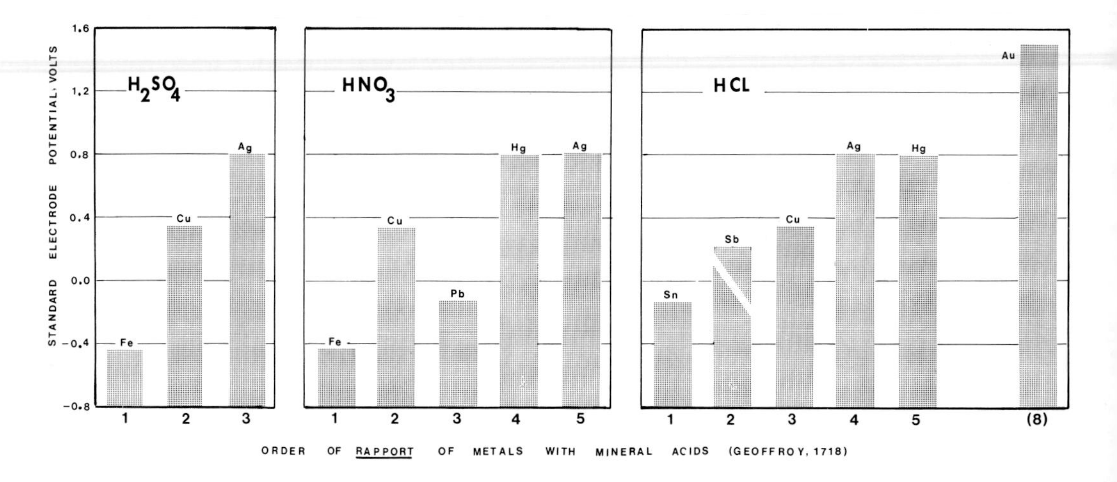

Figure 12.3
Geoffroy's table of the relative affinities between various salts, acids, and metals, ranked in order.

Figure 12.4
Modern values of the electrode potentials of some common metals plotted against the order of their affinity with the mineral acids as given by Geoffroy in 1718.

of course, long before the discovery of the role of electricity. Figure 12.4 shows how Geoffroy's ranking of the common metals in reaction with the common mineral acids relates to modern values of the single-electrode potentials. Another landmark in the scientific study of corrosion is K. F. Achard's *Recherches sur les Propriétés des Alliages Métalliques* (Berlin, 1788), which lists the response of nearly 900 alloys to four different corrosive environments.

Electrochemical corrosion and redeposition was the basis of that pretty chemical toy, the *Arbor Dianae,* the treelike growths of metallic silver produced from solutions of silver nitrate in contact with mercury or copper. In the 1680s, electrolytic corrosion was observed, though not explained, in the loss of iron rudder fixings on some ships of the British Navy the bottoms of which had been coated with lead sheets from the new rolling mill at Deptford.[14] Though there were speculations about possible effluvia emanating from the lead and corroding the iron, the phenomenon attracted little attention. A century later, in reporting his fine early laboratory study of ancient metal objects, George Pearson commented both on the fact that water would not corrode iron if air was excluded (a fact experimentally established by the French cutler, J. J. Perret, in 1772) and on the acceleration of the corrosion of iron that occurred when it was in contact with copper.[15] To quote: "The destruction of the iron swordguard by oxygen within the copper scabbard, and the preservation of the part of it not in contact with the copper, is a good example of the action of copper and water united in destroying iron, the copper remaining entire. The effect of copper upon the iron bolts and nails, in copper-bottomed ships, is a loss of the greatest magnitude." Altogether, it is odd that voltaic electricity was not discovered much earlier than it actually was, and that it needed the intervention of a frog's leg. Evidently, a phenomenon can be well known without provoking curiosity on a theoretical level.

All work on electricity between Galvani's discovery and the replacement of batteries by the magnetoelectric generator could be claimed under our rubric of Constructive Corrosion, for the sacrificial solution of an anode was then the only source of current. The first practical uses were in the almost simultaneous inventions of the telegraph and of electroplating and electroforming. Beginning around 1840, the latter quickly became of great importance in the arts, and it provided a popular hobby that introduced many youngsters into the wonders of electricity.[16] Appropriately, the first print from an electrotyped plate to be regularly published was one of the electrolytic cell itself which appeared in the January 1840 issue of Sturgeon's *Annals of Electricity* (figure 12.5).

The etched plate of the graphic artist with its intaglio lines is probably the best known use of localized corrosion. Prints made by the process first appear around 1500 A.D.[17] Leonardo da Vinci used it in 1504 and Albrecht Dürer in 1515, but there was a long prehistory in the decorative etching of metal surfaces, particularly those of steel arms and armor (figure 12.6), but also of copper alloys.[18] The first printed book on iron and steel is mainly on etching them.[19]

Although in Europe etching was mainly used to bite in patterns previously drawn by an artist, in the Middle East and Far East the etching was used almost entirely to reveal the texture of the metal itself, thereby giving surface indications of both gross heteroge-

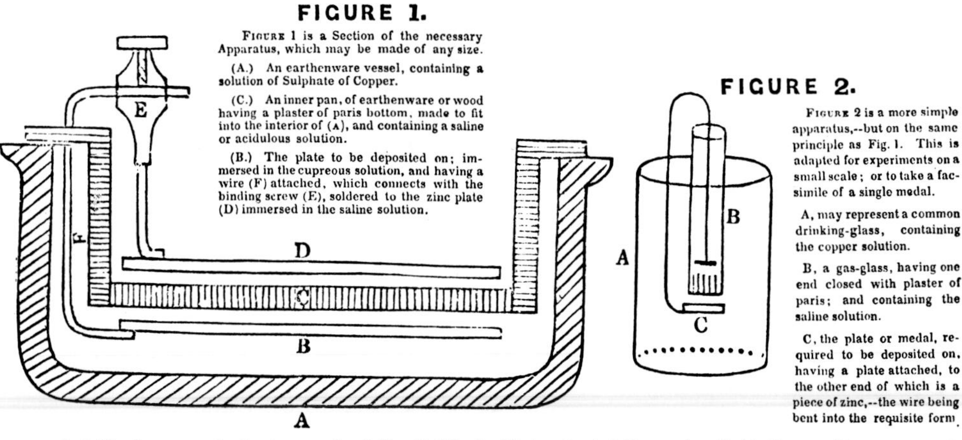

DESCRIPTION (AND ACCOMPANYING VIEW) OF THE APPARATUS.

FIGURE 1.

Figure 1 is a Section of the necessary Apparatus, which may be made of any size.

(A.) An earthenware vessel, containing a solution of Sulphate of Copper.

(C.) An inner pan, of earthenware or wood having a plaster of paris bottom, made to fit into the interior of (A), and containing a saline or acidulous solution.

(B.) The plate to be deposited on; immersed in the cupreous solution, and having a wire (F) attached, which connects with the binding screw (E), soldered to the zinc plate (D) immersed in the saline solution.

FIGURE 2.

Figure 2 is a more simple apparatus,--but on the same principle as Fig. 1. This is adapted for experiments on a small scale; or to take a fac-simile of a single medal.

A, may represent a common drinking-glass, containing the copper solution.

B, a gas-glass, having one end closed with plaster of paris; and containing the saline solution.

C, the plate or medal, required to be deposited on, having a plate attached, to the other end of which is a piece of zinc,--the wire being bent into the requisite form.

*** The above engraving has been produced (in relief) by the Electro-chemical Process described in the preceding pages;—and is the first result of that process appearing in print.

It is however by no means a fair specimen of what *may* be done by it.—The lines were originally cut in a sheet of soft lead, hastily, and without reference to ornamental beauty of execution. The want of a tool properly adapted to cut the lead, accurately, in the required manner, has made the lines less regular in formation than they would otherwise have been. The letters were punched into the lead, with types. The lead, thus prepared, was then put into the apparatus it illustrates, and a plate of copper deposited, the lines being in relief,—which is here printed off, accompanying the pages of the treatise.

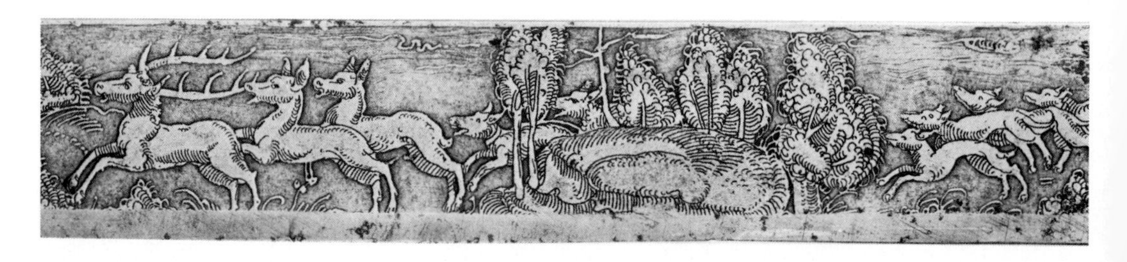

Figure 12.5
The first regularly published print from an electrotype plate, showing the cell used to produce the plate. From the *Annals of Electricity* no. 4 (4 January 1840). Courtesy of the Burndy Library.

Figure 12.6
Hunting sword with etched decoration. Made by Ambrosius Gemlich in 1540. Photo courtesy of the Trustees, The Wallace Collection, London.

neities resulting from casting and forging procedures as well as the finer structural features that depend upon crystalline separations of the kind that, when they were later studied under the microscope, provided the basis of today's materials science.[20] Although the internal structure of any material can be seen to some extent in the fine detail of fractured surfaces (and observations of these almost certainly prompted the first "atomic" or corpuscular theories of matter), the subtler details of microstructure can only be seen on carefully prepared sections that have been submitted to etching.

The discovery of carbon in steel occurred in 1774, after centuries of good steelmaking guided by the natural though mistaken belief that heating iron in a hot charcoal fire must purify it.[21] One of the many results of the stimulus given to European science by contact with Oriental materials occurred when a Swedish metallurgist came to look closely at the blackish residue responsible for the visible pattern on "Damascus" gunbarrels, which were finished by etching (figure 12.7). A few decades later the same etching technique was used to reveal crystal structure in metals for the first time, initially in a meteorite, then to make fancy bibelots of crystalline tinplate, and, almost fifty years later, for the first scientific studies of the microstructure of terrestrial steel (figure 12.8), from which stem modern metallography and much more.[22]

It is interesting to note that the discoverers of the meteorite and steel structures—Widmanstätten in 1813 and Sorby in 1864—both first published their results in the form of prints made directly from inked metal surfaces etched in relief (see figure 4.3), as Leonardo had proposed over three centuries before. By the 1880s, many engineers and chemists were studying etched metals under the microscope and relating the structures to problems of utility and understanding. Not the least of the consequences of the new method of study was a deeper understanding of the process of corrosion itself, though to this day a metallographer cannot help but feel that there is not enough attention paid to the microstructural aspects of the corrosion process.

Finally, the intentional coloring or patination of metals in jewelry, sculpture, and architecture is an enormous field with a large literature, much of which is worthless. Here more than in most areas an examination of objects in a museum laboratory is more revealing than reading about them in a library. Not all such patinations are simple sulphides or oxides. Particularly in need of further study are the Chinese black bronzes of the late Chou and Warring States periods;[23] the hot-forged high-tin bronzes of Soghdian Iran with their brownish black coating which seems to have been formed by oxidation prior to the final quench;[24] the fine black finish given to zinc alloy castings known as bidri ware, originating in the Hyderabad district of India (figure 12.9);[25] and, towering above all these in both technical ingenuity and beauty, the colored metals used by Japanese *tsuba* makers especially in the eighteenth and nineteenth centuries.[26] Prominent among the alloys used are *shakudo* and *shibuichi* which after pickling acquire, respectively, a beautiful warm purplish black color and a slightly frosty dark greenish brown (figure 12.10). These are alloys of copper with gold (ca. 5 percent) and silver (ca. 25–30 percent). Technically, if not historically, their ancestry includes the more gaudy finishes of metallic silver and gold that the metalworkers of South America

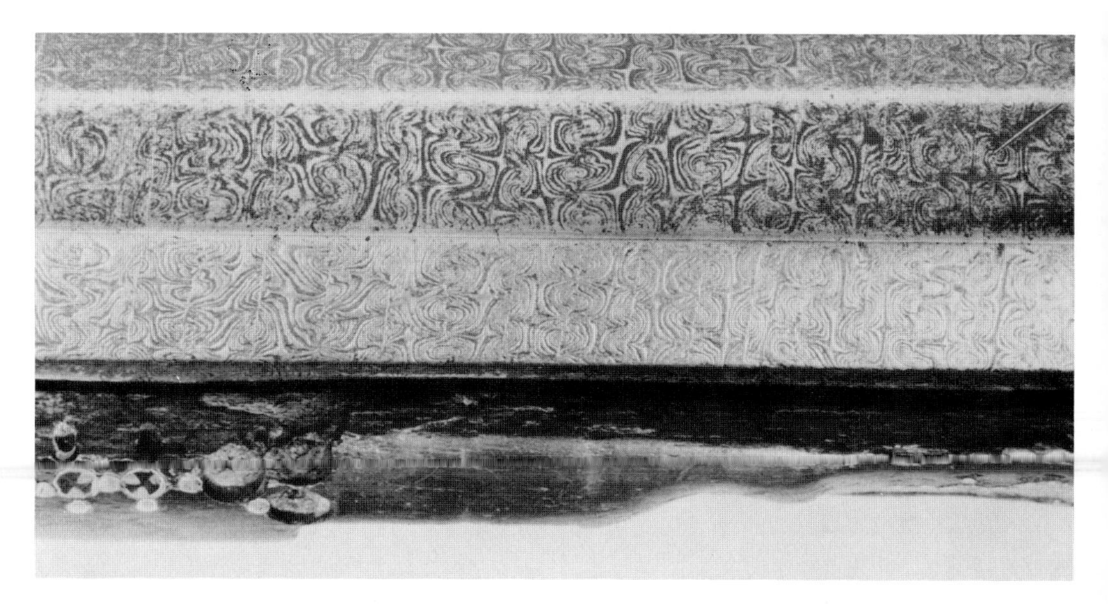

Figure 12.7
Etched gunbarrel with so-called Damascus texture. Turkish, eighteenth century. Photo courtesy of the Victoria and Albert Museum.

Figure 12.8
The first photograph of the microstructure of steel, made by Henry Clifton Sorby of Sheffield and shown at the British Association meeting in September 1864 but not published until 1887. The steel surface was polished with extreme care to avoid distortion, then etched in extremely dilute nitric acid to develop the structure. Photographed at a magnification of 9.

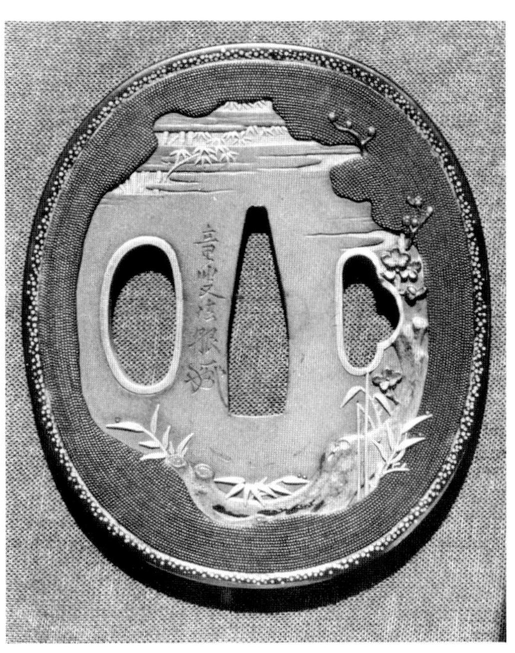

Figure 12.9
A small covered bowl of bidri metal inlaid with silver. Hyderabad, eighteenth century. Cast in sand molds, bidri metal is a zinc alloy with about 4 percent copper. It is given a rich warm black patina by pickling in a solution of ammonium chloride and saltpeter often containing copper.

Figure 12.10
Japanese swordguard (*tsuba*), nineteenth century. Several different alloys are combined by an intricate inlaying technique and given their different characteristic colors by a final treatment in a pickling solution. From the Museum of Fine Arts, Boston; photo by the author.

many centuries earlier had obtained by corroding more concentrated alloys.

All these corrodings, and many more, constitute an integral part of the story of man's discovery of the wonderful diversity of the properties of materials which he can enjoy, incorporate in his philosophies, and employ in innumerable devices.

Notes and References

1
E. R. Caley, "The Leyden Papyrus X," *Journal of Chemical Education* 3:1149–1166 (1926); C. S. Smith and J. Hawthorne, *Mappae Clavicula: A Little Key to Medieval Techniques* (Philadelphia, 1974).

2
Joseph Needham, *Science and Civilisation in China*, vol. 5, part 2, section A33, Alchemy and Chemistry (Cambridge, 1974). An important source that he does not cite is Book V "which treateth of Alchemy; shewing how Metals may be altered and transformed one into another" in G. B. della Porta, *Magiae Naturalis Libri Viginti* (Naples, 1589).

2a
R. Streblinger and W. Reif, "Das alchemistische Medaillon Leopold I," *Mitteilungen der Numismatischen Gesellschaft in Wein* 16:209–213 (1932).

3
Heather Lechtman, "Gilding of metals in Pre-Columbian Peru," in W. J. Young, ed., *Application of Science in the Examination of Works of Art* (Boston, 1973), pp. 38–52.

4
Jean Boizard, *Traité de Monoyes* (Paris, 1696); L. H. Cope, "Silver-surfaced ancient coins," in E. T. Hall and D. M. Metcalf, eds., *Methods of Chemical and Metallurgical Investigation of Ancient Coinage* (London, 1972), pp. 26–278. Similar techniques were used very early in Peru.

5
C. S. Smith, "An examination of the arsenic-rich coating on a bronze bull," in Young, *Application of Science*, pp. 93–102, H. McKerrell and R. F. Tylecote, "The working of copper-arsenic alloys," *Proceedings, Prehistoric Society* 38:209–218 (1972).

6
H. C. Beck, "Etched carnelian beads," *Antiquaries Journal* 13:384–398 (1933). [See also P. Francis, "Etched beads in Iran," *Ornament* 4(3):24–28 (1980).]

7
See particularly the sources given in translation in chapters 1, 3, and 4 in C. S. Smith, *Sources for the History of the Science of Steel, 1532–1786* (Cambridge, MA, 1968).

8
G. B. della Porta, *Magiae Naturalis Libri Viginti* (Naples, 1589). Quotation is from English translation (London, 1658), p. 188. Other good accounts of colored foils are B. Cellini, *Due tratate . . .* (Florence, 1568), and G. Smith, *Laboratory and School of Arts* (London, 1740).

9
Pliny, *Historia Naturalis,* XXXIV, 149; English translation by K. C. Bailey (London, 1932), vol. 2, pp. 60–61.

10
Cited by Needham (note 2), vol. 2, p. 267.

11
Gustav Alexander, *Herrengrunder Kupfergefässe* (Vienna, 1927). Early accounts of cement copper are: della Porta's *Magiae Naturalis,* book 4; E. Browne, *Philosophical Transactions* 5:1042–1044 (1670); and C. A. Schlüter, *Gründlicher Unterricht von Hütte-Werken* (Braunschweig, 1738), pp. 461–471.

12
Lazarus Ercker, *Beschreibung allerfurnemisten mineralischen Ertzt und Berckwerksarten* (Prague, 1574), English translation (Chicago, 1951), pp. 167–169, 223.

13
E. F. Geoffroy, *Mémoires de l'Académie Royale des Sciences* (Paris, 1718), pp. 202–212.

14
Samuel Pepys, "Naval minutes (1680–83)," *Publications of the Naval Record Society* 60:115 (1926); Thomas Hale, *The New Invention of Mill'd Lead for the Sheathing of Ships Againt the Worm* (London, 1691).

15
George Pearson, "Observations on ancient metallic arms," *Philosophical Transactions,* Read 6 June 1796, 57 pp.
16
For a brief history and bibliography of the beginnings of electrotyping, see item 190 in the bibliography.
17
Arthur M. Hind, *A History of Engraving and Etching* (Boston, 1923); L. Reti, "Leonardo da Vinci and the graphic arts," *Burlington Magazine* 113:189–195 (1971).
18
See chapter 9 in this volume.
19
Von Stahel und Eysen (Nuremberg, 1532). English translation in Smith, *Sources for the History of the Science of Steel,* pp. 1–19.
20
C. S. Smith, *A History of Metallography* (Chicago, 1960).
21
See chapter 2 in this volume.
22
Smith, *History of Metallography,* esp. chapters 12 and 13.
23
W. T. Chase and U. M. Franklin, "Early Chinese black mirrors and pattern-etched weapons," *Ars Orientalis* 11:216–258 (1979).
24
Pieter Meyers, Metropolitan Museum of Art, private communication.
25
A good illustrated account of bidri metal manufacture is given in Oppi Untracht, *Metal Techniques for Craftsmen* (Garden City, NY, 1968), pp. 138–149.
26
Morihiro Ogawa, in *Nippon-to: The Art Sword of Japan,* W. A. Compton, J. Homma, K. Sato, and M. Ogawa, eds. (New York, 1976); B. W. Robinson, *The Arts of the Japanese Sword* (London, 1970); W. A. Roberts-Austen, "Colours of metals and alloys considered in relation to their application to art," *Journal Society of Arts* 36:1137–1146 (1888), 41:1022–1043 (1893); W. Gowland, "Metals and metal working in old Japan," *Transactions Japan Society* (London) 13:20–100 (1915); E. Savage and C. S. Smith, "The techniques of the Japanese *tsuba* maker," *Ars Orientalis* 11:291–328 (1979).

13
A Highly Personal View of Science and Its History

One of Rainer Maria Rilke's poems runs somewhat to the effect that man need not sing to the angels of the glory of God. Angels know of this far better than does man, but they might be amazed on being told how man weaves, makes rope, shapes pots, smelts ore, and builds many beautiful and useful things. In much the same way, I, as a technologist awed by the great historians and revering their works, suggest in all due humility that they might find something of value were they to study the technical parts of their edifice in more detail—that the technical parts, in fact, constitute much of the prehistory of the whole marvellous structure of thought and civilization. Moreover, even "pure" science is much more than the astronomy, physics, and chemistry which have been the main preoccupation of the few historians in the past who have looked at science at all.

It is fashionable today to complain about the earlier science historian's overconcern with the history of ideas and the role of great men in formulating them. Many historians now seem to treat the history of science as little more than a branch of social history. Taking the development of scientific understanding of the world (surely one of the most magnificent achievements of man) as the central topic, one must of course pay attention to its social milieu both as a provider of patronage and as a source of encouragement and opposition. However, science, like any organism, has inner compulsions which are more central to its being than the environment in which it flourishes. Personally I believe that the internal substructure of science was formed very largely out of the earlier factual findings of technologists, and that it was in the main a successful technology that exposed the fallacy of classical closed systems of thought and forced theoreticians to relate to experimental evidence.

My own criticism of past approaches to the history of science is that they found too much of the logic of the product in the process by which it came about. To me, the critical stages have seemed to be more often irrational than logical. The mutant seeds of the most formal theories form in the mind of one individual, a mind shaped by experience that is more sensual than intellectual. Intellectual analysis can only follow discovery, and discovery makes more use of the aesthetic nature of the whole man than of his cerebral capacity—though the latter is, of course, essential for communication and for putting in memorable order as well as for setting the stage for larger synthesis.

Though a number of my papers have been accepted for publication or review in historical journals, I have had no training whatever as an historian: I came into the field simply by natural extension of my professional work as a metallurgist. Metallurgy has been in contact with the core of much human experience—witness the "bronze" and "iron" ages, and the influence of the special properties of metals on vast numbers of objects of utility or beauty.

It came as a surprise to me when in the early 1930s I began to read the major works in the history of science and found so little in them that related to my own experience. Neither economic history, based upon taken-for-granted changes in technology, nor main-line history of science, with its emphasis on the noble concepts of cosmology at one end and atomistic or elementary theory at the other, paid significant attention to the development of knowledge of the wonderfully diverse properties of materials as they were experienced in the research laboratory or used in craft or industrial operations. Science seemed to be concerned with matter, not materials. Eventually I came to see that most of the aspects of the behavior of materials that interested me were structure-dependent ones that were far too complex to be handled by the respectable mathematical methods of the physicist, which applied only to systems so defined as to exclude structural change above the level of the atom. Kinetics and other aspects of the mechanics of solids that depended only on mass and perfectly elastic behavior were part of the burgeoning of science in the seventeenth century; plasticity was not part of physics until well into the twentieth. It was only after I had published my narrowly focused *History of Metallography* in 1960 that I saw that it was a partial prehistory of the most flourishing branch of physics today, that of the solid state, which historians of science have not yet discovered.

The forefront of science seems now to be turning away from the search for simplistic, atomistic, analytical models (which since Newton have been associated with laudable rigor and which are as essential as a basis for understanding as are atoms for the existence of matter) toward real systems with their fascinating diversity and complexities. As my wife has pointed out, it is almost as if the Copernican Revolution has been undone, and things on man's own scale are returning as a topic central to scientific inquiry. Certain it is that complex things, whether composed of material atoms or social or intellectual units, do not spring full-panoplied into existence, but that they grow by locking into their structure some of the accidents of history. A knowledge of individual and collective history is as important for understanding complex inorganic materials as it is for understanding biology or human culture. There are some quite fascinating parallels between the complexities of real materials composed of imperfect aggregates of atoms and the interlocking aggregates of ideas and people which make up the sciences or society. Without in any way abandoning, in fact while exploiting to the fullest extent possible, all that has been learned by the main-line physical sciences, it is worthwhile looking again at the things that had to be pushed aside in their development, and to recover some of the holistic enjoyment of the world that motivated seventeenth-century figures such as Robert Hooke and René Descartes. Cartesian philosophy, though only qualitative, provided better structural models and carried greater insight into the origin of the physical properties of real materials than anything in the Newtonian philosophy that so triumphantly replaced it. The science of materials today is finding the appropriate levels of actual structure comparable to the purely imaginary corpuscles invoked by the Cartesians to explain all manner of properties. It also has revived the disgraced phlogiston of the eighteenth-century chemists (now the valence electron), and with a little stretch one can regard their particles of caloric as today's quanta, which are no less real than the atoms with which they enter into fixed or variable combination. One might also note that the discarded Aristotelean elements were a recognition of the states of matter (Earth, Water, Air, Fire = solid, liquid, gas, energy) and that the mixture of the qualities derived therefrom is by no means a foolish description of the physical properties of diverse substances, however far it may be from a chemical explanation.

In anything whatever, excepting only the smallest isolated particle, there is a hierarchy of balance between perfection and imperfection, between ordered aggregation and disorder. Every externally observed property depends upon a particular level of internal structural communication. The details of that structure reflect individual history, and are a record of it. Change follows an "S" curve based upon much the same structural sequence in human history as it does in materials. Both involve the rearrangement of the pattern of interactions between parts. The structural features that mark the beginning of change, the growth of the new form, and its internal and external adjustment at maturity, are quite similar in different systems. All involve the stages of the difficult formation of an interface between new and old, its rapid expansion until it meets restraint, and final adjustment. In the beginning, a small nucleus of the new structure forms, and if conditions are ripe it will grow at the expense of the larger structure of which it was initially a small aberration; at the end, deviant units submit to the demands of the larger organization and form local clusters where they will be least harmful, or most helpful, to it. In the beginning aesthetic creation suggests things that may, if widely adopted, cause disruptive change; at maturity, perceived needs force local change in a direction to promote greater stability of the whole. Is this history, or science, or the history of science?

The growth stage, whether of a crystal, an idea, or a social structure, is the easiest to follow, while the formation of an effective nucleus at the beginning is least clear. Change has to be preceded by metastability in the existing structure, as when the density of packing has been changed by external events or the parts have suffered internal change. The parts are held together by an overarching pattern of interlock inherited from the past, and they are ready for rearrangement if a sample of a new pattern is presented which better satisfies their new requirements and if a mechanism for producing change exists. Though eventually the change will be accompanied by relaxation at all levels, the new structure is at first an unwanted misfit in the existing one. It is a nisus, a feeling toward a structure that will have validity only in an environment that does not yet exist, and it is opposed by the entire structural inertia of the established interlock. It cannot be predicted ahead of time, and the first nucleus will not conform to accepted pattern. However, one can expect that it will form in regions of imperfection in the old structure, at places where the parts have compromised long-range disregistry with locally odd structures.

Small changes may occur within a tolerant maturely established structure without the disruption of the larger pattern. Slightly differing parts that had been randomly or incorrectly incorporated during rapid growth may move by diffusion (a continued sequence of local changes) to form clusters of preferred neighbors. Parts may also diffuse in from an external surface; and the parts themselves may change by internal transmutation such as by radioactive change in chemical atoms or by the capturing of an idea if the "atom" is human. An accumulation of these local catastrophic changes can in the end produce metastability in the larger system, for they make it ripe for reconstructive revolutionary change if an appropriate nucleus should appear. All change is catastrophic at some level, that is, it involves an irreversible change in the manner in which parts are connected, a switching of valencies.

The main difficulty in using this structural model in complex systems lies in the identification of the units when they are not truly atomic. Moreover, even simple units may participate in several superimposed patterns of interaction based upon different methods of communication, as, for example, magnetic domains differentiating regions within a single crystal lattice, the conflict of local and long-range loyalties in human society, and the different meanings of a single note in a musical composition.

This picture of structural change suggests that the logic of science can have little logic at the critical stages of its own growth. Kuhnian shifts of paradigm occur in an environment that has been rendered unsatisfactory by the introduction of some new components within an old superstructure. However, the initial suggestion of the new structural pattern must be a misfit, imaginatively suggested and difficult to accept. It will at first be indistinguishable from innumerable other imaginative or crazy suggestions that are quickly abandoned even in the minds of their originators. Not infrequently, a suggested pattern that is rejected when first proposed later turns out to be viable because the environment has changed. All things are interactive.

This emphasis on the structural conflicts at nucleation leads by a different route to the well-known fact that great scientific ideas form and spread because they show ordered interconnection between facts of nature that were previously known in a state of disorder or inferior order. The most important ideas always originated outside the accepted intellectual framework of their times and at first were opposed by its logic. Until very recently (when discovery became a visibly profitable business) the precursors of most scientific insights were empirical discoveries in technology, that is, things discovered and appreciated for their intrinsic interest without regard for their place in an intellectual order. (Without some level of appreciation they will not, of course, be remembered at all.) To carry the same principle one step further, the discovery of the materials, processes, and structures that comprise technology almost always arose out of aesthetic curiosity, out of the desire for decorative objects and not, as the popular phrase would have it, out of preconceived necessity.

Just as my own metallurgy contained within it much of the prehistory of scientific chemistry and physics, so does art contain most of the prehistory of metallurgy. Personally, I came to this view only after decades of laborious study of the literature of metallurgy without ever having found verbal records of the beginning of any significant aspect of it. Then I found early artifactual evidence for ingenious techniques in art museums; but I had frequented art museums for some years before the fact dawned upon me that discovery *is* art, not logic, and that new discoveries have to be cherished for reasons that are far more like love than purpose.

Historians are intellectuals and are committed to verbal communication. It is their professional responsibility to correct for the fact that verbal records leave out most of the human experience, that they are commonly intended to mislead and that they are almost always biased unconsciously. Nevertheless, with all the historian's care in interpretation, his conclusions cannot help but reflect his (or his patron's) interest and concept of order. Such selectivity is even more true in science, where the final intellectual triumph makes the messy underpinnings almost invisible.

Of course there are social influences on science, as on anything else, but these are less unique and less important than the inverse. I believe that the historian concerned with change would, at the present time, be better occupied with the neglected technology which underlies so much of the history of science and which perhaps more directly affects society. Changes in technology change how man communicates, how he feeds, clothes, houses, and amuses himself—and, most important, what he thinks about. Despite popular clichés, technology has been a fully human experience, with both sensual and intellectual attributes, working to fulfil the practical and aesthetic needs of society. At times its practitioners experience a rich interplay between their internal nature and the nature and meaning of things in general. This picture of the technologist is very different from the one that is current today of an almost evil, malformed creature, a destroyer of human values! I suggest that the sympathetic study of the technologist is far more important to the future than the study of the social dropouts popular with novelists and filmmakers today. The perspective that would come from more humanistic study of technology in history is desperately needed. The "atom" that is at the base of the whole structure of technological institutions is the individual technologist, with various motives, driven by glimpses of beauty or of profit or of service to mankind; sometimes a far-out discoverer, sometimes a repetitive craftsman, sometimes a deviser of ways of making more, sometimes an entrepreneur who seeks the inventions of others which he can adapt to perceivable needs; but he is always human and always contributing to the intellectual and social changes the study of which is the very business of the historian. However, the human characteristics of the individual technologist are not always incorporated in the megamachines (Lewis Mumford's term) that are constructed out of his work. This is a different story.

The aspects of science that in the past have interested historians have usually been the concerns of formal science itself after a lapse of some decades or centuries. Scientists, rightly proud of their achievement, have underestimated the contributions of the prior practical arts and have overestimated the practical value of theory, while historians have followed them in rarely noticing the practical man with his knowledge that matter is usually complicated enough to render other people's theories of less value in its management than his own sensually acquired experience. Modern instrumentation and automatic control of industrial processes is rapidly changing this, but the historian concerned with the past must understand it. I well remember an occasion in 1962 when in a remote Iranian village I asked a blacksmith famous for his superior penknives to tell me the difference between iron and steel. "What's the difference?" he replied. "What's the difference between an oak tree and a willow—they have different natures and one must adapt to them." He did not accept the suggestion that some material absorbed from the fire's charcoal might have something to do with it, and he would not have understood a word of any lecture I could have given him on diffusion, crystal structure, and phase transformations; yet he could make a good knife and I could not. (However, I was visiting him, not vice versa, and I was driven by curiosity, with the cost of my travel half way around the world paid for from the surplus of an industrial society that did have use for such knowledge.)

No scientific theory of matter could have arisen prior to the discovery of, and empirical experience with, the differing natures of diverse substances. Democritus's atoms were an extrapolation from the granular textures of stone, wood, and metal seen by the sculptor on his broken and carved surfaces. Aristotle's elements and elementary qualities were the states of matter as experienced in the workshop—today we call his earth, water, air, and fire, solid, liquid, gas, and energy. To read the list of the properties of matter to be explained in his *Meteorologica* is to walk with him down Hephaestos Street observing the various artisans manipulating their materials as they fashion useful or beautiful objects. The transmutations on which the alchemists later built their mystical theories began as real changes of tangible properties, first in the alloys and colorations used by the makers of jewelry and today of concern to the solid-state physicist.

The first break into modern scientific understanding of materials came, however, not with physics and the seventeenth-century Scientific Revolution (though Cartesian corpuscles were marvellously prescient), but a century later with analytical chemistry and the discovery of the many species of chemical element, of something that remained unchanged and recoverable through many transformations and combinations. This in turn was a direct outgrowth of the practical metallurgist's intimate experience with the many phase separations and partitions used in his smelting and refining operations and the precise quantitative methods of assay based upon them that he had developed. Having read the standard histories of chemistry available in the 1920s with their fuzzy background in alchemy and medicine, I was quite unprepared for stumbling across the precise clear descriptions of chemical processes in Lazarus Ercker's book on assaying, published in 1574. Here was a historical source I could understand. By 1935 I had learned for myself the historian's prime rule, which is the same as the scientist's, namely to work with real sources and to be fundamentally critical of all secondary works and accepted interpretations.

Since I am a poor linguist, I have devoted much of my historical labor to the translation of significant sources, working in collaboration with people like Martha Gnudi, Anneliese Sisco, and John Hawthorne, who knew their basic languages well and were willing to argue over the technical meaning of every phrase. Italian, Latin, French, German, Spanish, and Polish—I have sought records of the role of metals in theory and practice anywhere throughout the whole human record, with a most unprofessional disregard for the historian's usual concentration on one period and place.

I have come to the conclusion that everyone should *write* history based upon his own selection of sources that appear significant to him, but that no one should *read* it except to obtain general information in areas of peripheral concern. Oddly, only an amateur can be so detached. It takes time. In my own case, as an industrial metallurgist with no courses to teach and having neither occasion nor desire to make a name for myself in the academic world, I was able to follow my curiosity for twenty years without regard for the breadth of knowledge and professional discipline that someone seeking promotion in his department or membership in a professional historical society would have to have.

It was somewhat the same in my relationship with science. I missed the beginnings of calculus because I had been advanced too fast in grammar school, and I never caught up. Consequently, when I came to enter the

University of Birmingham my mathematical ability was inadequate to qualify for admission to the physics course which was my first choice, and I settled for metallurgy. Perhaps this was not too bad. Instead of becoming a third-rate physicist I became a second-rate metallurgist and, temporarily losing sight of the shining goal of theory, I spent enormous amounts of time familiarizing myself with the details of the diverse behavior of alloys. Even in grammar school I had had my own laboratory at home which began with a mortar, some flasks and test tubes, a Bunsen burner, and a few chemicals given me by my uncle, Herbert Vigurs, on my tenth birthday. He also discussed geology with me over his collection of fossils, and I remember his telling me about crystal cleavage and its "molecular" significance. I was fond of qualitative analysis and did innumerable experiments with all kinds of substances. Someone gave me an 1880-ish set of Plattner's blowpipe analysis equipment, the use of which could hardly have been better preparation for my later interest in the history of metallurgy, for the techniques represented in miniature and in logical form all the old methods at their apex. All of this gave me an intimate sensual feeling for the nature of substances and their reactions and behavior that no modern scientific training conveys, and that has been of great help in interpreting ancient technology from both artifactual and written records. I see retrospectively that I enjoy the down-to-earth diversity of experience rather more than the austere and exalted levels of exact theory, but most of all I enjoy the relationships between the two and would be dissatisfied with either alone.

Another strong formative influence in my childhood was playing with a Meccano set, composing from steel strips, angles, nuts, and bolts all kinds of machines and structures, both known and invented. This, and a superb teacher of geometry at Bishop Vesey's Grammar School, undoubtedly laid the basis for my work in structural principles in both science and everything else. At a very early age I became familiar with and fascinated by microscopy and photomicrography under the influence of the father of a sunday-school friend of mine. He was a highly skilled amateur of pond life and he used to exhibit at the converzationi of the Midland Institute held annually in the Town Hall at Birmingham, all in the best nineteenth-century tradition. A parental gift added a fine Watson metallurgical microscope to my home laboratory in 1923. I built all the auxiliary equipment for photography myself, even to the point of making sand castings, of aluminum alloy, in my home foundry. I disliked history in school and have had not even one university-level course in it. I started my professional life as a metallurgist, and my turn to history was a direct result of efforts better to understand my particular branch of science. I have done little that was not connected in some way with the understanding and use of materials. In 1931 I married Alice Kimball, then a student of English social history, and perhaps it was my unsuccessful attempt to turn her toward the history of science that helped spark my own interest in it. My first purchase of a rare book was John Webster's *Metallographia* (1671, London) which she found in one of her catalogs. No historian inspired me in my formative years, though I was moved by the historical writings of the great nineteenth-century English metallurgists Roberts-Austen and John Percy. Herbert Hoover's translation of Agricola's *De Re Metallica* was clearly the model for my first historical publication, the *Pirotechnia* of Vannoccio Biringuccio in 1942. I owe an immense debt to the Sterling Memorial Library at Yale, which I was able to use on weekends from 1928 to 1942, thanks to the shortened industrial workweek of the depression. In the postwar years the University of Chicago, all of it, was an environment that provided immense stimulation to me, although the historians there were not aware of my existence. Not until 1955–56, when I spent a year in London on a Guggenheim fellowship, was history a primary occupation for me, and I did not return to this blessed state until 1961, when I went to MIT as an Institute Professor with no assigned duties.

The only part of my work in science that has any chance of being of lasting consequence is that published between 1948 and 1955 on the equilibrium form and the topology of packing of the nonsymmetrical grains in a crystalline aggregate. It was an approach more appropriate to the nineteenth century than the atomistic study of crystal lattice symmetry that was attracting my more advanced scientific colleagues. I also found some basic relationships between the features that are observable in two-dimensional sections of structures such as are usually studied in the laboratory and their three-dimensional reality, and I have been fascinated by hierarchy in all systems ever since.

Quite late in life I began to find meaning in art, and through it a better understanding of much that had interested me earlier. (A Japanese sword at the Victoria and Albert Museum and its curator, B. W. Robinson, provided the first impetus for the change.) Many philosophers and historians have examined the relations between science and art in terms of underlying concepts and *Weltanschauungen*: I found it mainly in the techniques of making beautiful things and in the use, particularly in the art of the Far East, of quite subtle aspects of the behavior of matter to achieve aesthetically appealing effects. It has long been known that some chemical reactions and mechanical principles were used in toys and trinkets before they became the basis of "purposeful" applications of more social importance, but I contend that aesthetic curiosity is the very root of all discovery. It has led man to explore the natural world and to find things for enjoyment before they can be understood. Collectively, it led to ritual and society.

Yet discovery needs preparation. I believe that, although occasionally great insights into broad questions may come without specialization, in most cases a period of very deep immersion in a narrow area is a necessary preliminary. A general education if not supplemented by concentration in something specific is likely to produce mediocrity. One needs both a mass of facts to puzzle about and the experience of using the mind in both rigorous and imaginative modes.

In my general hierarchical view of the structure of things and of the understanding of things, it is permissible and proper to select any level one chooses as the center of attention so long as both the substructure and the superstructure are studied, namely, the internal interactions that give the thing itself and the external interactions that give it value or interest. Change involves catastrophic change on some level that belongs to neither the old nor the new, and change in the intellectual structure that is science involves shifts that are not at first intellectual. The history of science as an intellectual effort cannot be understood without studying the extrascientific environmental influences of society, but it also cannot be understood without more attention to nonintellectual precursors on a smaller scale. The history of technology is even more interactive with social history than is science, but even it begins with the personal experience of a man working with materials and mechanisms. He is the atom from which the whole great superstructure of ideas and social forms arise in both their continuity and their change.

For the future, I see the greatest need and the greatest opportunity in the development of nonverbal sources for history and of methods for their interpretation. The greater use of artifactual records from recent as well as archaeological periods would cause technology to take its proper place in historical research. In the history of the pure sciences, the richest opportunities lie in the area of solid-state physics. Not only is this the largest subfield of physics today, but it is changing the concept of what physics can be. The boundary conditions no longer need to be those of idealized atomism and simple symmetry, but include imperfections and real structures.

The history of technology and science should be a significant fraction of all the history that is taught in schools and universities. It throws light on all aspects of man's being, and it underlies all great social changes. To the scientist it shows the essential dependence of the monumental moments of sudden enlargement of understanding and applicable theory upon preceding empirical experience with structures, phenomena, and processes that were discovered more by art than logic. Such a history of technology and science would be understandable to general historians and would lead to less exclusion of its practitioners from the main body of historians. Its study would help those in all fields preparing to face problems in industry or government today. Though it is, of course, a specialization, connections with other fields are intrinsically involved.

The distinctions that separate science from technology or either of these from art are rather recent intellectual simplifications. The words imply modern definitions that are not fully appropriate to the past, and they cannot avoid dependence upon certain preselected types of record. I believe that historians would gain a better sense of the past if they would work more with things, not only seeking the cultural factors embodied in the form, decoration, and use as they were experienced by users in their heyday but also learning to read the evidence that is preserved in the internal structure of materials of the treatment to which they have been subject and which reflects the actual sensual experience of the men shaping them. Such "reading" involves the use of modern laboratory equipment and methods of interpretation that are more like those of the scientist than the traditional ones of the historian. Suitable techniques, only recently developed, are already extensively used in art museums as well as by archaeologists. Their wider use by historians would not only give humanists a better understanding of the scientist's view of the world, both past and present, but would provide many opportunities to give a humanistic context to scientific endeavor. The biggest opportunities undoubtedly lie in the field of the history of technology, but even in main-line history of science the repetition of classical experiments helps to reconstruct the full experience of the past. As Philip Morrison has said, Clio must increasingly repair to the laboratory.

Postscript (1980)

The foregoing was written specifically in response to a letter from the editor of *Annals of Science*, Ivor Grattan-Guinness, asking some historians of science to "discuss such questions as their initial and subsequent involvement with the subject, their views on its development during their career and its relationship with related disciplines, and their hopes (and fears) for its future." Since it is the nearest approach to an autobiographical statement that I am likely to write, I should perhaps rectify some omissions.

My home was not an intellectual environment, but neither was it anti-intellectual. My parents sent my brothers and me to a good school. They subscribed for us to *The Children's Encyclopedia*, edited by Arthur Mees, when it was first being issued in fortnightly parts. Its arrival was eagerly awaited, and I can trace my first awareness of most fields of knowledge to its well-written and highly diverse articles.

My turn toward more than a vague interest in science was at a lecture on physics with fine demonstrations of light and sound given to boys in the lower school, age about nine. I was excited by seeing a flame dancing to an organ pipe, with its multiple images spread out by Helmholtz's rotating mirror, but the first two years of school classes in science were devoted to chemistry not physics.

Because I failed to get good enough grades in the Oxford Senior Local Examinations in History or English—I don't recall which—I had a year of independent study between leaving school and entering the University. I took a job as lab assistant at King Edward's Grammar School in Camp Hill, and this together with my home laboratory kept up my manipulative skills. My second try at matriculation was successful, and I entered the University of Birmingham with advanced standing in 1921, to graduate in 1924. My twenty-first birthday came a week or two after my arrival in America.

I had almost no contact with music beyond the singing of the choir in the local Wesleyan Methodist Church, or with visual art beyond rare visits to the Birmingham Municipal Art Gallery. I was more baffled than inspired or exhilarated by the arts, but I accepted their importance, and puzzled perseverance eventually paid off. And, of course, like the man who found he had been speaking prose all his life, I had had, without being aware of it, innumerable aesthetic experiences in the laboratory and workshop. In 1972 in a letter to Robert Cahn I reminisced thus:

> In the long gone days when I was developing alloys I certainly came to have a very strong feeling of natural understanding, a feeling of how I would behave if I were a certain alloy, a sense of hardness and softness and conductivity and fusibility and deformability and brittleness—all in a curiously internal and quite literally sensual way, even before I had a sensual contact with the alloy itself.
>
> Even now, mention of an alloy first brings to mind the way it felt when it was hammered hot or cold on an anvil, how it looked as it flowed from a crucible, a harshness or otherwise to the touch. And all the work I did on interfaces really began with a combination of an aesthetic feeling for a balanced structure and a muscular feeling of the interfaces pulling against each other!
>
> Of course, before publishing anything I tried to put it in respectable scientific terminology and it was fun to do so, but the stage of *discovery* was entirely sensual and the mathematics was only necessary to be able to communicate with other people. Perhaps

discovery is always this way, though development rarely is.

My perception of the world is undoubtedly influenced by the fact that I am colorblind (a protonope—my peculiarities are quantitatively documented as subject "CSS" in a 1935 report by F. G. Pitt for the Medical Research Council, entitled "Characteristics of dichromatic vision"). I suspect that my scientific interest in structure is in large measure a consequence of my inability to react to the principal component of painting, while the hierarchy of texture in etchings and the graphic arts generally has always delighted me.

In those days few Englishmen went on to university-level education. Only one of my classmates at school did so, and I doubt if I would have gone had not my grammar-school science teacher, proudly brought home to tea one day, urged it on my father with a little diagram of forking paths impressed (to my mother's distress) on the damask tablecloth with a knife.

All my early interests were narrow, pretty much confined to experimental science. I had not a single University course that was not directly technical in content. However, I do remember a lecture before the student metallurgical society by W. R. Barclay, the Chief Metallurgist of the Henry Wiggin Company, producers of nickel alloys, and a little pamphlet of his entitled "A technical man's reading." Neither original interest nor purpose but simple puzzlement that people thought art and literature to be important led me to wider reading. Barclay led me to Herbert Spencer, the first broad view of knowledge I encountered, and for a brief period I even thought of studying philosophy. More important, though, was the delight in discovery of the value of a large library, so much more exciting than the small or purposeful collections of school or university departments. One can learn without being taught.

I emigrated to America partly perhaps because working in distant colonial parts of the globe was more or less a family tradition, but specifically because my eldest brother, living in France, went to buy cigarettes one day when I was visiting him, and I idly picked up a copy of the popular magazine *La Science et la Vie* for sale in the tobacconist's. I subscribed, and sometime in 1923 read an article on "Les Grandes Laboratoires Américaines." Its account of the Bell Telephone Laboratories sounded like the environment in which I wanted to work and planted the idea of coming to America, but immigration restrictions turned me into a graduate student at MIT instead. I suppose I was a contributor to what has come to be known as the brain drain, but neither the brain nor the drain amounted to very much. Both intellectually and socially I matured in the United States, and when I return to England I have no nostalgic feelings but enjoy it greatly as a sympathetic and admiring foreigner. Yet my years in Cambridge (1933–1934) and in London (1955–1956) were tremendously important to me.

After graduation with a D.Sc. in 1926 I remained at MIT as a research associate working not very productively in the x-ray laboratory under John Norton. I wanted a permanent academic position, but no suitable one appeared and I went to industry. A summer job in Waterbury, Connecticut, with the American Brass Company turned into a permanent one. In retrospect this was most fortunate; indeed I do not see how one can claim to be a metallurgist without some industrial experience.

My main job was alloy development, at first simply as an adjunct to the main laboratory responsible for routine analytical control of plant operations, but later (following a request from the parent company, Anaconda, for research on possible markets for copper in the steel industry) I was put on my own in charge of "The Copper Alloys Research Laboratory" with all of two assistants. I became intimately acquainted with the nature and preparation of a wide range of materials and the careful measurement of their mechanical and physical properties in relation to constitution and structure. A biographical note by "a former associate" (probably Daniel Hull of the American Brass Co.) in *Metal Progress* (February 1948, pp. 217–218) gives an external if uncritical view of my activities at this time. My fifteen years (1927–1942) in industry also gave me a sense of history, for it was a time when many old processes were being replaced. The book by Daniel Hull, *The Casting of Brass Ingots* (Cleveland: American Society for Metals, 1950), gives a wonderful picture of what it used to be like.

I still have vivid sensual memories of that time: The smell of burning lard oil. Streams of molten brass in the casting shop. Some of the last coke-fired pit furnaces in operation, and men drawing crucibles, skimming and pouring the metal. The magnificent row of rolling mills, all driven continuously by a Corliss engine with a huge flywheel and a shaft running the full length of the large shop. The dance and clangor of drop and screw presses; and a sympathetic feeling for the pressure in the extrusion presses. Men seizing red-hot snakes of copper, threading them curving back and forth through the wire rod mill at Ansonia. (To this day a frequent dream is of wandering through complex assemblies of industrial buildings full of such machines, in search of something I never find.)

All of this was nineteenth century, some of it much earlier. The older industrial processes, with all of the hardships imposed on the workers, gave them a rather fierce and passionate pride in their skill and an experience of participation almost as in a dance with very real aesthetic overtones. Seeing it I came to admire the skill and the real, if unintellectualized, knowledge possessed by men who work with furnaces, tools, and machines to make things. Much of this has disappeared in today's production lines: the computers experience what a man in an old rolling mill did. Creative participatory joy is felt only by the few people who develop and design the machinery and instrumental control for them, while their operation needs little of the coordination of the physical, muscular component with mental understanding that seems in earlier times to have made the skillful carrying out of repetitive work tolerable, even enjoyable, because of its challenge.

My respect for the workman's knowledge was given added depth during my visits to Iran in 1962 and 1966, arranged by T. A. Wertime, then Cultural Attaché at the U.S. Embassy in Tehran. Many preindustrial techniques could still be seen in active operation in their natural habitat, demonstrating close sympathy between the man and his material.

During the years in Connecticut I also learned something of the nature and necessity of management and of the economic factors behind the dominance of production over research. But more important for my later activities was the fact that, unlike the academic who thinks his ideas seven days a week, as an industrial employee I had weekends free, and these enabled me to exploit the Sterling Library at nearby Yale University and develop my historical interests.

World War II ended all this. For a year I served ineffectively in a desk job with the War Metallurgy Committee in Washington, but in March 1943 I went to Los Alamos to direct the work on metallurgy. This involved preparing the fissionable metal for the cores of the Hiroshima and Nagasaki bombs and shaping a great variety of other materials—metallic, inorganic, and organic—for nuclear experiments. The work was applied science at its best and involved interaction with all aspects of the project (see item 85 in the bibliography). Robert Oppenheimer was a superb project director and became a good friend as well as an occasionally critical boss. The physical environment of mountains, mesa, and forest provided an inspiring backdrop; the community was compact and was united in both technical and social matters by the shared sense of urgent purpose. It was a very different world from the Connecticut industrial town, and my intellectual horizons were tremendously widened as I found myself able to play a part in association with many of the greatest physicists of the time. Not only in the laboratory—I frequently went hiking in the mountains with Bethe, Fermi, Weisskopf, Teller, G. I. Taylor, and once even cajoled Johnny von Neumann into a short scramble up the Quemazon trail. I knew well Peierls, Segrè, Bloch, Rabi, Frisch, Feynman, Kistiakowsky, and others who have since become famous. On his rare visits to Los Alamos, I talked with Niels Bohr. All this served to intensify the intellectual slant of my mind that had been barely perceptible before, and it made the appeal of the academic world irresistible.

On 1 January 1946 I moved to the University of Chicago as founder and first director of the Institute for the Study of Metals. This was the first academic interdisciplinary research organization in America dealing with materials. A natural outgrowth of the close association of metallurgists with chemists and physicists on the Manhattan Project, it became somewhat of a model for later laboratories. Nowadays the heads of such organizations have to devote a major part of their time to fund raising, but in the postwar scientific euphoria and in the University milieu created by Robert Hutchins, funds came easily from industry, and I had plenty of time for my own research.

When in Chicago I was appointed by President Truman as one of the original nine members of the General Advisory Committee to the Atomic Energy Commission. At our first meeting Robert Oppenheimer was elected chairman, and he led it brilliantly until the expiration of this appointment some months after my own resignation in 1952. On 30 October 1949 I voted with enthusiasm for the Committee's famous (some say infamous) recommendation against a crash program to develop the hydrogen bomb. To participate in such discussions and decisions at that time was exciting, and it carried me into many aspects of government. I had glimpses of a possible alternative and more influential career, but the halls of academia have a stronger appeal for me than the corridors of power, and in retrospect I am glad that my distaste for the development of superweapons gave the nudge to get out.

Later I served briefly on the President's Science Advisory Committee at the time of its greatest influence under James Killian and George Kistiakowsky, but I resigned from this also as I realized again my temperamental incompatibility with such work. On the other hand I enjoyed service in Washington as a member of the Committee on Science and Public Policy of the National Academy of Sciences, and particularly a ten-year stint on the Council of the Smithsonian Institution (1966–1976).

In 1961 I came to MIT, not because I was in any way dissatisfied with the University of Chicago (indeed I still think it provides an almost ideal academic environment) but because I thought it was time for a change in emphasis and aim. Leo Szilard has put it well in one of his "Ten Commandments": "Do your work for six years; but in the seventh, go into solitude or among strangers, so that the recollection of your friends does not hinder you from being what you have become."

The move to Cambridge provided, if anything, somewhat greater professional stimulation to my wife than it did me. She became a fellow and eventually Dean of the Radcliffe Institute for Independent Study, and finished some articles and important books on the changing public roles of scientists.* My main appointment at MIT was as an Institute Professor, which carried no assigned duties, and although both the humanities and the metallurgy departments gave me nominal appointments, I somehow never found myself acting as an effective member of either. Yet MIT has been the environment admirably suited for the drawing together of my varied earlier experiences and has enabled (perhaps, by relative indifference, even encouraged) the final development of my "philosophy" in which structural change—that is, physics and history combined—is seen as the common factor uniting all my previous interests.

From beginning to end I have been a simple metallurgist using metals and their structure as a kind of inverted touchstone to assay all things.

---

*Alice Kimball Smith, *A Peril and a Hope* (Chicago, 1965); and with Charles Weiner, *Robert Oppenheimer, Letters and Recollections* (Cambridge, MA, 1980).

14
Structural Hierarchy in Science, Art, and History

Introduction

The author is a metallurgist. When, years ago, I became more than casually interested in the history of my profession, I searched the written records of the last eight centuries without even approaching the beginnings of important techniques, but eventually I found that the earliest evidence of knowledge of the nature and behavior of metals was provided by objects in art museums. Slowly I came to see that this was not a coincidence but a consequence of the very nature of discovery, for discovery derives from aesthetically motivated curiosity and is rarely a result of practical purposefulness.

Also, having spent many years seeking quantitative formulations of the structure of metals and trying to understand the ways in which the structures change with composition and with treatment, and the ways in which structure relates to useful properties, I have slowly come to realize that the analytical quantitative approach that I had been taught to regard as the only respectable one for a scientist is insufficient. Analytical atomism is beyond doubt an essential requisite for the understanding of things, and the achievements of the sciences during the last four centuries must rank with the greatest achievements of man at any time: yet, granting this, one still must acknowledge that the richest aspects of any large and complicated system arise in factors that cannot be measured easily, if at all. For these, the artist's approach, uncertain though it inevitably is, seems to find and convey more meaning. Some of the biological and engineering sciences are finding more and more inspiration from the arts.

It is important to note that the word "hierarchy" is used throughout this paper not in the sense of a rigid boss/slave relationship, with control passing unidirectionally from top to bottom, but even more as the inverse, for it is the interlock of the smaller parts that generates the larger overarching structures. My model is not the command tree of figure 14.1 but is more like the self-forming, locally diverse aggregate shown in the frontispiece. Though environment will strongly influence the behavior of small things within it, the environment itself comes from the interaction of individuals to produce ever-larger structural groupings. All things are interactive, both up and down the scale of clumpings. Though the larger groups necessarily incorporate the smaller ones, the aspects of hierarchy here considered are far more than mere inclusion, just as they are more than mere addition or mere dissection. One of my colleagues at MIT has pointed out that, since most complex structures arise without control from above, "anarchy" might be a better term than "hierarchy" for the subject of this paper. It would be good to avoid both terms for they are overloaded with political emotion, but I know of no better word than hierarchy to convey the idea of an interpenetrating sequence of structural levels.

The author's experience with the hierarchy of structural changes associated with the hardening of bronze, steel, and duralumin may seem like poor qualification for writing on the elusive subject of art, but I hope to show that it is not entirely irrelevant since style in a work of art is an aspect of its structure, and structures of all kinds have certain similarities regardless of what materials, ob-

This paper is dedicated to the memory of Lancelot Law Whyte in grateful acknowledgment of many stimulating conversations.

jects, societies, or concepts they relate to. The differences, and the problems, come in the identification of the significant units and the manner in which hierarchical entities successively present themselves as the size of the aggregate increases and as the scale of resolution and the character of what can be perceived changes with the method of observation.

The unique quality of a work of art depends on the manner in which its component parts are shaped and put together. It has style, but the style is unrecognizable except by comparison with other works having similar structure and overtones, an extension of an inevitable hierarchy. In chemistry, the characterization of a phase (whether crystalline or not) also depends on how the parts are put together and how this internal structure affects external contacts to build up materials on the scale at which we use them.

Structure provides a universal metaphor: the apparent mixing of metaphors throughout in this article is not entirely due to carelessness! *Everything* involves structural hierarchy, an alternation of external and internal, homogeneity and heterogeneity. Externally perceived quality (property) is dependent upon internal structure; *nothing* can be understood without looking not only at it in isolation on its own level but also at both its internal structure and the external relationships which simultaneously establish the larger structure and modify the smaller one. Most human misunderstanding arises less from differing points of view than from perceptions of different levels of significance. The world is a complex system, and our understanding of it comes, in science, from the matching of model structures with the physical structure of matter and, in art, from a perceived relationship between its physical structure and the levels of sensual and imagi-

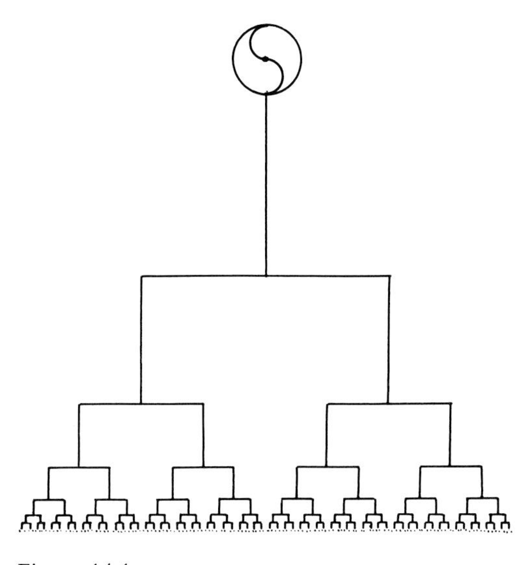

Figure 14.1
A tree of logic, representing the conventional view of hierarchy as a one-directional system of dominance.

native perception that are possible within the structure of our brain's workings. All is pattern matching, with the misfits, if they can interact, forming a superstructure of their own as in moiré patterns or beat notes.

As a rudimentary example, compare figures 14.2 and 14.3. The former represents the structural arrangement of atoms in an alloy such as brass, with two crystals of one phase having "atoms" in square array (the two being incompatible only because of their differing orientations), and a third crystal of a different structure based on triangles. More correctly, figure 14.2 is not the structure of brass but rather a sketch reflecting my idea of the structure, which itself already partially maps other conceptual and physical patterns and is rendered meaningful only by the possibility of establishing resonant matches with patterns in other minds.

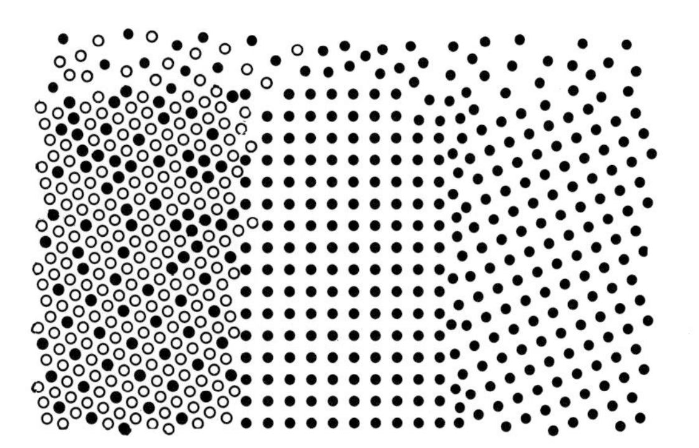

Figure 14.2
Diagram showing the arrangement of atoms in three crystals. Two are based on a square lattice and have a boundary between them resulting only from their different orientations, while the third crystal (on the left) is based on a triangular lattice and contains a ratio of three "white" atoms to each "black" atom, in regular ordered array at the bottom, disordered near the top. The disorder of the three or four atom layers at the top suggests the structure of a liquid.

Figure 14.3 is a detail of a recent etching by Tanaka Ryohei, who has skillfully used his needle and other devices to produce textural arrays in areas that define and depict walls, roofs, and other features that relate to each other to produce a meaningful picture on the scale of the whole (see detail). In both figures 14.2 and 14.3 there is a just-visible detail in local pattern which the eye quickly passes over to appreciate the larger areas characterized by its extension. These areas in their turn are seen as units in a new pattern grouped within a larger composition. The units are essential, but it is not the units that signify, it is the pattern formed by the repetition of their relationship. Moreover, the distinction between one area and the next in either figure 14.2 or 14.3 comes from the incompatibility of their microstructures. The mismatch need not be geometric (at least not on a visible scale), but may be based on any perceivable interaction in art or any physical interaction in matter—or, more commonly,

Figure 14.3
Detail of an etching, *House in Ohara,* by Tanaka Ryohei (1975). Compare with figure 14.2. The extension of different local patterns of juxtaposed fine detail is used to define areas representing different features in the overall design (right).

on many different interactions superimposed. Seurat performed magic with small patches of color. In ceramic glazes the aesthetic units are sometimes identical with the areas of a physical phase.

From the viewpoint of a physicist, the hierarchy in figure 14.2 stops with the symmetry of the crystal lattices. The metallurgist sees, on a larger but still small scale, the two-phase microstructure in a sample of 60/40 brass, while the engineer sees a lump of brass, useful to be shaped into naval hardware, a plumbing fixture, a telescope body, or other object to play its role within the still larger structure of society.

A work of art has failed if it does not further enlarge in the viewer's mind to encompass and to relate many overtones and different levels of personal, emotional, social, and cultural themes, some joined for the first time and uniquely by the sensing of a larger pattern of harmony between imagination and understanding.

All things have internal structure. Externally, a thing may have form, and possess considerable grace in the balance of its parts, but it cannot be said to have a style unless some aspect of the relationship of its parts appears in other objects and, by its replication, provides a basis for a perceptive eye to group them together. Original creations must inspire copies. Style is the recognition of a quality shared among many things; the quality, however, lies in structure on a smaller scale than that of the things possessing the quality.

Style in art is closely analogous to the relationship between internal structure and externally measurable property that in science distinguishes one chemical phase from another. Similar problems of characterization, boundary identification, and mechanisms of change exist in both fields. Bulk properties of matter such as density, color, conductivity, crystal structure, or vapor pressure by which a chemical phase is identified are not a property of any of the parts (though they would not exist without them) but rather are external characteristics depending on the pattern of interaction between the atomic nuclei, electrons, and energy quanta and the extension of this pattern by repetition throughout the entire volume of the phase concerned. The pattern and the property both disappear when there are too few parts, and they change when an interface into another distinguishable phase is crossed. So with styles of art. They cannot be seen from inside, although the structure on which they depend can be. Some styles, like some phases (for example, liquid or solid solutions), can tolerate considerable diversity in the shape and constitution of their parts, while others (like simple molecules and covalent crystals depending upon the precise symmetry of nearest-neighbor interaction) are intolerant of substitution: if different parts are introduced, the whole structure will adjust to a new form. Many important differences in the behavior of large systems arise from differences in the degree of tolerance or intolerance of neighbors within the smallest groupings that compose them.

In the aesthetic case the identifying uniformity corresponding to the physical-property test of a chemical phase is that of some psychological response triggered by the repetition of a detail of pattern or a color relationship. Though equilibrium is not involved, there is a kind of phase rule relating the number of distinguishable types of features in a painting to the number of kinds of just-visible units with which the artist works and the manner and density of their packing. Etchers and engravers use cross hatching and stippling almost like crystal lattices to distinguish areas. In figure 14.3 the eye immediately senses an individual quality in each of the areas that, on close inspection, can be seen to arise from arrays of lines that differ in width, uniformity, spacing, and orientation in a manner that is quite analogous to the way in which chemical phases are distinguished and given identity. Neither atoms nor lines are enough by themselves: It is their interplay that gives unity and character to the assembly even when the parts are invisible, and the assembly on one level—whether in matter or in the perception that underlies art—itself becomes a part in a larger aggregate with its own pattern of interplay.

Chinese landscape paintings, constructed as they usually are with a limited number of distinct types of brush stroke, show this particularly well. The literati painters of the Ming dynasty made use of this device very effectively, but it originated much earlier. Figure 14.4 is a landscape painting by the early Ch'ing painter Kung Hsien (ca. 1620–1698) and figure 14.4a a detail. Each of the areas, for example, those suggesting trees, rocks, sky, or mountains, has an immediately sensed distinctive quality which relates it to others of the same kind within the painting as a whole, but which arises in the repetition within it of a particular type of brush stroke and the manner in which the brush strokes relate to their immediate neighbors. In a physical system the maximum number of phases that can coexist in equilibrium at a given temperature depends upon the number of atomic species present and how closely they are pressed together. Whether a boundary can exist between one area and another depends upon cooperative hierarchical interaction, on whether the valencies (desire for resonant communication) of individual atoms are compatible with those of the aggregate of their neighbors. Ultimately, the possibility of continuity or separation between two regions depends upon the numerical topological requirements of space filling—in whatever space and by whatever means of connection is appropriate for the occasion. The requirement of uniformity within the connected region of each "phase" is unfulfilled when equilibrium has not been attained, that is, when the system retains a memory of some previous structure. In these, and rather generally in works of art, gradients in microsymmetry exist on a scale comparable to that of the granularity. However, too much disorder can soon cause a phase to lose its identity, and a proliferation of gradients of color or texture in a painting quickly diminishes its visual impact. In systems of any kind, effective diversity does not continue to increase with the complexity that accompanies a greatly increased number of units. Properties such as density, compressibility, or vapor pressure that are established by the interaction of a few atoms do not change much in larger aggregates, although others, such as strength and plasticity, are enormously influenced by imperfections and larger clusterings. In visual perception the angular resolution limit means that the number of things that can be perceived at any one time remains approximately constant while their absolute scale may be altered almost without limit depending on distance or on the use of external instrumentation. Sharp diversity seen as part of a region on one scale becomes mere texture on another and eventually becomes entirely irrelevant except as it contributes to some average property. Though one could sequentially resolve all the details in a

Figure 14.4
Landscape painting, *Wintry Mountains,* by the early Ch'ing painter Kung Hsien (ca. 1620–1698). 165 × 49 cm. Collection of John M. Crawford, Jr. Note the use of brush strokes almost as atoms to form, to relate, and to distinguish the larger features. Figure 14.4a (right) shows a detail of the painting.

large scene, their relationship is rendered fuzzy by the fortunate failure of memory.

As in a physical aggregate, areas in the Kung Hsien painting can be distinguished not only by the nature of the individual brush strokes (=atoms) but also by the density and orientation of their arrangements. The trees in the forest are separable from each other by the orientation of their leaves, just as individual microcrystals in a pure polycrystalline material are distinguished by lattice orientation alone.

Each tree is unique; yet the recognition of qualities shared by trees gives style to them as well as speciation. But style is hierarchical; it resides at all levels, or rather between any interrelatable levels. Considering the whole painting, its individuality and its style depend upon the distribution of and relationship between the styles of the smaller components. The painting itself can be said to be in the style of Kung Hsien because others by him are known with similar nuance of interplay between brushwork, pattern, content, and cultural overtones. On a coarser association, it is recognizable as the work of a scholar-artist of the early Ch'ing period, then as Chinese, and possibly on a universal scale as earthbound and distinguishable from the styles that have developed in the works of beings on other planets. The painting itself as an undivided whole has become the unit for successive stages in the larger aggregate. The same sort of relationship appears in inverse form as one goes down in scale. Just as the areas were at first distinguished because of the patterns of brush strokes within them, so the brush strokes are recognized by the just-visible streaks and patches of light and dark on the paper, and these in turn derive from patterns of carbon particles (visible under a low-power microscope) that were deposited by the ink under the competing capillarities of the fibers in the brush and in the paper, while these in their turn arise in the molecular and electronic structure of colloids and polymers, and ultimately (or rather *pro tem*) in the patterns of interlock, mediated via photons, between electrons and atomic nuclei. One could move from here to any region among the atoms of any other solid without noting very much difference—until one returned up the scale of molecule, cell, crystal, and aggregate. The painting is just at the level at which the widening viewpoints of the artist and the narrowing viewpoints of the scientist merge. It is the scale of human experience, from which thought and imagination take off, and to which they must return.

Aspects of style are more than relationships of areas and volumes. The repetition of color or texture or of purely linear geometric detail can produce an aesthetic response in a viewer's mind. Note in figure 14.4 the unity that comes from the repetition of certain directions, curvatures, and angles of bend or junction which do not themselves interlock physically but are connected only in the viewer's mind. (The Japanese artist Sesshu [1420–1506] used such geometric resonance most strikingly in his famous *Four Seasons* scroll.) Moreover, the resonant pattern formed between the trunks of the larger trees in the foreground of figure 14.4 is different from that of the copselike clumping of the smaller trees, but both serve to unify the whole on the larger scale of the painting. This extends still farther: though each tree is unique, recognition of a common dendritic quality relates them in the viewer's mind with all trees. Similarly, the individual houses in the painting are identified by the familiar closure of the pattern of lines, but the sight of them invokes the broader concepts of "housiness" in general.

Wang Wei, a painter of the late T'ang dynasty, delighted in emphasizing junction (figure 14.5). For reasons evidently both philosophic and aesthetic, he played on the similarity between the branching connectivity of fissures in nearby rocks, the wrinkles and river valleys in distant mountains, and the branches of trees, the last, of course, the model for them all (as well as for modern computer programming!).

When the underlying structure is being examined, the style of the whole is not visible, for this resides only in an external view of the whole. Bohr's famous principle of complementarity is, perhaps, nothing but a statement that things react on different levels. Our view is limited partly because we can neither see nor think of very much at one time, more fundamentally because the pattern of relationships is indeterminate until all the parts have been examined. George Kubler has said that style is like a rainbow. He is right because both style and bow depend on a constancy of relationships that exists on a scale below their own—the constant angle of refraction of light of different wavelengths in the case of the rainbow, and of a more complex mental refraction in the case of style. In both cases the phenomenon recedes as we approach the place where we thought we saw it, and is replaced by a previously hidden structure. Anthropologists can recognize a culture and analyze it, but are hard put to define it.

The eye finds beauty, even something analogous to style, in natural objects that have never felt the touch of a creative artist. In the last half century, as artists have moved away from culturally determined iconography to abstract or supposedly nonrepresentational art, they have unconsciously represented and have sometimes anticipated the natural forms that are revealed by advanced techniques in the laboratory. Actually, there are not many basic units of composition, and as large things merge into smaller ones, and vice versa, both nature and the eye favor much the same principles of assembly. Landscapes, whether real or imaginary and whatever their scale or origin, have recognizable "style" based upon repetition, relationship, selection, and adjustment. Matter, whether living or dead, when left to itself adjusts its resonances on all scales, and the resulting structures seem to be based upon much the same relationships as those which give, or perhaps which actually constitute, aesthetic satisfaction in the mind of the human observer.

For five centuries the advancing frontiers of science have been associated mainly with an increase in resolving power and the discovery of new levels of structure. In one direction we have found universes and galaxies, in the other we have found particles and subparticles, but having isolated them we still cannot understand their reassociation except in rudimentarily simple ways which omit those qualitative features that give the richness that provoked human interest in the first place. Analysis followed by logical synthesis does *not* reconstruct reality. It leaves out local historical accident and balance and all the cumulative complex consequences of individual history. A list of types of bricks used in the Hagia Sophia may help one to build an interesting brick wall, but it poorly suggests the great edifice from which they came.

Science in the past has been almost synonymous with distrust of the senses, but it seems to me that we are now ready to make

Figure 14.5
Detail of a handscroll, *Clearing after Snowstorm on Mountains and Rivers,* by Wang Wei (699–759). This is probably a sixteenth-century copy.

better use of the other properties of the human brain besides its capacity for logical rigor. After logical methods that require an exact identification and control of boundary conditions have yielded what they can, we should seek a bridge to the more sensual study of whole systems. This will require an acceptance by scientists of macroscopic imprecision in the application of microscopically precise laws, and an appreciation of the individuality that arises historically in any complex system but has been excluded by current analytical or statistical approaches. We are not interested equally in all possible systems that could be formed from the units we know: those that *do* exist must take priority. Nevertheless, although atomistic details are insufficient for full understanding, they cannot be ignored. All the established "facts" must be considered before imaginative interpretation can be indulged in. Atoms and entropy may be dull to anyone except a physicist or chemist, but without them the whole rich world of thing and thought would not exist and there would be no one to enjoy it.

Entity and Entitation

Nothing is a thing by itself: it takes meaning, indeed existence, only as it interacts with something else. Once this is admitted, hierarchy becomes inevitable in all systems excepting only those that are completely ordered or completely disordered, for these have no levels between the units and the whole assembly. There are two unavoidable dissymmetries—moving upward, where the inner structure meets the surface of whatever we are talking about, and moving downward, where it meets the surface of a part.

The very fact that something is recognized as an entity means that it has associated with it a distinguishable response to external probing—in physical terms, some measurable property; in more general situations, some recognizable quality. The hierarchical alternation between an externally observable quality, property, or trait and an internal structure that gives rise to it occurs at all levels and applies to all things. The interface, being both separation and junction, is always Janus-faced—that which is characterized as an entity is in some way more closely connected within than without. To define an entity, to separate it from the rest of the world, the interface closes upon itself, and hence, in sum but not necessarily everywhere, is concave inwards; connections converge and tighten inside, at the same time that they diverge and loosen outside it. Gertrude Stein once remarked that identity, based upon interaction, was much more interesting than entity, which marks completeness. Moreover, as Alice's whiting said to the snail, "The further off from England the nearer is to France."

Though all structures depend in the main on the repetition of relationships, there is always some hierarchical level in any natural, social, or aesthetic structure at which it can withstand the replacement of some of its parts by others. At what level are anomalies or misfits rejected, tolerated, or welcomed? In some structures the simple hierarchical concept of the whole selects and places the parts, leaving little possibility of interchange without destroying the whole. In others the parts rigidly fix the pattern but not the extent of the whole, as, for example, the lattices of hard polygons in a Moorish mosaic in which no random change is possible, except at the edges, without disruption—though color remains a variable to be played

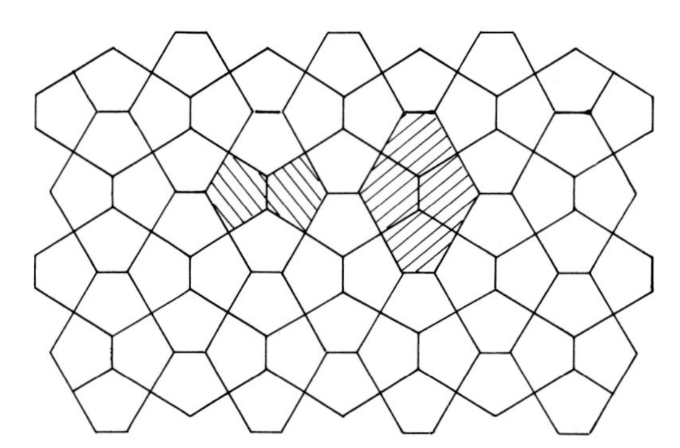

Figure 14.6
Space-filling arrays of pentagons in hexagonal and quadrilateral tesselations meeting, respectively, at trivalent or quadrivalent vertices.

with by adjustment and substitution within modules maintained at a higher level, as in pieces on the squares of a chess board, or atoms in alloy crystals. It is also common for misfits to compensate each other and to satisfy the balance of connectivities within larger modules. Thus pentagons, which when regular cannot fill space, do so when grouped in pairs to form quadrilaterals or in fours to compose hexagons (figure 14.6). Any extended array, however irregular or lacking in symmetry it may be, can always be subdivided into space-filling groups within boundaries that connect in quadrilateral or hexagonal array. The scale at which this can happen is an index of the hierarchical nature of the system. The balance or imbalance of internal and external connections determines both identity and the possibility of extensive aggregation.

When an extended array is viewed with decreasing resolution on larger and larger scale, the system becomes simplified by the merging of compatible connections, turning polygons into their duals (points with valence equal to the original number of sides) and groups of polygons into simpler polygons and eventually into points with the same residual valence. Thus up the hierarchical scale forever as local differences merge into global balance.

Such requirements are very simple applications of Euler's law relating the numbers of 0-, 1-, 2-, and 3-dimensional features in a simply connected group of polyhedra. Being topological, this does not relate in any way to shape, only to contacts and separations; but when they are combined with simple physical interactions, interesting geometric effects appear. Thus in a random soap froth in two dimensions (figure 14.7), the necessity of compensating for cells having

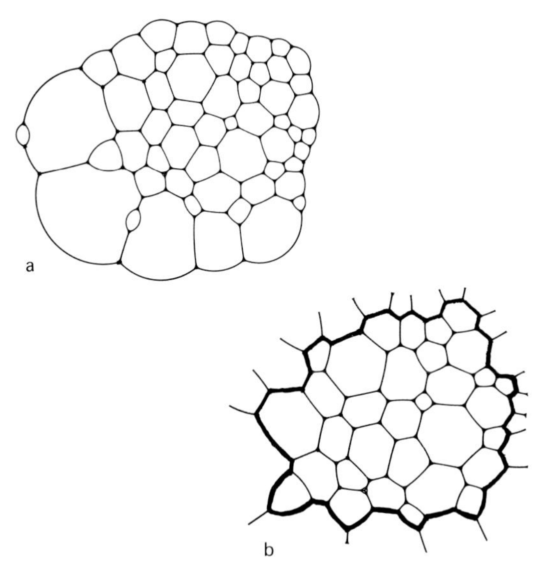

Figure 14.7
Soap bubbles in two-dimensional array. Since all corners are trivalent, the average bubble in a large aggregate must approximate six sides, and in an array of any size the departure from six is detectable by inspection of the boundary alone.

In an isolated set such as (a), with all vertices threefold and with no dangling edges, then, if P_n is the number of polygons with n sides and E_b the number of unshared edges constituting the boundary,

$$\sum (6 - n)P_n - E_b = 6.$$

Note that this is not the average but the sum of departures from hexagonality. Similarly, if a group of such polygons is excised from a larger array, as in (b), then

$$\sum (6 - n)P_n + E_o - E_i = 6,$$

where E_o and E_i are, respectively, the number of edges extending outward (severed) and inward from the boundary. Similar relations between internal and external connections apply in all systems, but they are only simple in two dimensions and when all vertices are restricted to the same valence. (For more detail on these relations see chapter 1 in this volume.)

more than six sides with those having less than six is joined to the surface-tension requirement for minimum boundary length and junctions at 120 degrees. The cell walls therefore have to be curved, and they can be rendered stable only by pressure differences between adjacent bubbles. This, in turn, causes slow diffusion of gas from small bubbles into their larger neighbors and introduces perpetual instability.

The limitations represented by Euler's law make it impossible to have extended simply connected aggregates of polygons in two dimensions with greater valence (number of neighbors) than six, or convex cells in three dimensions with more than fourteen: the number of neighbors needed to satisfy higher valencies cannot be assembled repetitively into a connected aggregate. However, a one-dimensional branching tree with no return loops can be of any extent and degree of complication. In three dimensions, such a tree can be externally superimposed on a two- or three-dimensional array to connect together any number of cells whatever. This, of course, is the basis of organism, for it provides a means of communication that transcends neighborhood: aggregation and communication are no longer limited to interactions only with, or via, nearest neighbors. The human mind similarly establishes relationships between parts of systems both internal and external to itself which do not necessarily have any physical interconnection. Thought is supposedly a kind of trial of patterns of interconnection of neurons as they already exist within the brain and as they are modified by selective interaction of the structure of the whole body with the outer world via the senses of sight, sound, touch, taste, and smell.

The distinction between an inorganic aggregate and a viable organization or organism lies partly in the fact that the latter contains an essentially one-dimensional treelike system maintaining communication to and from a center of some complexity. The communication system, being one-dimensional, cannot maintain itself unless it is immersed in a three-dimensional aggregate, but it often provides the means whereby the nature of the three-dimensional parts can be changed and their capacity for interaction with their neighbors thereby altered. Communication may be directed or broadcast, but there must be eventual feedback if the system is to be maintained.

Without both tension and compression and the balance between them nothing could exist, for it would either expand to infinity or shrink to nothing. Though tension can be one-dimensional, compression, as every engineer knows, has to be three-dimensional if pressure is to be contained or buckling avoided. The number of restraints that are needed to prevent buckling is, indeed, a mark of the dimensionality of the system. The part in compression can be internal within a skin stressed in tension, as in a soap bubble, or it can be external, as in a tetrahedral framework of rods supporting four stretched wires meeting at a central point or in the more complicated "tensegrity" structures of Buckminster Fuller and his followers. Even the brain needs a three-dimensional structure underlying the sometimes two-dimensional patterns of thought. A structural entity is one in which tension and compression are internally self-balanced. (Two-dimensional systems can achieve stability in a three-dimensional world by the balance between centripetal and centrifugal forces, as in planetary systems, the bolas, or the spinning electron—though note that the electron cannot interact with anything else without the intervention of a photon having an internal quantized structure that is at least three-dimensional.)

Few entities are completely isolated, without some residual connections to other entities comprising a larger aggregate within which either tension or compression or both form cell walls, simultaneously dividing and maintaining continuity. It is the stability of an overarching three-dimensional environment that allows special freedoms and local structures—for example, the holding of feebly bonded atoms within the chelated structure of some molecules, the tolerance of a crystal lattice for vacancies, dislocations, and substitutional atoms, and, most important of all, the balanced interlocking of separable molecules with homopolar bonds, hydrogen bonds, and van der Waals bonds that lies at the basis of biological replication. In solids, the bonds between atoms may be as sharply directed as the covalent bonds between tetrahedral carbon atoms in diamond, as specifically neighborly as in an ionic crystal, or as regionally compensating as in a metal.

Substructure

When more than one type of "atom" is involved, many different hierarchies become possible within the same overall lattice of principal connections. A part can be an unchangeable "atom" for some purposes and a variable organism for others.

This kind of structure is well illustrated by the simplest binary alloy incorporating equal numbers of two kinds of atoms, shown schematically in figure 14.8. Preference for dissimilar neighbors leads ideally to perfect order, figure 14.8b, and this is observed in crystals of many ionic compounds such as sodium chloride where the energy penalty

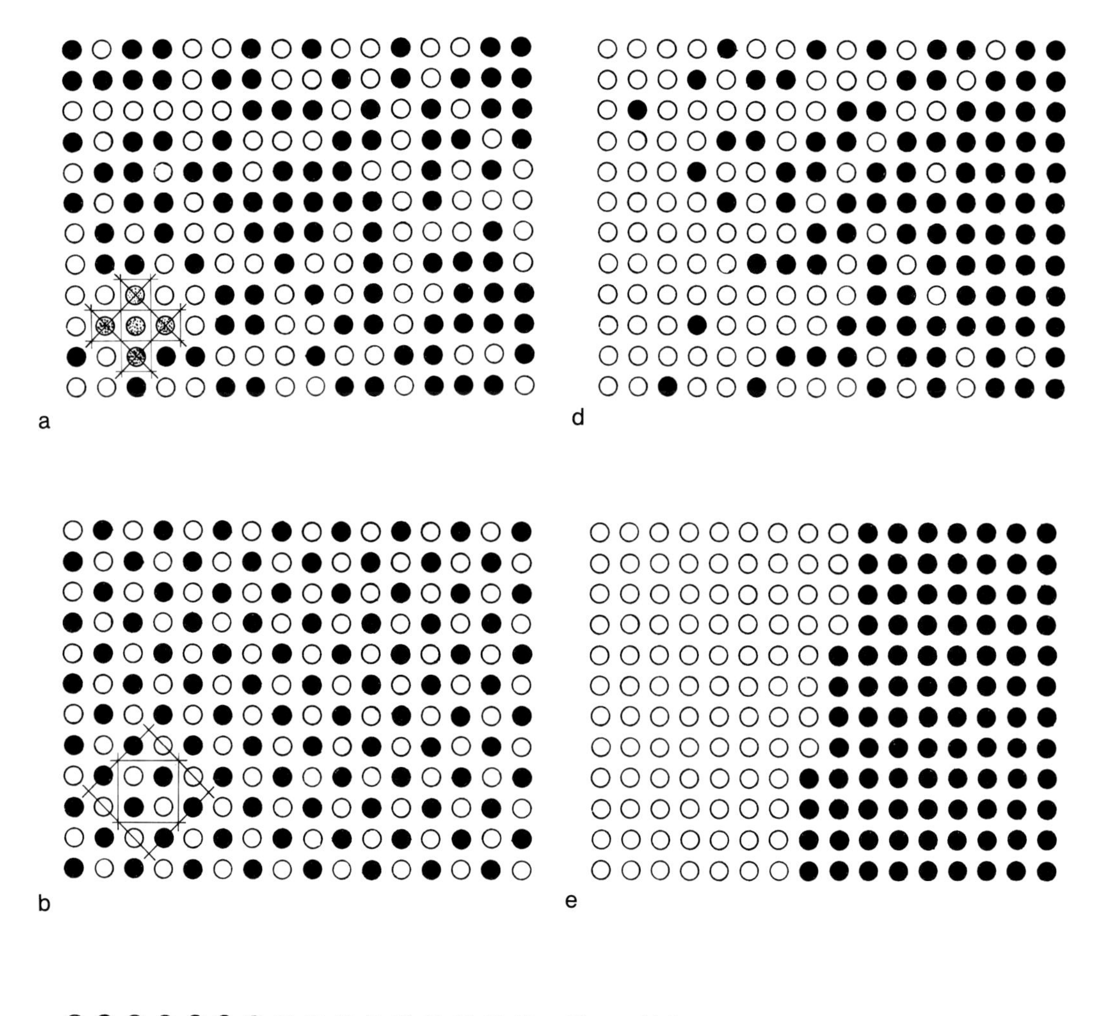

c

Figure 14.8
Lattice composed of two different kinds of atoms distributed randomly (a); in ordered array with each atom surrounded by the opposite kind (b); in two regions of order with an out-of-step boundary (c); with a diffuse gradient in composition (d); and with complete segregation (e). The basis of a boundary outlining structure on a higher hierarchical level is present in all but (a) and (b).

for wrong neighbors is high. With less dominant neighborly interaction there may be no order at high temperatures—the random solid solution of figure 14.8a—but on cooling order begins slowly and gradually increases. Here it is easy for local regions of order to be out of step, producing a zone of misfit (figure 14.8c), a domain boundary somewhat analogous to the orientation misfit of figure 14.1, but far less costly energetically. The same occurs with ferromagnetic interaction, where interlock of local magnetic polarity due to electron spin orientation defines regions that (except in the very smallest crystals) compensate each other in the whole. Changes into structures of this type, occurring within a dominant framework that is not significantly altered, are the second-order transformations of the thermodynamicist. They are called cooperative phenomena, but they could just as well be called anticooperative, for local cooperation produces larger-scale opposition: it depends on what scale is observed. Diffuse interfaces are common.

An adequate description of a system must include not only local neighbors and the average of the whole, but also a sense or measure of the hierarchical interlock between statistics and scale. Imperfections in local ideal arrangements inevitably arise as the structure on any one level is extended, and those imperfections point to and permit the existence of a higher level of structure. This continues until eventually, on a sufficiently large scale, *all* segregation, order, disorder, and diversity become invisible. My colleague John Cahn has called attention to the fact that even such a simple property as chemical composition has hierarchical aspects. What is the composition of a solution? For example, consider a crystal of beta brass, an alloy of copper and zinc containing about equal proportions of atoms of the two elements arranged in structures rather like those in figures 14.8a–d depending upon temperature. Obviously if one takes for analysis a sample the size of a single atom, the composition will be 100 percent of either copper or zinc, while a smaller sample might fail to contain any nucleus whatever. With a sample volume large enough to include two atoms, both could be of either element, or a one-to-one association of copper and zinc could come up, indistinguishable from the nineteenth-century ideal molecule of a compound CuZn, whatever degree of order or disorder might prevail on a larger scale. This hierarchical-statistical quality is quite general and applies to the component parts of anything whatever.

It should be noted that if there is more than one component in a structure, each of the smaller ones can, independently of the others, take its place in its own array with the characteristics of a solid, liquid, or gas. In a gas all things are miscible, but gaslike distribution can also occur of some atoms in a crystalline solid. An ordered framework of one component may or may not be accompanied by order in the others, although the smaller components cannot have more order than the larger ones (except on a different scale following segregation). In feldspar, for example, within a persisting lattice-network formed by the negative ions of oxygen, the major positive ions of silicon and aluminum and the smaller positive ions of potassium, sodium, and calcium can be substituted for each other either randomly or with each one in separately ordered lattice sites to an extent limited only by the thermal history, the size of the atom, and the requirement of electrical neutrality. Of course, there is strong interaction between atoms which, locally, can affect their neighbors just as an external pressure would. Some components are more easily substituted than others, and some subgroups form lattice complexes or radicals that have to be moved intact. Analogies to all this are easy to find in the structures of art and social institutions.

Figures 14.8 and 14.9 illustrate an important structural feature that can only exist within higher order, namely vacant lattice sites. The atoms adjacent to these vacancies are restrained by the normal interlock of their environment, but an atom and a hole can interchange positions easily with no immediate long-range opposition or effect. Lattice vacancies are as essential to the physical world as the crystals within which they form. They are the basis of diffusion in solids, for they allow movement of units by local interchange under thermal agitation without destruction of the overall pattern: the system can move toward either segregation or randomness depending on the statistical bias in the interchange of neighbors as vacancies move about. It should be noted that such diffusion affects only the units of the structure, not the superstructure itself. However, although diffusion may occur by easy stages within an existing framework, the resulting local changes in the nature and concentration of units may set up conditions that will feed a revolutionary reconstructive change once a nucleus appears.

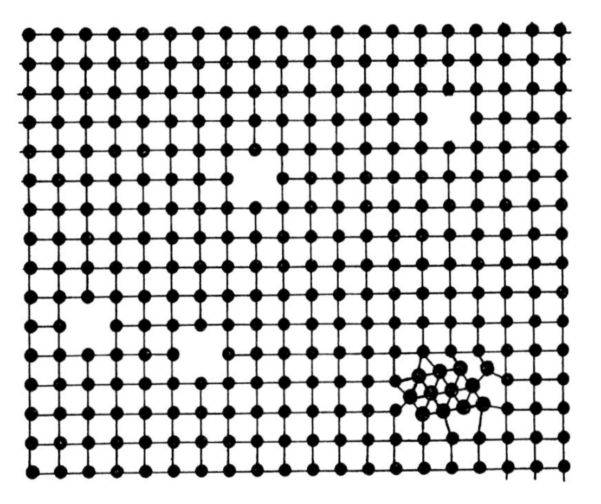

Figure 14.9
Diagram of a "square" lattice containing four places where atoms are missing without disturbing the larger arrangement, and a small region of misfit where the atoms are arranged in triangular array.

As in matter, so in the structure of ideas. "The thing that had happened in the heart" is a necessary preliminary to political revolution. In the history of science, one sees similar patterns of interlock followed by gradual change preceding the structurally reconstructive changes that constitute the paradigm shifts of Thomas Kuhn. Dalton's simple model of the molecule made nonstoichiometric compounds invisible to chemists for half a century, while physicists were unable to deal with the properties of real materials until after the discovery of disorder, which in turn was only possible after a period in which first atomic and then crystalline order had been found and overemphasized. In the last half-century departures from older norms of crystallographic symmetry and ideal electronic energy levels have been recognized, causing a complete revolution in the concept of materials and an explosive growth in the utility of solid-state physics.

Mechanisms of Change

This discussion is intended to help those familiar with the material sciences to see a little deeper into the quality of art, and those who enjoy art to understand a little more about the nature of substances. I have likened the recognition of style in a work of art or a school to the identification of a chemical phase, for both require the recognition of a more or less extended area characterized by certain qualities that inhere in the structural arrangements of parts on a lower hierarchical level. The differences lie mainly in the number of distinguishable components and in the character of the external observation.

The similarity between phase and style extends beyond mere existence; it can be seen even better in the mechanism of change that occurs to new forms when modifications of conditions have rendered an existing structure no longer the most appropriate one. Change of both phase and style occur by the regrouping of parts (which are usually far more stable than the whole) into a new pattern of interlock—a pattern which is fortuitous and highly localized when it first appears, but which provides the opportunity for self-discovery of the fact that a new grouping is preferable to the old, and is followed by an increase in the amount of the new at the expense of the old until the latter is consumed. Without a tiny sample of the new structure to serve suggestively as a nucleus, the old will remain forever happily metastable, for it has no means of knowing that change is desirable. Thus crystals of ice melt not because they are suddenly less stable at +0.01°C than they are at −0.01°C, but because the whole system benefits by the motion of the interface between ice and water. The interior of the ice does not disintegrate or change in any way: the volume of water simply increases at the expense of ice as long as the necessary heat is supplied from outside.

A system that is in equilibrium under given conditions has reached balance between the requirements of different levels, and a subsequent change in any level will eventually influence others, though at different rates. Because parts adjust faster than wholes, change moves structurally upward. In an equilibrated structure change will not begin unless there has been an externally induced change in conditions—something equivalent to the relative numbers of different species of units (the composition), or their intimacy (density, reciprocal of pressure), or the frequency or form of communication (temperature, number of quantum states or larger groupings of information). A change will affect at least four levels: that of the changed entity itself, that of its environment, that of its internal structure, and that of the medium of communication. A change in any of these levels may make a stable structure become metastable, ripe for change as soon as a nucleus appears to suggest a better arrangement. In an atomic aggregate different atoms do not, of course, form internally (except in the rare case of nuclear transmutation), but something akin to this occurs in more complex systems: the parts, which are themselves complex, can individually and locally undergo internal transformation. If enough similar units form as a result of suggestive communication, the replicated new units (which do not need to be identical, only reinforcingly communicative) will find each other and jointly respond by reformation when an appropriate nucleus of a better superstructure appears. This is the way ideas spread and topple governments, or, in science, Kuhnian paradigms succeed each other.

Not all changes, of course, involve the complete reorganization of the environment. There are many cases of small nonrevolutionary change that occur in a lower level of a preexisting structure without straining the local connections to the point of disruption. Inventions are improved by bright ideas in every mechanic's shop, and every member of a school of artists helps to round out the master's style in popularizing it. Such is the very essence of ecological gradualism, and this has strong implications for biological and social development, in that change can begin by gradual substitution and without discontinuity. Such change may stabilize the larger structure, or it may gradually set up conditions that make disruptive change probable. An example of the latter is the easy acceptance of technological change on an individual basis—the automobile, telephone, or dishwashing machine, for example—eventually interacting to make large social change necessary. This is analogous to the slow substitution of, say, chromium atoms for iron in a face-centered cubic lattice at 1000°C by diffusion, eventually making a body-centered cubic structure preferable, requiring only the appearance of a nucleus to produce it.

Any existing structure possesses structural inertia. Because of the reinforcing interactions of its parts, a structure resists change even after it has become thermodynamically, philosophically, or socially metastable. It came into existence because at some time in the past the parts had found stable patterns of overall interlock, and the interaction of any one pair of neighbors cannot now be altered without altering many adjacent ones. Anything but an isolated structureless particle or the nonresonant chaos of a gas has some history locked into it, and the more complex a thing is the more its present features depend upon the retention of unique configurations that resulted from the resolution of some historic conflicts during growth. Minerva's full-panoplied origin is a material impossibility, but the myth itself has permanence because of the long conceptual history behind it.

Simple structures allow few opportunities for the incorporation of alternative arrangements as they grow, and their history is without much interest. Conversely, the local groupings embodied in a complex structure both record past events and provide a unique framework on or within which future changes must occur. Though overarching order may erase some details, local differences that mesh into the larger structure will be preserved and built into the future.

Some time ago—appropriately on Memorial Day, 1970—I coined the word "funeous" for this aspect of structure, after the unfortunate character in Jorge Luis Borges's story "Funes the Memorious" who remembered everything. Different structures have widely differing degrees of "funicity."

Physics is largely afuneous. In general, both the analytical approach of the physical scientist and the averaging methods of the statistician achieve their exactness by the elimination of funeous aspects of the world. However, it is both impossible and unnecessary to study all structures that might have come about in the developing universe, and in the future science will inevitably pay more attention to those complex funeous structures that actually do exist. History, which produced the record, and physics, which analyzes it, must work hand in hand. Biological organisms and human cultures arising therefrom have a high density of funeous detail; indeed their very nature depends on the transfer of blocks of historically acquired pattern, not statistical exploration at each point of change in the development of each individual. The heredity-versus-environment argument can be applied to all things, for the past has given structures at various levels that cooperate to respond to present opportunities.

Man is at the scale at which structures built of chemical atoms have achieved something like the maximum significant degree of funicity, but his discovery of new means of preserving, replicating, and communicating blocks of thought patterns makes new levels possible. By using only the properties of pattern in electronic communication, thought can extend immeasurably far beyond the limits imposed by the aggregation of matter. (It does, however, need a material substrate on which the patterns can be formed.)

In the generation of new structures, that is, in change, the ordering of the parts has to pass through a period of explorative contacts and the gradual adjustment and locking in of the slow responses with the fast ones. Funicity, that is, historical diversity, is rooted in quantized interaction of structural units at various scales of interlock, though ultimately in the energy quantum. Quantum mechanics is the basis of all stability. Too much emphasis is commonly placed on the principle of uncertainty in quantum physics, which does not affect the result of an interaction, but only whether or not a particular interaction occurs within a given interval of time. Perhaps time itself is nothing but sequence in the hierarchy of structural inertia. The various quantum levels within the atom represent the patterns of resonant interlock between different ratios and densities of nucleons, electrons, and quanta, exactly as, on the next level of hierarchy, atoms in different ratios interlock to form a series of chemical compounds in Daltonian molecules. Larger cooperative structures such as crystal aggregates take longer to form because the response time to find if resonance has occurred is not based on the speed of light, but requires diffusion to relieve strain in cooperative configurations.

Once a structure has found itself, the integrity of the whole stabilizes the parts and resists their rearrangement. There is a close interplay between the numerical and the morphic aspects of any system. Structural inertia is at least as important as mass inertia, indeed it is the basis of inertial mass in everything larger than an electron. Macroscopic behavior predicated upon Newton's third law (action = reaction) occurs only when the interacting bodies are not stressed beyond the point where their internal structure ceases to respond elastically. (This is why physicists in the eighteenth and nineteenth centuries studied only the mechanics of elastic bodies and left plasticity to the potter, the pastry cook, and the metallurgist.) Reversibility applies only to binary interaction or to an elastic assembly. Consider the difference between the action of a hard steel punch and the reaction of a piece of soft silverware being chased. They are anything but equal and opposite except on the subatomic scale. The classification of actions as either elastic or inelastic is quite fundamental. All structures, and parts of structures, have a kind of elastic limit in their response to external conditions beyond which relaxation is to a pattern of connections that differs from the initial one. Without this there would be no historical individuality in complex structures.

Any change whatsoever involves a catastrophic change of neighbors at some level, while both above and below this level connections, though they may be strained, remain topologically unchanged. Atoms persist through changes of state and combination, while the local fury of a storm leaves the global atmospheric balance unchanged. Statesmen can talk of destroying government but not people, and the death of one man does not immediately change either society or atoms.

Nothing is unchangeable unless its environment is also unchanged. The only ultimate truth is, "It all depends." However, within a complex system not in equilibrium, local differences in the rate of loss of memory (that is, the erasing of structural features formed under conditions that no longer exist) will continue to cause internal change. A crystalline bar can be bent by an external force (figure 14.10a, b), but if it is elastic it recovers as soon as the force is removed: The strain caused a change only in the patterns of quantum/electron resonance forming the bonds, not in the topology of the connections between atoms and their neighbors. At a sufficiently high stress, irreversible plastic deformation becomes possible, but this of necessity involves a topological change; the

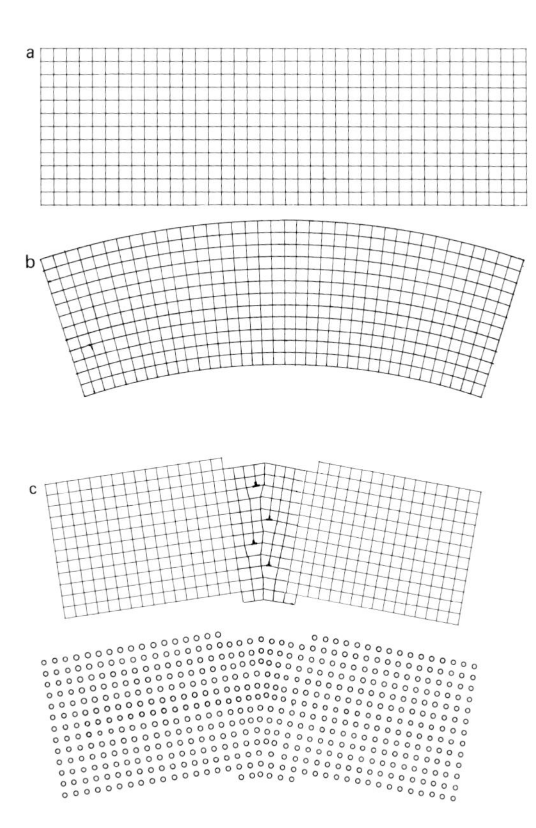

Figure 14.10
Diagrams of elastic and plastic deformation. The unstrained crystal (a) can be bent by an external force (b), but will return on removal of constraint. However, if local imperfections ("dislocations") are introduced, the crystal will remain bent while leaving most of the lattice relaxed (c).

production and movement of some local imperfection in the symmetry of the original crystal lattice. The imperfections (figure 14.10c), known to the solid-state physicist as lattice dislocations, have as much geometric reality as the groupings in the perfect lattice, but they can exist only in an extended environment of such perfection. They are hard to form, but, once formed, they are easy to move. In the simple two-dimensional case of figure 14.10, every "atom" has four neighbors except at the dislocations in figure 14.10c, where a single atom has three neighbors and all connections outside its immediate neighborhood are unchanged except for a regional change in direction. Thus a few places of concentrated misfit have absorbed nearly all of the applied strain, and the rest of the structure has been allowed to relax. The energy of the entire system has been decreased by replacing the sum of many small strains extending throughout a large region with the extreme strain of a local catastrophe. The bent bar no longer unbends on the release of the external force. The plasticity of metals, like the possibility of social change, depends entirely upon the generation and movement of imperfections within a predominantly ordered environment.

Both the nature of the parts and the nature of the communication between them are necessarily involved in the interactions that stabilize structures, and changes in either may render the superstructure metastable. Internal structural change may also be rendered desirable by an externally produced change in volume or in temperature, or if the atoms themselves are changed as a result of nuclear transmutation or diffusion from outside. However, even if a new array would have lower energy, it cannot form instantaneously, for the new structure must be nucleated in an environment that is opposed to it, and it must find a mechanism for growth.

A small group of units which, if isolated, could easily move into a new configuration cannot do so when it is embedded in a larger structure, for the initial coherence with the environment produces stresses at the interface that are equal and opposite to the displacive ones until some atoms switch their connections. The initial elasticity hinders the discovery of the possible advantages of any new structure. However, the advantages of the new phase increase in proportion to its volume, and there is some critical size beyond which the energy required for the interface is more than compensated by the volumetric gain. So with things in general, for a "thing" exists only in interaction with its environment, and it includes the interface through which it merges with or adjusts to its neighbors.

In the arts, once stylistic norms are established, slight departures are welcomed as improvements, but really original deviants are extinguished or at least ignored. At some point, either because of boredom or because of long-range changes in mental attitude (perhaps as a result of an intrusion of forms from another culture), social tolerance for at least some kinds of novelty will increase; the innovations will become numerous enough to reinforce each other and so overcome the conservative pressures of the status quo to form an effective nucleus. As more and more patrons accept the new form, more and more artists will produce works in it, and it becomes a recognizable style with its own opposition to further change.

Look at figure 14.9, bearing in mind that it is two-dimensional and overly simplistic. It depicts an area of a two-dimensional crystal based upon a square unit cell, each "atom" being joined to four others. It contains, however, a local region near the lower right corner in a six-connected triangular arrangement. Now, even if this should be a preferable structure and even if there should be no change in volume, its formation is opposed by the whole existing array of established connections between neighboring atoms, and both atom positions and a whole hierarchy of internal and external bond resonances must be disruptively changed to allow the formation of the six bonds needed for each atom in place of the original four. Once the particle has become large enough so that the energy associated with the disruption at the interface is less than the gain within the volume, the system will move in whatever direction favors the structure with lower free energy. This kind of interface is of necessity disordered, and atoms can pass into and out of it with little more restraint than if it were liquid. Its structure permits and its translation produces the desirable change of arrangement that could not occur within the uniform lattice of the adjacent material.

When a change of temperature, pressure, composition, or social climate has made a system ripe for change, nuclei of a more stable form will not appear everywhere, but only in a few places within the old structure that are for some local reason deformed or strained, or at an interface with an intrusive structure that serves to catalyze the change with a suggestion of a possible new order. In a physical system of crystals, new forms are most likely to appear in the zone of misfit where one phase or crystal impinges upon another. (Note the many such regions in the bubble raft shown in the frontispiece.) Socially, it is where individual freedom is greatest, usually at places where classes or cultures clash; aesthetically, it is where ideas least conform to the established values or where an existing style is impressed on a new material or a different technique. However, these same structural suggestions can exist locally without nucleating massive change as long as the dominant structure best matches the prevailing "thermodynamic" conditions, which are overall, not local.

One cannot overemphasize the fact that *everything*—meaning and value as well as appropriateness of individual human conduct or the energy state of an atom—depends upon the interaction of the thing itself and its environment. The drive for both stability and change is the minimization of free energy in a physical system and, in a social system, something like unhappiness or dissatisfaction—both summed, not averaged or individually measured. The mechanism of change is by transfer across the interface, each atom making its own choice, and quickly being brought back if not supported by compatible choices on the part of its neighbors. Not infrequently, such short-range improvement delays an overall readjustment that would be better—the metastable phase or the tragedy of the commons. Is loyalty to a country or belief in one system of moral values admirable when it denies equal value to others?

In a large system local changes may be nucleated in many locations, and patches of the new, more stable, structure will continue to spread until eventually they impinge on and interfere with each other. Once the difficult stage of nucleation has been passed, patches of the new form will grow by accretion at a rate depending only on how quickly the parts at the interface can rearrange themselves. This is the central rising slope of the now-familiar "S" curve of growth (see figure 11.1). The slowing and stopping of growth comes from the depletion of material to be changed into the new form or from the concentration of rejected parts that cannot be removed by diffusion or otherwise. So with salt in water that is freezing or, less permanently, heat in almost all reactions and conservative ideas in a radical society.

Change, then, begins slowly, uncertainly, and in places that are highly dependent upon local circumstances because the nuclei necessarily are misfits in the existing structure or orthodoxy. The nuclei are unpredictable (except perhaps by the Witch of Endor of whom Banquo demanded, "If you can look into the seeds of time, and say which grains will grow and which will not, Speak then to me") because no system can by itself know ahead of time what, if any, new structure can supplant it. Nuclei do form, however, in those regions of the old structure that are least contented. A phase change is analogous to a political revolution; not the destruction of all individuals but the rearrangement of most of them into a new pattern of interaction. A revolution, driven by the injustices of the old regime, needs its formative nucleus, and its growth (which occurs via an interface of high disorder) is slowed by the need to accommodate or eliminate dissenters. Eventually, however, the structure that began as the highly creative work of a successful innovator becomes an ideology, and as it spreads it becomes indoctrination, not creation.

At the top end of the "S" curve of growth where the new structure has in the main replaced the old, there follows a period of ripening, of adjustment in response to smaller energy changes. Personal preferences begin to modify a dominant but overly simplistic ideology; necessity at last becomes the mother of invention and mechanics can improve mechanisms; a school of lesser artists refines the creator's insights and methods; local regions of stable minor phases are formed by the segregation of "impurity" atoms and the breaking away of clusters of them from the dominant older matrix; and large particles of the stable phases absorb smaller ones to decrease interfacial area. The system slowly moves toward a stable maturity that will be upset once again only if something from outside changes the nature or density of the component parts or their means of communication. While at the beginning a small nuclear region expands to create large change in an unstable structure, at the end the stability of the new larger structure dominates the smaller details. There is individual injustice in both cases. The conflict of large regions of order produces local conditions of disorder that can neither extend nor vanish for their dissymmetry is in local equilibrium with the overarching stability of the larger environment: They shrink but cannot disappear. This contrasts with the beginning, where it was the environment that was unstable and the interface moved outward from the misfit. What was once an innovative individual thing has become just one part of a larger structure which now claims the individuality.

The total effect of these overlapping three stages in growth forms the "S" curve. Both beginning and end are gradual and both depend on virtually unpredictable local conditions. Though it is easy to follow the growth of a new style or a new phase, for it is simply a quantitative change, it is virtually impossible to observe its beginning or even retrospectively to trace back to it. The creation of a form cannot be observed until it exists, and, in practice, it is not usually noticed until after it has existed for some time. The only certainty is that the beginning of a new structure must be a misfit with the old. New political ideas, new aesthetic forms, or new scientific theories inevitably seem crazy in the framework within which they appear. The converse, however, is not true: it must not be assumed that all misfits have the germ of the future in them! In physical systems most fluctuations are transient, and in human societies the overwhelming majority of crazy ideas remain crazy forever, in any environment.

Any historian who has tried to trace the origin of an important concept or invention knows how difficult it is to do so, for even the people most concerned are unaware of the full consequences of what they are doing. Even in their own time, the records of the beginning stages of any art, technology, language, or other component of culture are few in number and hard to detect, and even when a change has proved itself by incorporation in the social structure, its origins are cherished only in retrospective, perhaps overly mature, societies. On the other hand, the period of growth leaves many records and is driven by the very type of social consensus that had earlier opposed it. Change is in one direction only, and people at the interface must choose between old and new, being in contact with both. The period of maturity again has many locally diverse features, unpredictable because they result from the interaction of previously unconnected events, from the accumulation of rejected or unexpected factors, and from the reconciliation of larger conflicting domains. Even the most careful design or advanced systems analysis can rarely include all important consequences of a new activity. "Technology Assessment" cannot hope to anticipate all the side effects of a new technology, though it may be able to detect them early, to avoid catastrophe if they are harmful, and to obtain greater benefits otherwise. The aesthetic imagination of the artist has little role to play in the middle of the "S" curve but is essential at both ends, as it suggests new structures that are capable of expanding or new arrangements that best resolve local tensions arising in the conflict of larger stable structures. If a means of replication exists, these new structural arrangements may in their turn become units in a structure of society on a still higher hierarchical level. The fact that the social consequences of art are not easily seen at first makes it easy for new concepts in art to spread on a personal level.

This model of nucleation, growth, and mature adjustment, simplistic though it is, may illuminate some anthropological problems. It is only after mass acceptance and interlock with other aspects of behavior that a trait can be regarded as characterizing a culture.

As we have seen, the interactive quality of structural inertia makes the formation of a new structure extremely difficult. If it should occur once, the chance of a similar nucleus occurring independently at another place remains not a bit greater than it was before. But a new structure can, of course, spread into other parts of the old by the example of direct contact or by word and diagram carry just the idea to catalyze comparable change in other receptive areas. The stimulus has no power to force change but merely offers an example of a pattern the response to which depends entirely upon local conditions. The difficulties of "modernization" programs and the transfer of technology to "underdeveloped" countries show how different levels of a structure may be transferred at different rates, producing curious hierarchical disregistry. In fact, except by conquest, only those parts of a culture or civilization that are capable of being reintegrated into a different larger pattern are transferable. Cultures are the hierarchical result of collective selection—a phrase I owe to Avraham Wachman.

Personally, I believe that human communication in some form is behind nearly all cases of seemingly independent simultaneous invention, the multiple appearance of ideas or concepts that are outside the preexisting framework—things like the first smelting of ores, for example, and perhaps first ideas of kingship and of God. That the route of communication does not always remain visible is a consequence of the selective unreadiness of the intermediate terrain. An ice crystal moving over a moist windowpane will leave no structural record except in those places below freezing temperature, and travelers' tales are usually disbelieved.

The changes involved in the refinement of a mature system are different. The adjustment of the larger structure tends to force a pattern upon the localities of conflict, and rather similar structural resolutions may occur in many different places. These are the "reverse salients" of economic history, and sometimes they may themselves interact unexpectedly to form really revolutionary restructuring on a new higher level of hierarchical structure.

All changes, however, do not involve misfit at the higher level. There are many pairs of structures that are so related topologically that it is possible to pass between them via continuous distortion without upsetting the overall connectivity. Liquids or solutions are of this type. Particularly interesting is the critical point where a liquid and a gas merge into each other—but only at a specific combination of composition, temperature, and pressure. Partially miscible liquids that differ in composition can do the same, for the solubility of each in the other increases with temperature, and unless something else like vaporization sets in, they will eventually merge. In crystalline solids such merging can occur between isomorphous solid solutions in which atoms are substitutable for each other without change of the lattice except in size. Diffusion, made possible by the random mobility of vacant lattice sites, enables the components to become randomly distributed at high temperatures, but at lower temperatures the system has fewer thermal quanta to be incorporated, and the different atoms assert their preferences for like or unlike neighbors and begin to associate into either an alternating order on the scale of the unit lattice cell or larger regions of segregated clumping (figure 14.8). If the overall structure is tolerant, the interface between the structures can be diffuse and gradual, the interface energy being distributed over many units with the local strain never high enough to destroy topological continuity. The boundaries that can be drawn to divide such systems into discrete space-filling cells do not mark a disruption, but only a change in the second derivative of some property. The parts organize into an aggregation of cells of intermediate size each with exactly compensating positive and negative deviations from the average. These cells must, of course, ultimately fill space, but individually they do not necessarily have to be similar in size or to have any symmetry. Symmetry, indeed, has been grossly overemphasized in both art and science: its main value is in giving meaning to its absence, dissymmetry, without which there could be no hierarchy. A random aggregate of soap bubbles is a more generally applicable analogy than is a crystal lattice. Many structures come from the adjustment of boundaries and the internal rearrangement of groups of parts within an already existing structure. The shaping of valleys and mountains from the self-intensifying cooperative interaction between initial minute hints of a depression, the formation of cities and countries into cultural regions, the development of social institutions that inspire loyalty or moral commitment, or the enjoyment of games or art—all of these form as a result of cumulative regional weak interaction and are quite different from patterns resulting from rigidly specified identical-neighbor interlock. They are a consequence of a long-range search for community reinforcement, though they only form after the overriding short-range demand has been satisfied.

On Art

The importance of the cumulative nuance as distinct from a brutally clear and simple statement is what much of art is about. Socially and culturally, the aggregation of nuances in diverse utilitarian things, daily custom, and ritual gives richness without disruption, for the forces are gentle, not overly specific, and nearly invisible. Since the outcome is statistically determined and requires no basic change of connectivity while forming local clumps, they form easily and everywhere and do not require the disordered sites for nucleation of a first-order change (see the gradient at the left in figure 14.2).

In art generally the hierarchy rarely arises from such well-defined "crystallinity" as in figure 14.3, which was chosen to illustrate the analogy with the chemical phase. The requirement is simply that parts of some kind be perceptible and relatable to each other by the operations of the mind as the eye scans and compares, noting connections, invariances, symmetries, deletions, and modifications while mentally changing scale and orientation, and relating the current sensual input to other forms and concepts from recent experience or from more deeply embedded memory. Aesthetic pleasure seems to come as a kind of moiré pattern emerging on a higher level from the superposition of sensed and remembered images, somewhat as the experience of the third dimension arises from binocular vision. It is what is left over when what is expected has been canceled out. It seems always to involve some interaction between what is immediately visible and features on scales both above and below this.

Beauty can be perceived in a simple line (except a hard straight one) and it seems to arise from the relationships that the eye detects between the curvatures of different parts of it and the variety in less obvious, more than one-dimensional qualities such as width, density, and sharpness of definition. Curvature can be uniform throughout as in a circle, or smoothly varying as in other conic sections, or in most natural forms where surface tension and viscosity serve to reconcile the geometric requirements of locally effective forces with distant larger ones. Other interesting lines may be complex, with reversals of curvature or abrupt divisions into segments. The eye is repulsed by complexity if no order is detected, but it can be delighted by repetition, translation, rotation, reflection, magnification, and other simple variations of the parts. As more levels of hierarchy can be constructed from the simple initial components, the richer becomes the experience.

The retina of the eye receives information in two dimensions, but the mind interprets it in three or more. Features in the object in two or three dimensions can be suggested by outline or by internal shading, or both. Areas of pure color derive their effect from an unresolved but real substructure of wavelike photons producing a selective texture in the array of three types of color-sensitive rods and cones in the retina, much as the response to texture in a surface without color arises in just-resolvable granulation of the primary image. Some of the more enjoyable surfaces (for example, the grain of a fine mahogany table top or a Japanese sword) have an interplay between pattern and texture which, though two-dimensional, suggests the unseen internal three-dimensional array.

As every painter knows, gradients of color or shading can distinguish areas without exactly defining them, and juxtaposition or repetition of such features are as effective visually in suggesting hierarchical relationships as are well-defined ones; indeed the very fuzziness invites the mind to search the field, the memory, and the imagination for similar but not exactly matching patterns and experience.

The style of a painting or other object lies in the way in which its parts (both physical and iconographic) are related to each other, with repetitions or symmetries and contrasts that give a sense of the whole object presently in view and suggest a larger structural and cultural wholeness. The object itself is doubtless content to exist as a mass of atoms terminating in an external irregular surface, but the human mind, which learned in infancy to deduce a three-dimensional world from the changes in time of two-dimensional visual images and the sense of touch, notes similarities and differences between the various curvatures, junctions, shadings, and textures and then operates upon them to generate from the inconsistencies other levels in the hierarchy of matter and space and to extend into quite different kinds of worlds. The object, as it is experienced by a human being, is just one level in a hierarchical set, at the point where a series of crystalline and molecular and atomic structures reaching from below merge into a set of conceptual and cultural structures stretching both up and down.

Notice how all these factors contribute to the beauty of a piece of jade (figure 14.11). Overall, there is the approximate gross symmetry, but the many departures from the shape of the plane-surfaced Euclidean polyhedron that was first sawed from the boulder of nephrite give richness by inviting the detection of differences between one part of the outline or surface and another. The subtle curvatures arising from the soft-backed abrasive finishing techniques cause the sharp lines of junction to fair beautifully into more gently curved surfaces. Though basically simple and geometric, all of the features require and suggest something of a higher dimension. To this are added the local, fuzzily defined regions of shading and color within the stone itself that increase the geometric relational possibilities, the slightly matte finish that gently disperses and softens the reflection of light from the surface, and the transparency and granularity of the stone that gives a deep three-dimensional sense of texture leading inward with decreasing clarity but no limit, while also recording something of the history of the earth and (via the deduced technology of manufacture and ceremonial purpose of the blade) something of man's history and aspirations. Every one of these features, by provoking the eye, the imagination, and the memory to find some

Figure 14.11
Ceremonial axe of nephrite, cut and polished by the use of abrasives. Chinese, Shang (?) dynasty. Photo courtesy of the Fogg Art Museum, Harvard University.

resonances, facilitates the establishment of a pattern of interlocking connectivity like that between atoms making crystals but in a richer, less geometric mode, and characterizing many more levels of the hierarchy of existence.

As David Pye has shown in his fine little book *The Nature and Art of Workmanship* (Cambridge, 1968), the resonances arising in workmanship are often very subtle. The fact that the material itself guides the tool differently in different processes of working introduces changes in the overall relationship of curvatures. The smooth curves of surfaces approaching the edge of a jade axe that come about from innumerable abrasive particles moving against a slightly yielding and mechanically unconstrained backing would seem incongruous if other surfaces or outlines were present that had come from cleavage or from the geometric motions of a machine. These could be produced easily enough, but the eye would not establish larger resonances among them.

In the above discussion, I have dealt mainly with the nature of things that, like inorganic matter, are composed of parts that are rather simply aggregative, possessing some degree of interchangeability despite their complexity. A biologist would be more concerned with organisms in which a treelike branching linear interconnection of specialized parts is superimposed on problems of cellular space-filling. In complex biological organisms as in social institutions, the differences between the parts assume an importance comparable to the similarities, and integration becomes more a matter of functional interdependence than of simple structural interlock. There is a recognizable center of communication. The fast, almost volumetric communication via elastic strain in a crystal aggregate is replaced in biological organisms by the slower and less directed but more specific communication resulting from the broadcast diffusion of molecular units and the directed impulses of a nerve cell. Communication involves both emanation and absorption, and complex messages can pass only between complex structures. In both crystals and organisms resonant feedback is necessary, but the stability of the organism depends on different and complex communications between specialized parts. Without internal diversity and functional communication it would simply be an aggregate. One should note, however, that the functional interdependence is hierarchical, defining an entity by the interaction of its parts, differing from that in an aggregate mainly in distance and specificity. The units that become organisms are more individual, more polarized, and the boundaries defining them more distinct than in inorganic aggregates.

In a society, the specialized activities of hunting, ceremony, family life, cooking, or the production of food, tools, or art represent different aspects of the individuality of their practitioners and products, but they nevertheless are interdependent within the culture, and each will develop in time some characteristic that marks the culture as well as the particular purpose. In the archaeological record the style thus arising can most easily be seen in objects that are duplicated by mass production (not necessarily using machines) and are accepted in the environment of most members of the culture. In the fine arts a work is too individual to be a style until its essential aspects have been replicated, and new forms will be subversive until they have found enough copiers and patrons to break through the conservatism of the old institutions and set their own style. An aggregate in which order and diversity are combined necessarily imposes differing status on its parts. In society there are two kinds of elite: the conservative managers involved in the functioning of an existing hierarchy and the radical intellectuals who are at the apex of a new structure that is about to develop. Neither is a common man. Neither can work by logical rote, for both need aesthetic intuition.

Though my illustrative diagrams (as my own experience) have concerned structures familiar to the solid-state physicist and metallurgist, my examination of their hierarchical nature has led me away from science toward art, in which I have no professionalism whatever. While science has traditionally looked as exactly as possible at one level at a time in relative isolation, art seems to be in its very essence hierarchical. The viewing of art involves both the immediate recognition of simple patterns of color and resonance and symmetry in the parts, the slower recognition of less exact and conflicting relationships and of similarities persisting through changes of scale, orientation, and content, and the still slower perception (or perhaps subliminal absorption without perception) of allusions and cultural overtones. It is likely to involve a number of shifts of gestalt and a slow modification and renewed appreciation of detail. The creative artist seeks new discoveries in unexplored areas or new viewpoints in old ones where he senses relationships that have remained hidden. He makes suggestions rather than demonstrating precise conclusions. Depending on the state of his environment his work may nucleate a prompt revolution in seeing, it may simply cooperate with other factors to produce a gradual change, or it may fail to establish any external resonance whatever and be forgotten.

All structures need units and some means of resonant interaction between the units that affects their being. The world is the sum of all structural interactions, and in a way it knows no parts. Yet reacting within it are nucleons, photons, and atoms selectively associating with each other, and human minds endeavoring to evolve structures to match. By thought we can select a level and a means of probing it. In time we might be able to probe in succession its workings on all levels, but we can never hope to think or sense them simultaneously. Despite this we can learn something. The nature of both the parts available and the means of communication within a system may change, and the means of externally probing it may differ, but patterns of interaction seem to have some consistency. Structure, most easily understood when presented visually, has much of the character of a universal metaphor. A field of view comprising relatively clear interrelations between parts in the middle and progressively more indistinct but nevertheless essential interactions toward the fringes can be applied at almost any level in almost any medium. For different purposes, different levels and different components will be chosen.

Nothing can be understood without at least a simplified glance at levels that are above and below the one of major interest. In both science and art the center of the limited perception of the human mind can be placed anywhere. In science the boundary of concern has traditionally been sharply defined to exclude unknowns, while in art the boundary is fuzzy and undefined: it can be anywhere, nowhere, or everywhere. The future seems to lie with a more extensive science, but it will have to be a multilevel science that, eschewing mysticism but not metaphor, will be able to pass continuously with a controllable focus and precision into the field of art.

There is no absolute scale, but being human, man should not, even if he could, wish to avoid emphasis on things at his own scale. It is a very interesting scale, being at about the point of maximum significant complexity based upon the properties of the chemical atom. But regardless of scale, the meaning of everything lies in interaction, in the cumulative and changing history of associations on many levels which change at different rates. The structure that exists at any given moment is funeous, the product and the record of past associations and interactions, and it is also the framework within which future changes must commence.

Change is always unpleasant at the level most involved. The bitter pill that causes it has to be sugar-coated, and it has to be small in order to be swallowed. This is why art on an individual scale has so often been the precursor of large technological change—and why art on a social scale, which we so desperately need today, is so hard to experiment with.

Acknowledgments

Discussions with John Cahn, David Hawkins, Philip Morrison, and Victor Weisskopf have helped to shape the structural view of the world presented here. The conference on structure arranged by my colleagues Heather Lechtman, John Cahn, and Arthur Steinberg to mark my "retirement" in 1968 provided the impetus to consolidate the less scientific aspects of the structural principles that began to develop in my mind with studies of microstructure and x-ray diffraction in the twenties: the relationships with art came after a search for records of early metallurgy had made me a frequenter of museums, especially the Asiatic section of the Victoria and Albert Museum, the Fogg Museum, the Freer Gallery, the Museum of Fine Arts, and the Metropolitan Museum of Art, whose curatorial staffs have been both tolerant and helpful. Most recently, membership in the informal group of people in the Cambridge area who call themselves the Philomorphs has served to enliven for me the problem of structure.

Bibliography

It is impractical to give detailed references to the prior literature. In addition to D'Arcy W. Thompson's *Growth and Form* (Cambridge: At the University Press, 1917), which marks the beginning of broadly integrated structural thinking, the following have proved particularly helpful:

Rudolf Arnheim, ed., *Entropy and Art* (Berkeley: Univ. of Calif. Press, 1971).

H. S. M. Coxeter, *Regular Polytopes* (New York: Macmillan Co., 1963).

Gyorgy Kepes, ed., *Structure in Art and Science* (New York: George Braziller, 1965).

Gyorgy Kepes, ed., *Module, Proportion, System, Rhythm* (New York: George Braziller, 1966).

Wolfgang Köhler, *Gestalt Psychology* (New York: H. Liveright, 1929).

George Kubler, *The Shape of Time* (New Haven: Yale Univ. Press, 1962).

Arthur Loeb, *Space Structures, Their Harmony and Counterpoint* (Reading, MA: Addison-Wesley, 1976).

Howard H. Pattee, ed., *Hierarchy Theory: the Challenge of Complex Systems* (New York: George Braziller, 1973).

Herbert A. Simon, "The Architecture of Complexity," *Proceedings of the American Philosophical Society*, 106:467–482 (1962).

Peter S. Stevens, *Patterns in Nature* (Boston: Little, Brown, 1974).

Paul A. Weiss, *Life, Order, and Understanding* (Austin: Univ. of Texas Press, 1970).

Hermann Weyl, *Symmetry* (Princeton: Princeton Univ. Press, 1952).

Lancelot L. Whyte, *Accent on Form* (New York: Harper & Row, 1954).

Lancelot L. Whyte, *Aspects of Form* (London: Humphries, 1951).

Lancelot L. Whyte, A. Wilson, and D. Wilson, eds., *Hierarchical Structures* (New York: American Elsevier, 1969). Contains a good bibliographical essay.

It will be clear that the author believes the nature of hierarchy to be most easily accessible to human understanding via visual patterns and material structures. References to mathematical treatments of group theory and algebraic topology (in which views paralleling those given in this paper are doubtless to be found) are therefore omitted. For similar reasons, the copious literature on systems analysis and that on structuralism in anthropology and linguistics is not cited despite some similarities in both problems and solutions.

Bibliography of the Work of Cyril Stanley Smith

1925–1929

1
The oxy-acetylene welding of copper, *The Metal Industry* (New York) 23:360–361 (1925).

2
(with C. R. Hayward) The action of hydrogen on hot solid copper, *Journal of the Institute of Metals* 36:211–230 (1926).

3
Note on cathodic disintegration as a method of etching specimens for metallography, *Journal of the Institute of Metals* 38:133–135 (1927).

4
Note on the crystal structure of copper-gold alloys, *Mining and Metallurgy* 9:458–459 (1928).

5
The alpha phase boundary of the copper-silicon system, *Journal of the Institute of Metals* 40:359–373 (1928).

6
[Cadmium-zinc and lead-tin-cadmium solders (discussion)], *Transactions of the American Institute of Mining and Metallurgical Engineers* (henceforth *TAIME*) 78:362–366 (1928).

7
The constitution of the copper-silicon system, *TAIME* 83:414–439 (1929).

1930–1934

8
An air-hardening copper-cobalt alloy, *Mining and Metallurgy* 11:213–215 (1930).

9
Thermal conductivity of copper alloys, *TAIME* 89:84–105 (1930).

10
Alpha phase boundary of the ternary system copper-silicon-manganese, *TAIME* 89:164–193 (1930).

11
Thermal conductivity of copper alloys, parts II and III, *TAIME* 93:176–184 (1931).

12
Specimen holder for inverted microscope, *Mining and Metallurgy* 12:323 (1931).

13
[Internal oxidation of copper alloys (discussion)], *Journal of the Institute of Metals* 46:49–51 (1931).

14
The equilibrium diagram of the copper-rich copper silver alloys, *TAIME* 99:101–114 (1932).

15
[Internal oxidation of copper alloys (discussion)], *Mining and Metallurgy* 13:481 (1932).

16
The interconversion of atomic, weight and volume percentages in binary and ternary systems, *AIME Contribution* no. 60 (1933).

17
(with W. E. Lindlief) A micrographic study of the decomposition of the beta phase in the copper-aluminum system, *TAIME* 104:69–105 (1933).

18
(with E. W. Palmer) The precipitation hardening of copper steels, *TAIME* 105:133–164 (1933).

19
(with C. H. Lorig) Effect of copper in malleable iron, *Transactions of the American Foundrymen's Association* 42:211–226 (1934).

1935–1939

20
(with E. W. Palmer) Some effects of copper in malleable iron, *TAIME* 116:363–382 (1935).

21
Magnetic susceptibility of some copper alloys of gamma brass structure, *Physics* 6(1):47–52 (1935).

22
The relation between the thermal and electrical conductivities of copper alloys, *Physical Review* 48:166–167 (1935).

23
(with E. W. Palmer) Thermal and electrical conductivity of copper alloys, *TAIME* 117:225–245 (1935).

24
[On polarisation in the metallurgical microscope (discussion)], *TAIME* 117:153–155 (1935).

25
Methods of illuminating Brinell impressions, *Mining and Metallurgy* 17:206 (1936).

26
(with C. S. Taylor) Thermal and electrical conductivities of aluminum alloys, *TAIME* 124:287–298 (1937).

27
Copper alloys containing sulphur, selenium, and tellurium, *TAIME* 128:325–334 (1938).

28
[Precipitation and ferromagnetism in copper-iron alloys (discussion)], in *Age Hardening of Metals,* ASM Seminar Report (Cleveland: American Society for Metals, 1939), pp. 186–187.

29
[Mechanism of precipitation from solid solution (discussion)], in *Age Hardening of Metals,* ASM Seminar Report (Cleveland: American Society for Metals, 1939), pp. 425–426.

30
Mechanical properties of copper and its alloys at low temperatures—a review, *Proceedings of the American Society for the Testing of Metals* (henceforth *PASTM*) 39:642–648 (1939).

31
Constitution of alloys of copper with aluminum, cadmium, chromium, silicon, silver and zinc-tin; physical constants of copper; properties of commercially pure copper; articles in *Metals Handbook* (Cleveland: American Society for Metals, 1939), pp. 1341–1396.

32
An early testing machine, *ASTM Bulletin* 100:13 (1939).

1940–1944

33
Constitution and microstructure of copper-rich silicon-copper alloys, *TAIME* 137:313–329 (1940).

34
A simple method of thermal analysis permitting quantitative measurements of specific and latent heats, *TAIME* 137:236–244 (1940).

35
Biringuccio's *Pirotechnia*—a neglected Italian metallurgical classic, *Mining and Metallurgy* 21:189–192 (1940).

36
Proportional limit tests on copper alloys, *PASTM* 40:864–884 (1940).

37
Structure and ferromagnetism of cold-worked copper containing iron, *Physical Review* 57:642 (1940).

38
[Action of hydrogen, steam and carbon monoxide on copper (discussion)], *TAIME* 137:297–299 (1940).

39
The development of powder metallurgy, in *Report of the Powder Metallurgy Conference* (Cambridge: MIT, 1940), pp. 9–14.

40
Seventeenth-century English brass making, *Metal Industry* (London) 56:285–287 (1940).

41
Of Typecasting in the Sixteenth Century, New Haven, CT: printed by Carl Rollins for the Columbiad Club of Connecticut, 1941.

42
Nonferrous metals, in Lionel S. Marks, ed., *Mechanical Engineers' Handbook* (New York: Interscience, 4th ed., 1941), pp. 623–671.

43
Metallurgy of the Renaissance, *Technology Review* 43:155–157, 174–177 (1941).

44
(with R. W. VanWagner) The tensile properties of some copper alloys, *PASTM* 41:825–848 (1941).

45
(with A. R. Anderson) Fatigue tests on some copper alloys, *PASTM* 41:849–857 (1941).

46
(with M. T. Gnudi) *The Pirotechnia of Vannoccio Biringuccio*, New York: American Institute of Mining and Metallurgical Engineers and Yale University Press, 1942 (paperback edition, 1966, by The MIT Press).

47
(with C. Zener and H. Clarke) Effect of cold work and annealing upon internal friction of alpha brass, *TAIME* 147:90–95 (1942).

48
Magnetic studies on the precipitation of iron in alpha and beta brass, *TAIME* 147:111–121 (1942).

49
(with W. R. Hibbard, Jr.) The constitution of copper-rich copper-silicon-manganese alloys, *TAIME* 147:222–225 (1942).
50
The early development of powder metallurgy, in J. Wulff, ed., *Powder Metallurgy* (Cleveland: American Society for Metals, 1942), pp. 4–17.
51
(with E. W. Palmer) Effect of some mill variables on earing of brass, *TAIME* 147:164–181 (1942).
52
Hardness changes accompanying the ordering of beta brass, *TAIME* 152:144–148 (1943).
53
[On impurities from circulating scrap (discussion)], *TAIME* 152:142 (1943).

1945–1949

54
(with S. K. Allison et al.) Atomic energy and its implications, *Metal Progress* 49:761–779 (1946).
55
Metals in modern society, *Mining and Metallurgy* 27:541–543 (1946).
56
(with J. E. Burke) The formation of uranium hydride, *Journal of the American Chemical Society* 69:2500–2502 (1947).
57
Institute for the Study of Metals, The University of Chicago, *Scientific Monthly* 65:489–492 (1947).
58
Trends in metallurgical research, in *Yearbook of the American Iron & Steel Institute* (New York: AISI, 1947), pp. 529–538.
59
Grains, phases, and interfaces: an interpretation of microstructure, *TAIME* 175:15–51 (1948).
60
(with A. G. Sisco) *Bergwerck und Probierbuchlein*, New York: American Institute of Mining and Metallurgical Engineers, 1949.
61
[Effect of a dispersed phase on grain growth in Al-Mn alloys (discussion)], *TAIME* 185:312–313 (1949).
62
(with M. Farnsworth and J. L. Rodda) Metallographic examination of a sample of metallic zinc from ancient Athens, *Hesperia* (Athens: American School of Classical Studies) Supplement 8:126–131 (1949).
63
(with D. F. Clifton) Microsampling and microanalysis of metals, *Review of Scientific Instruments* 20:583–586 (1949).
64
On blowing bubbles for Bragg's dynamic crystal model (Letter to the Editor), *Journal of Applied Physics* 20:631 (1949).
65
Solid nuclei in liquid metals, *TAIME* 185:204 (1949).
66
(with K. K. Ikeuye) Studies of interface energies in some aluminum and copper alloys, *TAIME* 185:762–768 (1949).

1950–1954

67
(with E. W. Palmer) Alloys of copper and iron, *TAIME* 188:1486–1499 (1950).
68
A decade of metallurgical science, *Metal Progress* 58:479–483 (1950).
69
(with C. C. Wang) Undercooling of minor liquid phases in binary alloys, *TAIME* 188:136–138 (1950).
70
(with A. G. Sisco) *Lazarus Ercker's Treatise on Ores and Assaying*, translated from the German edition of 1580, Chicago: University of Chicago Press, 1951.
71
(with L. H. Beck) Copper-zinc constitution diagram, redetermined in the vicinity of the beta phase by means of quantitative metallography, *TAIME* 194:1079–1083 (1952).
72
Grain shapes and other metallurgical applications of topology, *Metal Interfaces*, ASM Seminar Re-

port (Cleveland: American Society for Metals, 1952), pp. 65–108.

73
Interfaces between crystals, in *L'Etat Solide*, Report of the 19th Solvay Conference on Physics (Brussels, 1952), pp. 11–53.

74
Interphase interfaces, in W. Shockley et al., eds., *Imperfections in Nearly Perfect Crystals* (New York: Wiley, 1952), pp. 377–401.

75
Pure and applied science in American metallurgy, *Metal Progress* 61:51–54 (1952).

76
(with W. M. Williams) A study of grain shape in an aluminum alloy and other applications of stereoscopic microradiography, *TAIME* 194:755–765 (1952).

77
Review of R. J. Forbes's *Metallurgy in Antiquity: A Notebook for Archaeologists and Technologists, Isis* 43:283–285 (1952).

78
(with R. Gomer) *Structure and Properties of Solid Surfaces,* Chicago: University of Chicago Press, 1953.

79
Further notes on the shape of metal grains: space-filling polyhedra with unlimited sharing of corners and faces, *Acta Metallurgica* 1:295–300 (1953).

80
(with L. Guttman) Measurement of internal boundaries in three-dimensional structures by random sectioning, *TAIME* 197:81–87 (1953).

81
Microstructure, *Transactions of the American Society for Metals* 45:533–575 (1953).

82
Pure and applied science in metallurgy, *American Iron & Steel Institute, Regional Technical Meetings,* 1953, pp. 11–20.

83
Uranium, in C. H. Mathewson, ed., *Modern Uses of Nonferrous Metals* (New York: American Institute of Mining and Metallurgical Engineers, 1953), pp. 464–477.

84
The artist looks at the scientist's world, Foreword to the exhibition catalogue, *The Renaissance Society at the University of Chicago*, 1954.

85
Metallurgy at Los Alamos, 1943–45, *Metal Progress* 65:81–89 (1954).

86
The microstructure of polycrystalline materials, *Transactions of Chalmers University of Technology* (Gothenburg, Sweden) No. 152:1–49 (1954).

87
Properties of plutonium metal, *Physical Review* 94:1068–1069 (1954).

88
The shape of things, *Scientific American* 190:58–64 (January 1954).

1955–1959

89
A quarter century of metallurgical science, *Metal Progress* 68:137–140 (1955).

90
A sixteenth-century decimal system of weights, *Isis* 46:354–357 (1955).

91
[Speculation on the structure of shock fronts in crystalline materials (discussion)], in *Symposium on Phase Transformations in Metals* (London Institute of Metals, 1955), pp. 291–293.

92
(with A. G. Sisco) *Réaumur's Memoirs on Iron and Steel*, Chicago: University of Chicago Press, 1956.

93
(with H. Hu) The formation of low-energy interfaces during grain growth in alpha and alpha-beta brasses, *Acta Metallurgica* 4:638–646 (1956).

94
[On measuring particle size (discussion)], in *Proceedings of the Third International Conference on Reactivity of Solids* (Madrid, 1956), pp. 442–444.

95
The structure of metals as seen under the microscope, *Proceedings of the Royal Institution of Great Britain* 36:404–417 (1956).

96
Decorative etching and the science of metals, *Endeavour* 16:199–208 (1957).

97
A metallographic examination of some Japanese sword blades, *Quaderno II del Centro per la Storia della Metallurgia A.I.M.* (Milan, 1957), pp. 41–68.

98
(with R. J. Forbes) Metallurgy and assaying, in C. Singer et al., eds., *A History of Technology* (New York: Oxford University Press, 1957), vol. III, pp. 27–71.

99
Metallurgy and metal physics, *Physics Today* 10:10–11 (August 1957).

100
Metallographic studies of metals after explosive shock, *TAIME* 212:574–589 (1958).

101
Alloys of copper, nickel, and tantalum, *TAIME* 215:905–909 (1959).

102
[Corrosion of an ancient bronze and an ancient silver alloy (discussion)], in T. N. Rhodin, ed., *Physical Metallurgy of Stress Corrosion Fracture* (New York: Interscience, 1959), pp. 293–294.

103
The crystallization of metals and metallurgists under fatigue, with some remarks on the energy of engineers and the purity of physicists, in G. M. Rassweiler and W. L. Grube, eds., *Internal Stresses and Fatigue in Metals* (Amsterdam: Elsevier, 1959), pp. 431–436.

104
The development of ideas on the structure of metals, in M. Clagett, ed., *Critical Problems in the History of Science* (Madison, WI: University of Wisconsin Press, 1959), pp. 467–498.

105
(with M. Langsdorf) Science and art, *Bulletin of the Atomic Scientists* 15:50–51 (1959).

106
Methods of making chain mail (14th to 18th centuries): a metallographic note, *Technology and Culture* 1:60–67, 289–291 (1959).

1960–1964

107
A History of Metallography, Chicago: University of Chicago Press, 1960.

108
Some devices for quantitative metallography, *TAIME* 218:58–62 (1960).

109
Chronology of metals and metalworking to 1900 A.D., in *Metals Handbook* (New York: American Society for Metals, 8th ed., 1961), vol. 1, p. 43 (also included in the article "Metallurgy" in the *Encyclopaedia Britannica,* 1961).

110
(with S. Jovanovic) Elastic modulus of amorphous nickel films (Letter to the Editor), *Journal of Applied Physics* 32:121–122 (1961).

111
(with C. M. Fowler) Further metallographic studies on metals after explosive shock, in P. G. Shewmon and V. F. Zackay, eds., *Response of Metals to High Velocity Deformation* (New York: Interscience, 1961), pp. 309–341.

112
The interaction of science and practice in the history of metallurgy, *Technology and Culture* 2:357–367 (1961).

113
(with A. Sisco) Iron and steel in 1627: the *Fidelle Ouverture de l'Art de Serrurier* of Mathurin Jousse, *Technology and Culture* 2:131–145 (1961).

114
(with M. C. Donnelly) Notes on a Romanesque reliquary, *Gazette des Beaux-Arts,* July 1961, pp. 109–119.

115
Plutonium metallurgy at Los Alamos, 1943–1945, in A. S. Coffinberry and W. N. Miner, eds., *The Metal Plutonium* (Chicago: University of Chicago Press, 1961), pp. 26–35.

116
Science and art in the history of metallurgy, *Midway* (University of Chicago Press) no. 7:84–107 (1961).

117
Note on the history of the Widmanstätten structure, *Geochimica et Cosmochimica Acta* 26:971–972 (1962).

118
(with D. P. Spitzer) A simple method of measuring liquid interfacial tensions, especially at high temperatures, with measurements of the surface tension of tellurium, *Journal of Physical Chemistry* 66:946–947 (1962).

119
[On the formation of pearlite (discussion)], in V. F. Zackay and H. J. Aaronson, eds., *Decomposition of Austenite by Diffusional Processes* (New York: Interscience, 1962), pp. 237–243.

120
Four Outstanding Researches in Metallurgical History, Philadelphia: American Society for Testing and Materials, 1963.

121
(with J. G. Hawthorne) *On Divers Arts: The Treatise of Theophilus*, Chicago: University of Chicago Press, 1963 (paperback edition, 1979, by Dover Publications).

122
Note on a Japanese magic mirror, *Archives of the Chinese Art Society of America* 17:23–25 (1963).

123
Some important books in the history of metallurgy, *Metals Review* 36 (1963).

124
Review of N. Barnard's *Bronze Casting and Bronze Alloys in Ancient China, Harvard Journal of Asiatic Studies* 24:284–286 (1963).

125
The discovery of carbon in steel, *Technology and Culture* 5:149–175 (1964).

126
Granulating iron in Filarete's smelter, *Technology and Culture* 5:386–390 (1964).

127
Metallography: its history and aims; lesson 1 in a course series prepared by the Metals Engineering Institute (Metals Park, OH: American Society for Metals, 1964).

128
Metallography 1963 (Review of Special Report 80), *Journal of the Iron & Steel Institute* 202:857–858 (1964).

129
Some elementary principles of polycrystalline microstructure, *Metallurgical Reviews* 9:1–48 (1964).

130
Sorby centennial symposium on the history of metallurgy, *Journal of Metals* 16:46–48 (1964).

131
Structure, substructure, and superstructure, in G. Kepes, ed., *Structure in Art and Science* (New York: Braziller, 1964), pp. 29–41 (also in *Reviews of Modern Physics* 36:524–532, 1964).

1965–1969

132
(editor) *The Sorby Centennial Symposium on the History of Metallurgy*, New York: Gordon and Breach, 1965.

133
Materials and the development of civilization and science, *Science* 148:908–917 (1965).

134
A speculation on the origin of Fahrenheit's temperature scale, *Isis* 56:66–69 (1965).

135
Fahrenheit's temperature scale: postscript to a speculation, *Isis* 56:209–210 (1965).

136
The prehistory of solid-state physics, *Physics Today* 18(12):18–30 (1965).

137
Production de fer à la fenderie de Saugus aux alentours de 1660, *Revue d'Histoire de la Sidérurgie* 7:7–15 (1966).

138
Iron from the slitting mill at Saugus, Publications in the Humanities (MIT Department of Humanities, 1966), no. 75, pp. 1–11.

139
(with M. Stuiver) Radiocarbon dating of ancient mortar and plaster, in *Proceedings of the VI International Conference on Radiocarbon and Tritium Dating* (U.S. Atomic Energy Commission, Division of Technical Information, 1966), pp. 338–343.

140
The cover design: a jeweler's shop (1533), *Technology and Culture* 8:207–209 (1967).

141
Metallurgy in the seventeenth and eighteenth centuries; Mining and metallurgical production, 1800–1880; Metallurgy: science and practice before 1900; chapters in M. Kranzberg and C. W. Pursell, Jr., eds., *Technology in Western Civilization* (Madison, WI: University of Wisconsin Press, 1967), vol. I.

142
A historical view of one area of applied science—metallurgy, in *Applied Science and Technological Progress* (A Report to the Committee on Science and Astronautics, U.S. House of Representatives, by the National Academy of Sciences, 1967), pp. 57–71.

143
The texture of matter as viewed by artisan, philosopher, and scientist in the seventeenth and eighteenth centuries, in *Atoms, Blacksmiths, and Crystals* (Los Angeles: William Andrews Clark Memorial Library, University of California, 1967), pp. 1–31.

144
Sectioned textures in the decorative arts, in H. Elias, ed., *Stereology* (New York: Springer-Verlag, 1967), pp. 33–46.

145
On the nature of iron, in D. de Menil, ed., *Made of Iron* (Houston: University of St. Thomas Art Department, 1967), pp. 29–42.

146
Materials, *Scientific American* 217:69–79 (September 1967).

147
Historical perspectives in science and engineering, in *Man in Science,* Volume II of *The Human Condition* (Durham, NH: University of New Hampshire Press, 1967), pp. 95–107.

148
The interpretation of microstructures of metallic artifacts, in W. J. Young, ed., *Application of Science in Examination of Works of Art* (Boston: Museum of Fine Arts, 1967), pp. 20–52.

149
Crafts: forerunners of science, *Science* 156:1438–1439 (1967).

150
(editor) *Sources for the History of the Science of Steel, 1532–1786*, Cambridge, MA: Society for the History of Technology and MIT Press, 1968.

151
Note on grapie, steel making, and the origin of the blast furnace in Europe, in *Actes du XIe Congres International d'Histoire des Sciences* (Warsaw, 1968), vol. 1, p. 161.

152
Metallographic study of early artifacts made from native copper, in *Actes du XIe Congres International d'Histoire des Sciences* (Warsaw, 1968), vol. 6, pp. 237–252.

153
Matter versus materials: a historical view, *Science* 162:637–644 (1968).

154
The early history of casting, molds, and the science of solidification, in W. W. Mullins and M. C. Shaw, eds., *Metal Transformations* (New York: Gordon and Breach, 1968), pp. 3–51.

155
Simplicity and complexity, *Science and Technology* May:60–65 (1968).

156
Review of F. Szabadvary's *History of Analytical Chemistry, Isis* 60:553–554 (1969).

157
Porcelain and plutonism, in C. J. Schneer, ed., *Toward a History of Geology* (Cambridge, MA: MIT Press, 1969), pp. 317–338.

158
Structural hierarchy in inorganic systems, in L. L. Whyte, A. Wilson, and D. Wilson, eds., *Hierarchical Structures* (New York: American Elsevier, 1969), pp. 61–85.

1970–1974

159
Art, technology and science: notes on their historical interaction, in D. Roller, ed., *Perspectives in the History of Science and Technology* (Norman, OK: University of Oklahoma Press, 1971), pp. 129–165 (preprinted in *Technology and Culture* 11:493–549, 1970).

160
Foreword and discussion, in S. Doeringer, D. Mitten, and A. Steinberg, eds., *Art and Technology: A Symposium on Classical Bronzes* (Cambridge, MA: MIT Press, 1970), pp. v–vii, 51–56.

161
The techniques of the Luristan smith, in R. H. Brill, ed., *Science and Archaeology* (Cambridge, MA: MIT Press, 1971), pp. 32–54.

162
A. A. Barba; N. T. Belaiew; V. Biringuccio; W. Hume-Rothery; Z. Jeffries; P. D. Merica;

A. Sauveur; H. C. Sorby; Theophilus; articles in the *Dictionary of Scientific Biography* (New York: Scribners, 1970–1976).

163
A note on the steel [in an Iranian axe in the Fogg Art Museum], in D. G. Mitten, J. G. Pedley, and J. A. Scott, eds., *Studies Presented to George M. A. Hanfmann* (Cambridge, MA: Fogg Art Museum, 1971), pp. 206–209.

164
Metallography: how it started and where it's going, *Proceedings, Fourth Annual Technical Meeting, International Microstructural Analysis Society,* 1971, pp. i–ix (reprinted in *Metallography* 8:231–243, 1975).

165
Metallographic examination of some fragments of Cretan bronze armor from Afrati, in H. Hoffmann and A. E. Raubitschek, *Early Cretan Armorers* (Cambridge, MA: Fogg Art Museum, 1972), Appendix III, pp. 54–56.

166
Metallurgical footnotes to the history of art, *Proceedings of the American Philosophical Society* 116:97–135 (1972).

167
Architectural shapes of hot-rolled iron, 1753, *Technology and Culture* 13:59–65 (1972).

168
Thermal expansivity as a method of analysing coins, in E. T. Hall and D. M. Metcalf, eds., *Methods of Chemical and Metallurgical Investigation of Ancient Coinage* (London: Royal Numismatic Society Special Publication No. 8, 1972), pp. 305–306.

169
An examination of the arsenic-rich coating on a bronze bull from Horoztepe, in W. J. Young, ed., *Application of Science in Examination of Works of Art* (Boston: Museum of Fine Arts, 1973), pp. 96–102.

170
Bronze technology in the east: a metallurgical study of early Thai bronzes, with some speculations on the cultural transmission of technology, in M. Teich and R. Young, eds., *Changing Perspectives in the History of Science, Essays in Honour of Joseph Needham* (London: Heinemann, 1973), pp. 21–32.

171
Intuition, *Journal of Metals* 25(2):29 (1973).

172
(with John Hawthorne) *Mappae Clavicula, A Little Key to the World of Medieval Techniques,* Philadelphia: American Philosophical Society, 1974.

173
Aligning small metallographic specimens, *Metallography* 7:169 (1974).

174
Historical notes on the coloring of metals, in A. Bishay, ed., *Recent Advances in Science & Technology of Materials* (New York: Plenum, 1974), vol. III, pp. 157–167.

175
Reflections on technology and the decorative arts in the nineteenth century, in I. M. G. Quimby and P. A. Earl, eds., *Technological Innovation and the Decorative Arts* (Charlottesville, VA: University of Virginia Press, 1974), pp. 1–64.

176
Art, science and structure in the history of metals, *Proceedings of International Symposium on Fifty Years of Metallurgy* (Varanasi, India, 1973), pp. 1–10 (reprinted in *Transactions of the Indian Institute of Metals* 27(2):11–15, 1974).

177
Bergman's "equivalents": a further comment, *Isis* 65:393–394 (1974).

1975–1979

178
(with B. Wallraff) A seventeenth century manuscript: "Notabilia in essays of oars and mettals," *Historical Metallurgy* 8:76–87 (1975).

179
A seventeenth-century octonary arithmetic, *Isis* 66:390–394 (1975).

180
Metallurgy as a human experience, *Metallurgical Transactions* 6A:603–623 (1975) (reprinted as a booklet by the American Society for Metals, 1977).

181
Aesthetic curiosity—the root of invention, *New York Times*, 25 August 1975, section 2, pp. 1, 23.

182
(editor, with W. Rozanski) *Officina Ferraria, A Polish Poem of 1612 Describing the Noble Craft of Ironwork* by Walenty Rozdzienski, Cambridge, MA: Society for the History of Technology and MIT Press, 1976.

183
On art, invention, and technology, *Technology Review* 78(7):2–7 (1976) (reprinted in *Leonardo* 10:144–147, 1977).

184
Speculations on the corrosion of ancient metals, *Archaeometry* 18:114–116 (1976).

185
Review of N. Barnard's *Metallurgical Remains in Ancient China, Technology and Culture* 18:80–86 (1977).

186
A highly personal view of science and its history, *Annals of Science* 34:49–56 (1977).

187
Report of an examination of the "Drake" Plate, in *The Plate of Brass Reexamined* (Bancroft Library, University of California, 1977), appendix E.

188
Some constructive corrodings, in B. F. Brown et al., eds., *Corrosion and Metal Artifacts,* special report 479 (Washington, D.C.: National Bureau of Standards, 1977), pp. 143–153.

189
Structural hierarchy in science, art, and history, in J. Wechsler, ed., *On Aesthetics in Science* (Cambridge, MA: MIT Press, 1978), pp. 9–53.

190
The beginnings of industrial electrometallurgy, in G. Dubpernell and J. H. Westbrook, eds., *Selected Topics in the History of Electrochemistry* (Princeton, NJ: Electrochemical Society, 1978), pp. 360–405.

191
Speculations on dimensionality, valence, and aggregation, in D. V. Brisson, ed., *Hypergraphics: Visualizing Complex Relationships in Art, Science and Technology* (Boulder, CO: Westview Press, 1979), pp. 23–48.

192
Remarks on the discovery of techniques and on sources for the study of history, in G. Bugliarello and G. B. Doner, eds., *History and Philosophy of Technology* (Urbana, IL: University of Illinois Press, 1979), pp. 31–37.

193
(with E. Savage) The techniques of the Japanese *tsuba* maker, *Ars Orientalis* 11:291–328 (1979).

194
Preface [On early technologies], in D. Schmandt-Besserat, ed., *Early Technologies* (Malibu, CA: Undena Publications, 1979), pp. 3–6.

195
(with M. Kranzberg) Materials in history and society, *Materials Science and Engineering* 37:1–39 (1979). This special issue was also published as a book: M. Cohen, ed., *Materials Science and Engineering: Its Evolution, Practice and Prospects* (Lausanne: Elsevier-Sequoia, 1979).

1980–

196
From Art to Science: Seventy-two Objects Illustrating the Nature of Discovery, Cambridge, MA: MIT Press, 1980.

197
The physical scientist's place in art history, archaeology and history, in J. Frierman, ed., *Medieval Ceramics and the Physical Sciences* (Berkeley, CA: Lawrence Berkeley Laboratory, in press).

198
On the peculiar association of books and technology, in D. Kuhner and T. Rizzo, eds., *Bibliotheca De Re Metallica: The Herbert Hoover Collection on Mining and Metallurgy* (Claremont, CA: Claremont College, 1980).

199
Metallurgy as a link between technology and pure science, in *Vortrage, Internationales Symposium zur Geschichte des Bergbaus und Hüttenwesen* (Freiburg: Bergakademie for ICHOTEC, 1980), pp. 433–442.

Sources

1
"Grain shapes and other metallurgical applications of topology" was prepared for an American Society for Metals seminar in October 1951 and was included in the report *Metal Interfaces* (Cleveland: American Society for Metals, 1952), pp. 65–108.

2
"The discovery of carbon in steel" first appeared in *Technology and Culture* 5:149–175 (1964).

3
"Structure, substructure, and superstructure" was a contribution to *Structure in Art and Science* (New York: Braziller, 1964), part of the *Vision + Value* series edited by Gyorgy Kepes. It also appeared, minus a few figures, in the Robert Oppenheimer festschrift issue of *Reviews of Modern Physics* 36:524–532 (1964).

4
"The interpretation of microstructures of metallic artifacts" was a contribution to a seminar organized by William J. Young at the Museum of Fine Arts in Boston. It was included in the volume *Application of Science in Examination of Works of Art* (Boston: Museum of Fine Arts, 1967), pp. 20–52.

5
"Matter versus materials: A historical view" is based on the 1967 Sarton Lecture, delivered at the AAAS meeting in New York in December of that year. The present version was first published in *Science* 162:637–644 (1968).

6
"The early history of casting, molds, and the science of solidification" was prepared for the Second Buhl International Conference on Materials, held in 1966, and it appeared in the informal proceedings volume edited by W. W. Mullins and M. C. Shaw, *Metal Transformations* (New York: Gordon and Breach, 1968), pp. 3–51.

7
"Porcelain and plutonism" was prepared for the 1967 New Hampshire Interdisciplinary Conference on the History of Geology and appeared in the proceedings volume edited by Cecil J. Schneer, *Toward a History of Geology* (Cambridge, MA: The MIT Press, 1969), pp. 317–338.

8
"Art, technology, and science: Notes on their historical interaction" is based on a paper read at the 1969 University of Oklahoma Symposium on the History of Science and Technology and appeared in the proceedings volume edited by Duane Roller, *Perspectives in the History of Science and Technology* (Norman, OK: University of Oklahoma Press, 1971), pp. 129–165. The paper was preprinted in *Technology and Culture* 11:493–549 (1970).

9
"Metallurgical footnotes to the history of art" was the 1971 Penrose Memorial Lecture and was published in the *Proceedings of the American Philosophical Society* 116:97–135 (1972).

10
"Reflections on technology and the decorative arts in the nineteenth century" was the keynote address at the 1973 Winterthur Conference and appeared in the conference report edited by Ian M. G. Quimby and Polly Anne Earl, *Technological Innovation and the Decorative Arts* (Charlottesville, VA: The University of Virginia Press, 1974), pp. 1–64.

11
"On art, invention, and technology" incorporates part of the text of a short article in *The New York Times*, section 2 of 25 August 1975. The present version was prepared for the Philip Morrison festschrift issue of *Technology Review* 78(7):2–7 (1976). It was reprinted in *Leonardo* 10:144–147 (1977).

12
"Some constructive corrodings" was prepared for a 1976 seminar sponsored by the National Bureau of Standards to bring together conservators, archaeologists, and corrosion scientists. It was included in the proceedings volume edited by B. F. Brown et al., *Corrosion and Metal Artifacts* (Washington, D.C.: National Bureau of Standards, 1977), pp. 143–153.

13
"A highly personal view of science and its history" first appeared in *Annals of Science* 34:49–56 (1977).

14
"Structural hierarchy in science, art, and history" is an extension of notes made for an informal talk in the series on "Style" organized by the Cambridge Archaeological Seminar in the spring of 1974 and for one of the Danz lectures at the University of Washington in March 1975. The present version was prepared for inclusion in Judith Wechsler's collection *On Aesthetics in Science* (Cambridge, MA: The MIT Press, 1978), pp. 9–53.

Index